# A Primer for Mathematics Competitions

# MATHEMATICS TEXTS FROM OXFORD UNIVERSITY PRESS

David Acheson: *From Calculus to Chaos: An introduction to dynamics*
Norman L. Biggs: *Discrete Mathematics, second edition*
Bisseling: *Parallel Scientific Computation*
Cameron: *Introduction to Algebra*
A.W. Chatters and C.R. Hajarnavis: *An Introductory Course in
  Commutative Algebra*
René Cori and Daniel Lascar: *Mathematical Logic: A Course with
  Exercises, Part 1*
René Cori and Daniel Lascar: *Mathematical Logic: A Course with
  Exercises, Part 2*
Davidson: *Turbulence*
D'Inverno: *Introducing Einstein's Relativity*
Garthwaite, Jollife, and Jones: *Statistical Inference*
Geoffrey Grimmett and Dominic Welsh: *Probability: An Introduction*
G.R. Grimmett and D.R. Stirzaker: *Probability and Random Processes,
  third edition*
G.R. Grimmett and D.R. Stirzaker: *One Thousand Exercises in
  Probability, second edition*
G.H. Hardy and E.M. Wright: *An Introduction to the Theory of Numbers*
John Heilbron: *Geometry Civilized*
Hilborn: *Chaos and Nonlinear Dynamics*
Raymond Hill: *A First Course in Coding Theory*
D.W. Jordan and P. Smith: *Non Linear Ordinary Differential Equations*
Richard Kaye and Robert Wilson: *Linear Algebra*
J.K. Lindsey: *Introduction to Applied Statistics: A modelling approach,
  second edition*
Mary Lunn: *A First Course in Mechanics*
Jiři Matoušek and Jaroslav Nešetřil: *Invitation to Discrete Mathematics*
Tristan Needham: *Visual Complex Analysis*
John Ockendon, Sam Howison: *Applied Partial Differential Equations*
H.A. Priestley: *Introduction to Complex Analysis, second edition*
H.A. Priestley: *Introduction to Integration*
Roe: *Elementary Geometry*
Ian Stewart and David Hall: *The Foundations of Mathematics*
W.A. Sutherland: *Introduction to Metric and Topological Spaces*
Adrian F. Tuck: *Atmospheric Turbulence*
Dominic Welsh: *Codes and Cryptography*
Robert A. Wilson: *Graphs, Colourings and the Four Colour Theorem*
Alexander Zawaira and Gavin Hitchcock: *A Primer for Mathematics
  Competitions*

# A Primer for Mathematics Competitions

Alexander Zawaira
and
Gavin Hitchcock

OXFORD
UNIVERSITY PRESS

# OXFORD
UNIVERSITY PRESS

Great Clarendon Street, Oxford OX2 6DP

Oxford University Press is a department of the University of Oxford.
It furthers the University's objective of excellence in research, scholarship,
and education by publishing worldwide in

Oxford New York

Auckland  Cape Town  Dar es Salaam  Hong Kong  Karachi
Kuala Lumpur  Madrid  Melbourne  Mexico City  Nairobi
New Delhi  Shanghai  Taipei  Toronto
With offices in
Argentina  Austria  Brazil  Chile  Czech Republic  France  Greece
Guatemala  Hungary  Italy  Japan  South Korea  Poland  Portugal
Singapore  Switzerland  Thailand  Turkey  Ukraine  Vietnam

Oxford is a registered trade mark of Oxford University Press
in the UK and in certain other countries

Published in the United States
by Oxford University Press Inc., New York

ISBN 978-0-19-953988-8

Printed and bound in Great Britain by
CPI Antony Rowe, Chippenham and Eastbourne

# Contents

# Preface

### What is a Mathematics Olympiad?

The original meaning of the word 'Olympiad' was the period of four years reckoned from one celebration of the Olympic Games to the next, by which the ancient Greeks computed time. The year 776 BCE was taken as the first year of the first Olympiad. In modern times the word has become associated with the regular celebration of the Olympic Games revived in 1896. During the course of the twentieth century it came to refer also to other specialist competitions held on a regular basis, e.g. Chess Olympiad, Chemistry Olympiad, etc. One major difference from the Olympic Games is the general rule for Mathematical Olympiads, that contestants be high school students, not yet registered at a tertiary educational institution, and under 19 years of age. The reason behind this is that a primary objective of the World Federation of National Mathematics Competitions (WFNMC) is to stimulate interest in and enjoyment of the subject in high schools. This objective is achieved in a number of different ways. Firstly, mathematics competitions and Olympiads are a powerful way of convincing young people that mathematics is a lively and attractive subject. Secondly, there is ample evidence that such competitions are remarkably effective in identifying future mathematicians at an early stage, and helping them to discover and develop their gift. Thirdly, in mounting high-profile and often glamorous competitions, public awareness of mathematics is promoted, the morale of mathematics teachers is boosted, and the quality of mathematics education is improved. Fourthly, in a more subtle way, Mathematics Olympiads provide extremely valuable (and sadly rare) common ground for university mathematicians and high school teachers and students of mathematics to come together and share ideas on how the subject might be taught more effectively.

Hungary is a small country in Eastern Europe that has produced far more than its share of the world's top mathematicians in the last 60 years. This distinction has been attributed to the long-standing and important role of mathematics competitions in Hungary's education system. From such beginnings, the mathematics competitions movement has developed dramatically all over the world in recent decades, united in a common cause by the WFNMC.

A Mathematical Olympiad, then, whether provincial, national or international, is a form of regular competition in which young people pit their mathematical prowess and wits against each other in solving challenging and attractive problems of a mathematical sort.

## What kinds of problems feature in mathematics competitions?

The word usually used to distinguish Olympiad-style problems from the standard or 'drill' exercises in most textbooks is 'non-routine'. Such a problem will provide an appreciable challenge – it will provoke you to think, to explore, to wonder... This means that, while not much more actual background mathematical knowledge is demanded, yet the solution requires something quite different from the routine application of memorized techniques – it requires creative insight, drawing connections not explicitly given, seeing things in novel ways, thinking with both imaginative clarity and logical persistence. If the word 'problem' has a negative ring for you, think of 'puzzles', crosswords, riddles, 'Sudoku', etc. Most people love to play with, or chew on, a juicy puzzle that is neither too hard for them to attempt, nor so trivial that it is uninteresting and fails to tempt. For the mathematician, problems are everything. Paul Halmos (widely respected mathematician, author and teacher) says that the *heart of mathematics is problems*: the mathematician's main reason for existence is to solve problems. In Halmos' view (echoing Georg Polya, one of the greatest mathematicians and most influential mathematics educators of the twentieth century) a teacher does a grave disservice to students if he or she does not get them to grapple with significant problems and learn how to express the solutions in clear verbal and written form. In 1899, at the Second International Congress of Mathematicians (a four-yearly meeting inaugurated about the same time as the modern Olympics), David Hilbert, perhaps the greatest mathematician of his time, gave a talk in which he presented ten of his personal selection of the twenty-three most significant unsolved problems in mathematics at that time. He prefaced his talk with the assertion that the healthy development of any branch of mathematics depends upon an abundant supply of significant, challenging unsolved problems. And the (ongoing) story of the challenge and the solving of those twenty-three famous 'Hilbert's problems' is almost synonymous with the story of mathematics into the twenty-first century.

## Why do young people enter for Mathematics Olympiads?

Of course, there are usually mixed motives for competing in an Olympiad. There is the sheer challenge of any contest – of measuring one's abilities in a given discipline against those of other people. If you have developed a love of mathematics, there is the immense joy of tackling and solving beautiful problems. But there is another motive that brings many young people to train and compete in mathematics competitions: prowess in these competitions is often used as a basis for selecting people for places, bursaries and scholarships, to pursue studies in certain areas related to mathematics. Especially in recent years, the number of university-entry candidates with top grades in examinations like 'Ordinary Level', 'General Certificate of Education', and 'Advanced Level' has grown to the point where many universities and scholarship selection boards are looking for something else (for example, the International Baccalaureate, the Scholastic Aptitude Test, the 'Pre-U' Examination, essays, special entrance examinations and interviews) to test the true mettle of their candidates. Selection and participation in an international, national or even provincial Mathematics Olympiad generally carries a lot more weight in one's CV than top grades in more routine curricular examinations. This motive for Olympiad participation is not unlike aiming for a sports scholarship on the basis of a particular skill as a means to the end of acquiring a wider college or university education. There's nothing wrong with such 'ulterior' motives; but beware! – if you don't enjoy what you are training for, you may not be sufficiently enthused to stay the course when the going gets tough. Many young people embrace Mathematical Olympiad training programmes and aspire to compete in Olympiads because they share with dedicated athletes the exhilaration of discovering their innate gifts, developing their potential to its limits, and facing and overcoming well-timed challenges. Many of them have had a tantalizing taste of what mathematicians like to call 'the real thing', to distinguish it from its boring, mind-numbing caricature, parodied by the phrase 'doing sums'. It (the real thing) is the incomparable joy of doing mathematics, of solving attractive problems, discovering and proving theorems, exploring unexpected connections and fascinating analogies in an abstract, purely intellectual world of ravishing beauty. Naturally, young people also enjoy participation in competitions because of the fun and fellowship they can experience with kindred spirits. The ancient (thoroughly Olympic) ideal of true 'sportsmanship' is under threat in many ways today, but it represents something noble and worth protecting at all costs; in our mathematical arena we may dub it 'mathsmanship'. Winning medals is not everything, rival competitors can become close friends, competition is not cut-throat! As the football or rugby lovers say, it's the 'game' we all ultimately care for and serve – and our game is mathematics.

## What benefits does Mathematical Olympiad training bring?

One obvious answer is that for the mathematically-talented learner this is the very best way to learn the subject. We'll say more about this below. But another valid answer is: it brings with it a whole range of important thinking skills with wide application. One mathematician, speaking at a Mathematics Olympiad medal presentation ceremony, considered how an employer might assess and encourage desirable mental and physical qualities in employees. While employers naturally want their employees to be fit and healthy, so that they can execute their duties more efficiently, it would be absurd of the employer to expect the employee to jog 10 km each day, turn somersaults or lift 100 kg weights, although these activities do keep one fit. Analogously, most employers would like their employees to be sharp, quick-witted, and to think independently, so that, when the situation calls for it, they will make wise decisions that have no adverse effects on the company and its business. The speaker claimed that youthful participation in mathematics competitions offers just the right training, without having to go to the expensive and time-consuming extremes of professional training and specialization.

## What careers are appropriate for people with mathematical aptitude?

There is of course the highly rewarding vocation of high school teaching, and the academic worlds of mathematics and science. Beyond this, actuarial science is a popular field; so too is engineering, in a variety of forms: aeronautical, chemical, electrical, etc. More recently, weather prediction and epidemiology have become important and sophisticated sciences; and the burgeoning information sciences obviously need and attract mathematicians. Indeed, mathematical gifts and training are valuable assets in research and development in numerous technical and industrial concerns, where imaginative speculation and logical thinking combine fruitfully. Bankers, also, have begun to show a keen interest in mathematically talented young people, as the new and exciting discipline of financial mathematics has grown rapidly to prominence. Its patron saint should surely be Sir Isaac Newton, one of the greatest mathematicians that ever lived, who, though a poor investor, was Master of the Mint in London!

We hope that the portrait we have painted of what a Mathematical Olympiad is, and why you (or your children or your students) might want to train and enter for one, is an attractive one. If your interest has been aroused, you are ready for the crucial question.

## How will this book help to train you?

To acquire the mental equipment and attain the problem-solving agility characteristic of a successful Olympiad 'mathlete' is not easy. But, if you have the basic aptitudes, a good coach and good resources (like this book), you can rise to the challenge. As with all skills, it will take time and hard work. You will need lots of patience and commitment, like any athlete, sports-person or musician seeking to excel. Behind the apparently effortless performance – the grace and fluency of style and artistry – are the long, sometimes arduous hours of practice, coaching, rehearsal and constant striving to improve. Any one of a wide variety of skills developed gradually, painstakingly, during practice may make all the difference on the big day. Think of this repertoire of skills as a richly-stocked toolbox full of techniques ready to be fished out and used at the proper time.

Thus, an Olympiad contestant should have his or her own mental 'toolkit' of ideas, techniques and theorems. Take geometry, for example: in most international and even national contests, knowledge of theorems like Ceva's, Menelaus' or Ptolemy's is either assumed, or can substantially ease the solving of a problem. Similarly, in problems dealing with algebraic inequalities, the Cauchy–Schwarz inequality can often move mountains with ease! And in number theory it might be congruence arithmetic or Fermat's Little Theorem that rises splendidly to the occasion.

The chapters of this book offer you a set of 'Toolchests' for achieving competency in what we think are eight of the most important topics in the essential content of Olympiad mathematics. These Toolchests also contain incidentally most of the mathematics that is relevant to O level and A level, or similar school-leaving examinations. (Calculus is a notable exception, as it is traditionally excluded from Olympiad content. It makes a couple of brief appearances in this book, though always with an alternative method supplied.) This means that, as a considerable bonus, you can pick up in compact form the core mathematics of the standard school curriculum, consistently motivated and spiced with tempting problems! We think that this is actually the best way for a gifted person to learn it.

It's time to get to work, but first, a few words about the structure of the book and the relationships between the theory, examples, problems, and solutions. Although there is naturally some interdependence between the first eight Toolchests (with cross-referencing to indicate where you might dip into another Toolchest for some help), they can be studied independently and in any order. At the start of each Toolchest is a list of objectives outlining its focus and scope. Some Toolchests and many sections begin with one or more motivating 'appetizer' problems, followed by exposition of theory – a little at a time, liberally sprinkled with worked examples and climaxed by the solution to the motivating problem(s). Then at the end of each Toolchest come some relevant problems, followed by their solutions. Toolchest 9 consists of a mixture of problems (and their solutions) of all kinds and of varying difficulty, for you to practise your skills on, and to develop new skills in areas not directly covered in the previous eight Toolchests.

Our aim in this 'Primer' is not to equip you for the Himalayan heights of the IMO (International Mathematical Olympiad) but for the intermediate challenge of national Olympiads – say, of Welsh mountains, and moderately challenging Swiss Alps, North American Rockies, Peruvian Andes, South African Drakensberg, etc. There are other training resources to take you further; we recommend some at the end of the book. We have attempted to take you almost from ground-level, with unusually full and systematic coverage for an Olympiad training resource. We have not just handed out theorems and formulas – we have shown you how they are derived and how they are connected with other things you may already know. But the emphasis is on making you a problem-solver; therefore, in providing you with the necessary tools, we have consciously struck what we hope is a helpful balance between encyclopaedic, logically-ordered exposition, on the one hand, and practical application to solving problems, on the other hand. We want you to understand the tools and how they relate to each other, but more than that, we want you to be able to *use* the tools. Thus you also must aim for a balance in your use of this book.

## How can you achieve the right balance of theory, problems, and solutions in your training?

With this book we believe that we have supplied you, the Olympiad trainee, with a healthy abundance of significant and challenging problems (in David Hilbert's words), carefully ordered in such a way that your climb is not too steep at any point. Our primary aim is to get you to grapple, or wrestle, with these problems, and thus to develop your problem-solving muscle and agility. For this, you need the appropriate tools, and of course the theory and the examples are there in the Toolchests. But memorization of theory by itself is not sufficient to solve problems. The solutions are there too. But we urge you not to allow the presence of solutions a few pages away to distract you from your grappling. Only when you have thought hard and long enough about the problem to appreciate your predicament, and your need of a good idea, should you refer to the solutions, and then only for some direction. You may also turn back to the theory and examples, seeking a fruitful idea. Armed with some new insight, turn back courageously to your problem, to grapple again and make some progress before resorting again to the solution and the theory. It is possible that, without the comforting proximity of a solution, like a safety rope while you climb a steep rock face, you may not have the nerve to begin the climb. That's why the solutions are there. But we want you to learn to climb the rock by yourself, not to be hauled up passively by the rope. We want you to experience, more and more, the indescribable joy of finding the way, the solution, by yourself. And there are often many quite different routes to the solution of a problem.

Our solutions are therefore not to be taken as definitive, as the only way, or even the best way. We hope, however, that our solutions are 'good' solutions, in the sense that they will seed your mind with possibilities for applying to other problems, and will assist you (taking Paul Halmos' wise advice) to express your own solutions as clearly, elegantly and concisely as possible, using both words and symbols well. Georg Polya listed four steps in problem-solving: (1) understand the problem, (2) make a plan, (3) carry out your plan, (4) look back! – and he insisted that the most important step is that last step, which comes *after* you have solved the problem. Look back! Reflect on your solution! How did you get there? Check every step. What blocked you at various points? What was the breakthrough? What were the key ideas? What gave you the clues to finding them? Re-visit the theory in the light of this application. Is there another way to solve the problem? How does your solution differ from our solution? Which method is more elegant? Did you use all the given data/conditions? Can you shorten or simplify the solution? Can this solution be used to solve another related problem?

Happy problem-solving!
Alexander Zawaira and Gavin Hitchcock
December 2007

## Acknowledgements

We both wish to acknowledge the important contribution made by Professor John Webb of the University of Cape Town in stimulating an insatiable appetite for non-routine mathematical problem solving in high schools throughout Southern Africa. Alexander is one of the many who have been set on a journey of discovery by John Webb's pioneering of mathematics competitions in the region. Professor Webb and his colleagues and successors have encouraged our efforts north of the Limpopo, and The South African *Mathematical Digest* has provided the inspiration for Zimbabwe's *Zimaths Magazine*, which may (just!) survive its eleventh year in an extremely challenging environment.

We would also like to pay tribute to Richard Knottenbelt, mentor and encourager of many budding mathematicians, who has long continued to make important contributions in the teaching of mathematics (and chess) in Zimbabwean high schools, and particularly in Masvingo Province, from which a significantly number of Mathematics Olympiad medallists have come, including Alexander.

We acknowledge (on behalf of many) a debt to a number of other people who have worked hard and provided inspiration for helping, against enormous odds, to sustain Zimbabwean mathematical talent search activities and Pan-African Olympiad training and participation. Professor Alastair Stewart has ably chaired the Zimbabwe Mathematics Development Project (ZIMATHS) which took the preparation of this book under its wing some years ago. Erica Keogh and Thomas Masiwa, in particular, have worked hard at mounting and preserving the standards of the Zimbabwe Mathematical Olympiad, from which many of the problems in this book are taken. Although many of the problems in this book may not have appeared in book form previously, we make no claim for the originality of most. The creators of good problems (problems which are both challenging and beautiful, and have solutions which are also beautiful) are rare and greatly to be honoured, but are usually anonymous. We salute them.

**From Alexander Zawaira:** I am grateful to Gavin Hitchcock and Erica Keogh who took me on board the Zimbabwe Mathematics Olympiad

Steering Committee (ZMOSC, later to become ZIMATHS). The collaboration between Gavin and myself, which led to the production of this book, began with my joining ZMOSC. Upon obtaining a Beit Trust Fellowship to pursue a DPhil in molecular biophysics at Oxford University, it was Gavin's suggestion that I should seek to join Linacre College, where I was made to feel very much at home.

I am grateful to Professors Peter Savill and Endre Suli of Linacre College (University of Oxford) for taking an interest in the collection of problems and solutions that formed the basis for this book. Special thanks to Endre for making the time to go through the manuscript and recommending it for publication to Oxford University Press (OUP). Jessica Churchman of OUP took a keen interest at the early stages of the project, and others of the staff at OUP have willingly and ably assisted us in the later stages, in particular Dewi Jackson.

I have gained much from the advice and constructive criticism provided by Dinoj Surendran who read the work in an early manuscript form. I also wish to thank Archibald Karumbidza, my friend and fellow scientist, with whom I have had many fruitful and stimulating mathematical discussions.

My father, Thomas Zawaira, was supportive of my interest in mathematics and my curiosity with the natural world. I wish to register my appreciation for that support.

I have come a long way with Enoch Muchemwa who has watched this work evolve from mere ideas to a tangible reality. I wish to thank Enoch for unreservedly and unconditionally sharing his talents as an artist. The many discussions I have had with Gesina Human in Cape Town (South Africa) have yielded some of the most interesting free-hand illustrations in this book. I wish to thank her for sharing her talent as an artist. I also wish to thank Prince Dube who, 14 years ago, suggested that if one created and collected good mathematical problems, a few at a time, a good book can eventually be written.

I also wish to thank some of my high school and university teachers who made invaluable contributions to my intellectual growth: Mr J. Chivhungwa, Mr R. Munangwa, Mr A. K. Madhuvu, Mrs N. Muringani, Professor I. Sithole, Sister De Pace and Mr S. Tererai.

Alexander Zawaira, Cape Town, December 2007

**From Gavin Hitchcock:** I have appreciated my collaboration with Alexander, who commissioned me to take on board his 'baby', and, through long years of patient prodding from him, to help to bring it to maturity. The list of colleagues and graduate students in the Department of Mathematics at the University of Zimbabwe who have been involved with the various stages of typesetting and checking of this book (known in the Department as the Big Book) is too long for individual mention, but they are due our thanks. Thanks to OUP staff for their patient assistance. Finally, I wish to pay tribute my wife Rachel for her forbearance and encouragement along the way.

Gavin Hitchcock, Harare, December 2007

# 1 Geometry

*"Gentlemen. . .g-g-g-gentlemen, I agree with Archimedes-in-a-tub, we ought to be careful what we ask from Zeus these days. Last time we prayed for a miracle in geometry and we wound-up with Euclid and his insufferable pile of very thick thirteen volumes! Our popularity in schools took a dip thereafter"*

By the end of this topic you should be able to:

  (i)  Prove and apply the results of basic geometry.
 (ii)  Prove and apply the common base theorem of triangles.
(iii)  Prove and apply the triangle inequality.
 (iv)  Prove and apply the angle bisector theorem of a triangle.
  (v)  Prove and apply Heron's formula.
 (vi)  Extend Heron's formula to cyclic quadrilaterals.
(vii)  Prove and apply Ceva's theorem.
(viii)  Prove and apply Menelaus' theorem.
 (ix)  Prove and apply the intersecting chords theorem.
  (x)  Prove and apply Ptolemy's theorem.
 (xi)  Prove and apply Hippocrates' theorem.

The term 'geometry' is a fusion of two Greek words – *geos* and *metron*. The former means 'earth' and the latter 'measure', suggesting that geometry may have begun with problems about land distribution, reclamation (from flooding), and inheritance. Compare words such as geography, geology, from Greek *graphos* = drawing/writing, *logos* = word, reason or knowledge.

The subject was elevated to a pure deductive system from about 400 BCE, by the ancient Greeks especially, but also by cultures such as Babylonians, Indians and Chinese. The earliest 'geometry' was mainly a collection of techniques for the solution of practical problems like measuring fields, granaries, or water cisterns. Many of the most ancient Babylonian, Egyptian and Chinese problems preserved in surviving documents are concerned with the calculation of lengths, areas, and volumes. These early mathematicians invented technical terms for the most important geometrical concepts, and these names were wisely taken from everyday language. For example, the Egyptians called a rectangle a 'four-sided field', and a rectangular solid a 'four-sided container'. The Babylonians called a rectangle 'length-breadth', or 'field'.

Around 430 BCE, at the height of social and political struggles in the Greek Empire, there arose a group of critical thinkers called the 'sophists', less hampered by tradition than any previous group of learned persons. These people appreciated problems of a mathematical nature as part of a philosophical investigation of the natural and moral worlds and so set the stage for the development of a rigorous discipline of mathematics concerned with understanding rather than utility. One of these sophists was the Ionian philosopher Hippocrates of Chios (not the famous early physician whose name is preserved in the medical students' Hippocratic oath). The work of Hippocrates of Chios represents a high degree of perfection in mathematical reasoning, and he is famous for his work in an 'impractical' but theoretically valuable subject, the moon-shaped figures or crescents, bounded by two or more circular arcs, which we call 'lunes'. You will meet some of his results about lunes in this toolchest. Hippocrates probably explored these in connection with the great problem of 'squaring the circle' – the quest to construct (with compass and straight edge alone) a square provable by deductive reasoning to be equal in area to a given circle.

Hippocrates' work indicates that the mathematicians of the 'Golden Age' of Greece already had an ordered system of plane geometry, in which the principle of logical deduction from one statement to another (*apagoge*) had been fully grasped. This principle is thought to have originated especially with two men whose lives are shrouded in legend and celebrated in colourful tales: Thales and Pythagoras. Both are said to have travelled widely in the sophisticated civilisations of Mesopotamia, Persia and Egypt, and no doubt drew from already highly-developed mathematical traditions. Perhaps the best known theorem, and first significant *proven* theorem in mathematics, is what we call *Pythagoras' theorem* (see page 8), but this theorem was known to the ancient Babylonians, and was proved by the Chinese at about the same time Pythagoras and his disciples proved it. It is this distillation and culmination of all ancient geometry in Greece, and its development by the disciples of Thales and Pythagoras, that we deal with in this Toolchest. The ideas were first logically systematised and recorded in the historic *Elements of Euclid* around 300 BCE, but significant later developments took place in Alexandria and in the work of Arabic mathematicians.

Most of the basic ideas in geometry are included in school syllabi, but the fundamental notion of *proof* – so important to the ancient Greek (and other) founders of deductive mathematics, and so central to modern mathematics at the professional and research level – is often badly neglected. We will discuss many of these basic concepts in this first Toolchest, and (both to be true to the spirit of our subject, and to train the reader in the art of deductive reasoning) we will also give many proofs. Thus, you will find many logical deductions of one result from another. However, we will not attempt to present a complete deductive development in strict logical order from a few basic postulates, as Euclid does. While this is of great theoretical interest to those concerned with the logical structure of mathematics, it can be tedious to someone wanting primarily to acquire tools and skills for problem-solving. It is this we aim to provide in this book, with as much freshness and excitement and as little tedium as possible.

We can divide the geometry of interest to us roughly into two parts: the geometry of straight lines and polygons, and the geometry of circles and circular arcs. We also consider the sine and cosine rules as tools in elementary geometry, rather than trigonometry (Toolchest 5), simply because these will be familiar to many of our readers, and connections with them will be welcomed.

## 1.1  Brief reminder of basic geometry

Here, as tempting appetizers, are two typical problems of the kind you should be able to solve when you have worked through this section. You are invited to try them as soon as you like. You may find them too hard for now, but by the end of this section they should not look difficult to you. Their solutions are given at the end of the section.

**Appetizer Problem 1:**  *A square ABCD is drawn so that its side CD is tangent at E to the circle ABE. The ratio of the area of the circle ABE to the area of the square ABCD is:*

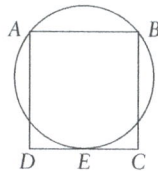

(A) $25\pi : 64$  (B) $5\pi : 8$  (C) $\sqrt{5}\pi : 2\sqrt{2}$  (D) $5 : 8$  (E) Cannot be determined

**Appetizer Problem 2:**  *A circle is inscribed in a 3-4-5 triangle, as shown in the diagram. What is the radius of the circle?*

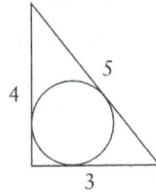

(A) 1  (B) $\frac{\sqrt{3}}{2}$  (C) $\sqrt{3}$  (D) $\frac{4}{3}$  (E) $\frac{2}{3}$

Building on the experience of solving the previous problem, we can solve this seemingly more difficult (because more general) problem.

**Appetizer Problem 3:**  *Consider a triangle with sides a, b and c such that $a^2 + b^2 = c^2$. The inscribed circle has radius r whilst the escribed (or circumscribed) circle has radius R. Find the ratio of r to R.*

(A) $\frac{2ab}{c(a+b+c)}$  (B) $\frac{ab}{c(a+b+c)}$  (C) $\frac{2ab}{a(a+b+c)}$  (D) $\frac{2ab}{b(a+b+c)}$  (E) $\frac{2ab}{a+b+c}$

### 1.1.1   Geometry of straight lines

#### Angles on a straight line

A **straight angle** is, by definition, the angle measuring two right angles, that is, 180°. When two angles lie beside each other and have a common vertex, we say they are **adjacent** to each other. Thus, in Figure 1 below, angles $P_1\hat{Q}_1R_1$ and $R_1\hat{Q}_1S_1$ are adjacent.

When a straight line stands on another straight line, two adjacent angles are formed and the sum of the two adjacent angles is two right angles. This follows from the definition above. Thus, in Figure 2 below, $P_2\hat{Q}_2S_2 + S_2\hat{Q}_2R_2 = 180°$.

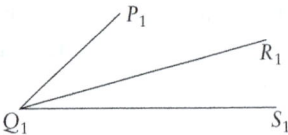

Figure 1

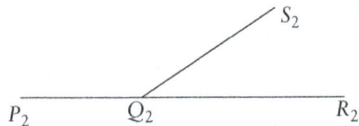

Figure 2

#### Vertically opposite angles

When two straight lines intersect, they will form four angles. Two angles opposite each other are said to be **vertically opposite**. Thus, in the figure below:

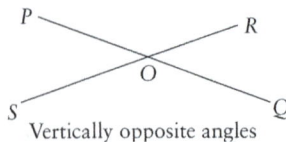

Vertically opposite angles

angles $P\hat{O}S$ and $R\hat{O}Q$ are 'vertically opposite', and so are angles $P\hat{O}R$ and $S\hat{O}Q$. What special properties do pairs of vertically opposite angles have? In the figure above, we see that

$$P\hat{O}S + Q\hat{O}S = 180° \quad \text{(or two right angles)} \tag{1.1}$$

and that

$$R\hat{O}Q + Q\hat{O}S = 180° \quad \text{(or two right angles);} \tag{1.2}$$

subtracting (1.2) from (1.1),

$$P\hat{O}S = R\hat{O}Q, \tag{1.3}$$

that is, vertically opposite angles are equal.

### Angles meeting at a point

A complete revolution measures, by definition, four right angles, that is, 360°.

When a number of lines meet at a point they will form the same number of angles, and the sum of the angles at a point is four right angles. Thus, in the figure below:

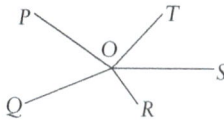

$$P\hat{O}Q + Q\hat{O}R + R\hat{O}S + S\hat{O}T + T\hat{O}P = 360° \quad \text{(or 4 right angles).}$$

### Parallel lines

If lines in a plane never meet, however far they are produced (i.e. extended) we say they are **parallel**. It is equivalent to say that any third line cutting the two lines makes equal **corresponding** angles at the two intersections. A line cutting a pair of parallel lines is called a **transversal**. The figure below shows pairs of corresponding angles (also called F-angles) for two transversals:

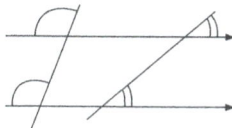

The figure below shows pairs of **alternate** angles (also called Z-angles) for two transversals. Alternate angles are equal, because the opposite angle of one is corresponding angle to the other.

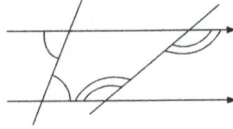

The figure below shows pairs of **allied**, or **co-interior** angles. Since the adjacent angle to angle $\alpha_1$ is alternate angle to the allied angle $\alpha_2$, it is clear that the sum of two allied angles is two right angles:

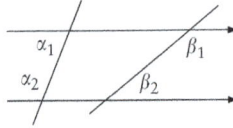

It follows from our definition of parallel lines that if a third line cutting two lines makes the sum of two allied angles on one side *less than two right angles*, then the two lines are not parallel, and must meet on that side. This property was used by Euclid (in his notorious *fifth axiom*, or *parallel axiom*) to characterize the notion of parallel lines without resorting to the idea of infinite or indefinite extension of the lines. A consequence of the parallel axiom in Euclidean geometry is this: if two lines are parallel, then 'the distance between the two lines is always the same', where by 'distance' we mean the length of the perpendicular drawn from any point on one line to the other line.

### 1.1.2 Geometry of polygons

A **polygon** is, roughly speaking, any plane figure with straight sides. However, we have to be more precise than this: $n$ points $A_1, A_2, \ldots, A_n$ form a polygon $A_1 A_2 \ldots A_n$ if the $n$ line segments $A_1 A_2, A_2 A_3, \ldots, A_{n-1} A_n, A_n A_1$ are disjoint except for the obvious common endpoints. The word 'polygon' actually comes from the Greek for 'many-angled': *poly* = many, *gonos* = angle. The former word features in many other English words: polygamy, polyglot, polyhedra, polytechnic, etc. The latter word *gonos* derives from the Greek word for knee, for obvious reasons! (It is interesting that the name *England* derives from *Angle-land* – the country of the Angles, so called because the people migrated there from a bent-leg-shaped region where modern Germany is.)

A **triangle** is a three-sided polygon (also called a 3-gon, since it has three angles too) and a **quadrilateral** is a four-sided polygon (also called a 4-gon). In general, an $n$-sided polygon is called an $n$-gon. The other polygons (5-gon, 6-gon, etc.) are similarly named using Greek words that tell us the number of sides (or angles), e.g. *penta* for five, *hexa* for six, etc. Hence we have **pentagon, hexagon, heptagon, octagon, nonagon, decagon**, etc. A **regular** polygon has all sides equal and all angles equal. NB: it is not enough to say all angles are equal (consider a non-square rectangle) nor to say all sides are equal (consider a 'flopped square', or rhombus). However, for triangles (only), each of these demands implies the other: if the angles are equal so are the sides, and *vice versa*. That is, a triangle is **equiangular** if and only if it is **equilateral**. This will be demonstrated at the end of the next section, on the geometry of the triangle.

### Interior and exterior angles of a polygon

We start off with the most basic polygon – the triangle.

(1) *The angle sum of a triangle is two right angles.*
Given any triangle $PQR$, produce $QR$ to a point $S$ and draw $RT$ parallel to $QP$. Using the lettering shown in the figure below,

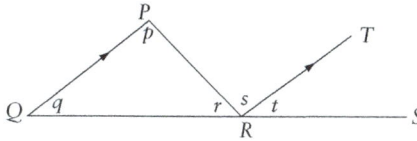

$$p = s \quad \text{(alternate angles)} \tag{1.4}$$
$$q = t \quad \text{(corresponding angles).} \tag{1.5}$$

Now,

$$r + p + q = r + s + t, \text{ from (1.4) and (1.5)}$$
$$= 2 \text{ right angles, since } QRS \text{ is a straight line.}$$

(2) *An exterior angle of a triangle is equal to the sum of the interior opposite angles.*
Given any triangle $PQR$ with $QR$ produced to $S$, we see from the diagram above that

$$P\hat{R}S + P\hat{R}Q = 2 \text{ right angles} \quad (QRS \text{ is a straight line});$$
$$\text{and} \quad p + q + P\hat{R}Q = 2 \text{ right angles} \quad \text{(angle sum of triangle } PQR);$$
$$\text{hence} \quad P\hat{R}S = p + q.$$

Note that this can also be shown by drawing the extra line $RT$ parallel to $QP$ as in the figure above, and observing that the equations (1.4) and (1.5) above yield $P\hat{R}S = s + t = p + q$.

(3) *The sum of the interior angles of any n-sided polygon is $(2n - 4)$ right angles, or $(n - 2)180°$. Hence each interior angle of a regular n-gon is $\pi(1 - \frac{2}{n})$ radians, or $180(1 - \frac{2}{n})$ degrees.*
This result is easy to prove for **convex** polygons, that is, polygons 'with no dents'. Given any convex polygon $ABCDE\ldots$ with $n$ sides as shown in the figure below, join the vertices $A, B, C, \ldots$ to some point $O$ inside the polygon.

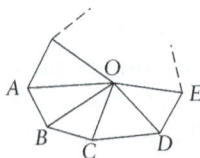

Now, by construction and using convexity, the interior of the polygon is dissected into $n$ triangles, the total sum of whose angles is $2n$ right angles. But the sum of the angles around the point $O$ is four right angles, hence the sum of the interior angles of the polygon, at the vertices $A, B, C, \ldots$, is $2n - 4$ right angles. Moreover, if it is regular, each angle is $(2 - \frac{4}{n})\frac{\pi}{2} = \pi(1 - \frac{2}{n})$ radians, or $180(1 - \frac{2}{n})$ degrees. (For discussion of radian and degree measure, see Toolchest 5.)

(4) *The sum of the exterior angles of any convex polygon (taken anticlockwise, say) is four right angles.*

Consider the $n$-sided polygon in the figure below, with each side produced:

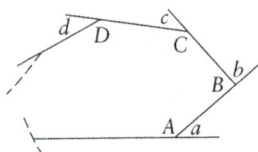

The interior angles of the polygon are $A, B, C, \ldots$, and the exterior angles are $a, b, c, \ldots$, and each corresponding (adjacent) pair sums to two right angles, so that $A + a = 180°$, $B + b = 180°$, etc. Since there are $n$ sides and $n$ vertices, we have:

$$(A + a) + (B + b) + \cdots = 180n$$
$$(A + B + C + \cdots) + (a + b + c + \cdots) = 180n$$
$$180(n - 2) + (a + b + c + \cdots) = 180n, \text{ by (3) above, hence}$$
$$(a + b + c + \cdots) = 180n - 180(n - 2) = 360°.$$

You may also see this by imagining yourself walking around the polygon anticlockwise, turning successively through each exterior angle. When you have gone right around to where you started, you will have turned through one complete revolution – 360°. Of course this result is also true for the exterior angles taken clockwise.

### 1.1.3  Geometry of the fundamental polygon – the triangle

#### Pythagoras' theorem

*In any right-angled triangle ABC, the square on the hypotenuse (long side, opposite the right angle) is equal in area to the sum of the squares on the other two sides.* That is, in the figure below,

$$AB^2 = BC^2 + AC^2, \text{ or } c^2 = a^2 + b^2.$$

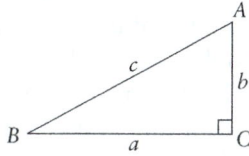

This great theorem was known to the Greeks about 500 BCE, and the Chinese about the same time, as a proven result. It was also known to the ancient Babylonians and perhaps the Egyptians, but it seems unlikely that they had a proof.

**Proof:**    This proof uses algebra. Let $PQRS$ be a square of side $a + b$ units and let $W$ be a point on PQ such that $PW = a$ units and $WQ = b$ units. Similarly for points $X$, $Y$, $Z$ on $QR$, $RS$, $SP$. The four lines joining these points give a square of side $c$, because the angles are all right angles. (To see that $X\hat{Y}Z$ is a right angle, observe that the three angles at $Y$ stand on a line, so sum to two right angles; but two of them are equal to the acute angles $\hat{X}$ and $\hat{Y}$ in the triangle $XYZ$, whose three angles also sum to two right angles. Hence $X\hat{Y}Z = X\hat{R}Y$, a given right angle.)

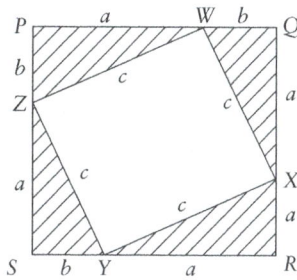

Proof of Pythagoras' theorem

Now we can find the area of the square $PQRS$ in two different ways:

1. area of $PQRS = (a + b)(a + b) = a^2 + 2ab + b^2$;

2. area of $PQRS$ = area of square $WXZY$ + area of four right
   triangles;

   $$= c^2 + 4\left(\frac{1}{2}ab\right).$$

Thus   $c^2 + 2ab = a^2 + 2ab + b^2$,

   so that   $c^2 = a^2 + b^2$.

◇

**Remark:** This algebraic method of proof, although modern students seem to find it easiest to understand, is not the method used to prove this theorem in Euclid's *Elements*. There, his first proof is based on purely deductive geometric reasoning, using Congruence Criterion (2) given at the end of this subsection, and his second proof uses the ideas of similarity which we deal with in Section 1.2. Problems 50 and 51 at the end of this Toolchest challenge you to rediscover Euclid's proofs. None of these three methods was the original approach (according to the best available evidence) by the Pythagoreans and Chinese. The first proofs arose from concrete manipulations and picturings of triangles and squares, and these should really be the first proofs that children encounter in their mathematics education. The following diagram 'says it all', but notice that the assumption that those three unshaded squares really are squares depends on the proposition that the angles of a triangle sum to a straight angle (180°), and the Euclidean postulate that all right angles are equal.

## Pythagorean and Chinese dissection proof

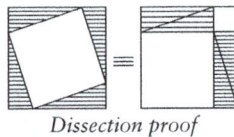

*Dissection proof*

## Converse of Pythagoras' theorem

*If, in any triangle ABC, the square on one side is equal in area to the sum of the squares on the other two sides, then the angle opposite the first side is a right angle.* That is, in Figure (a) below,

$$\text{If } c^2 = a^2 + b^2, \text{ then } \hat{C} = A\hat{C}B = 90°.$$

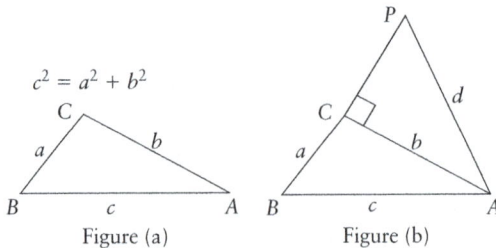

Figure (a)          Figure (b)

**Proof:** (This is close to Euclid's proof of his Proposition 1.48.) Construct another triangle $APC$ with $PC = BC = a$ and $A\hat{C}P = 90°$. Now the two triangles $ABC$ and $APC$ are congruent (that is, all sides and angles are the same), for $PC = BC$, $AC$ is common, and $AP = AB$ by the following argument. If $AP = d$, then:

$$d^2 = PC^2 + AC^2 \quad \text{(by Pythagoras)}$$
$$= a^2 + b^2 = c^2,$$

therefore   $d = c.$

Therefore, in the two congruent triangles, $A\hat{C}B = A\hat{C}P = 90°.$     ◇

At the end of this subsection, after introducing the sine and cosine rules, we shall use them to clarify exactly when we can be sure two triangles are congruent.

**The sine rule (version 1)**

*In any triangle ABC (acute or obtuse angled):*

$$\frac{a}{\sin A} = \frac{b}{\sin B} = \frac{c}{\sin C}.$$

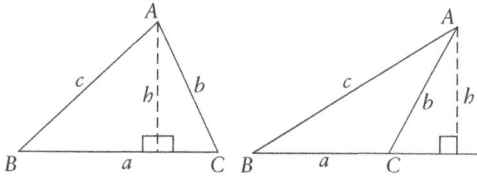

**Proof:**   Draw the perpendicular from $A$ to $BC$ (produced if necessary). In the figures above,

$$\sin B = \frac{h}{c} \tag{1.6}$$

while, in the left-hand figure,

$$\sin C = \frac{h}{b} \tag{1.7}$$

and, in the right-hand figure,

$$\sin(180 - C) = \frac{h}{b},$$

therefore    $\sin C = \dfrac{h}{b}.$

Therefore, whether acute or obtuse angled, we have from (1.6) and (1.7):

$$h = c \sin B \text{ and } h = b \sin C,$$
$$\text{so} \quad c \sin B = b \sin C,$$
$$\text{hence} \quad \frac{b}{\sin B} = \frac{c}{\sin C}.$$

Similarly, by drawing a perpendicular from $C$ to $AB$,

$$\frac{a}{\sin A} = \frac{b}{\sin B}.$$

◇

## The cosine rule

*In any triangle ABC (acute- or obtuse-angled):*

$$c^2 = a^2 + b^2 - 2ab \cos C.$$

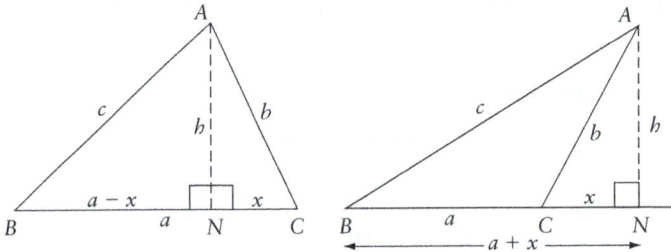

**Proof:**   Draw the perpendicular $AN$ from $A$ to $BC$ (produced if necessary) and let $CN = x$. In the left-hand figure above (with $\hat{C}$ acute),

$$
\begin{aligned}
c^2 &= (a - x)^2 + h^2 \quad \text{(Pythagoras)} \\
&= a^2 - 2ax + x^2 + h^2 \\
&= a^2 - 2ax + b^2 \quad \text{(in } \triangle ACN,\ x^2 + h^2 = b^2\text{)} \\
&= a^2 + b^2 - 2ab \cos C \quad \text{(in } \triangle ACN,\ x = b \cos C\text{)}.
\end{aligned}
$$

In the right-hand figure above (with $\hat{C}$ obtuse),

$$
\begin{aligned}
c^2 &= (a + x)^2 + h^2 \quad \text{(Pythagoras)} \\
&= a^2 + 2ax + x^2 + h^2 \\
&= a^2 + 2ax + b^2 \quad \text{(in } \triangle ACN,\ x^2 + h^2 = b^2\text{)} \\
&= a^2 + b^2 + 2a(-b \cos C) \quad \text{(in } \triangle ACN, \\
&\qquad x = b \cos A\hat{C}N = b \cos(180 - C) = b(-\cos C)) \\
&= a^2 + b^2 - 2ab \cos C.
\end{aligned}
$$

◇

Similarly,  $b^2 = a^2 + c^2 - 2ac \cos B$,  and  $a^2 = b^2 + c^2 - 2bc \cos A$. Some immediate consequences of the sine and cosine rules are:

- *An* **equiangular** *triangle must be* **equilateral**, *and conversely.*
- *An* **isosceles** *triangle (one with two sides equal) must have the two base angles equal, and conversely.*
- *If the sides of a triangle are given, this determines the angles.*

Recall that two triangles are called **congruent** if all corresponding sides and angles are equal. (Imagine them as coinciding when either is cut out and placed on top of the other.) From the previous result we have the first criterion for congruence (which we already used in proving the converse to Pythagoras' theorem):

**Congruence criterion (1): ('SSS')**   *The two triangles ABC and A'B'C' are congruent if corresponding sides are equal; that is, if a = a', b = b', c = c'.*

**Congruence criterion (2): ('SAS')**   *The two triangles ABC and A'B'C' are congruent if two corresponding sides are equal and the included angles are equal; that is, if b = b', c = c' and A = A'.*
    That $a = a'$ too follows from:

$$a^2 = b^2 + c^2 - 2bc\cos A = b'^2 + c'^2 - 2b'c'\cos A' = a'^2.$$

**Congruence criterion (3):**   *The two triangles ABC and A'B'C' are congruent if each has a right angle and two corresponding sides are equal.*

    In view of congruence criterion (2) we need only consider the case when $A = A' = 90°$ and $a = a'$, $c = c'$. (That is, the triangles have right angle, hypotenuse and a side in common.) That $b = b'$ too follows from Pythagoras' theorem:

$$b^2 = c^2 - a^2 = c'^2 - a'^2 = b'^2.$$

Before going on, convince yourself that for two triangles to have 'two sides and an angle in common' is in general insufficient for congruence. For example, construct two distinct triangles each having two sides of length 1 and one angle of 45°.

## 1.1.4   Geometry of circles and circular arcs

### Arcs and chords

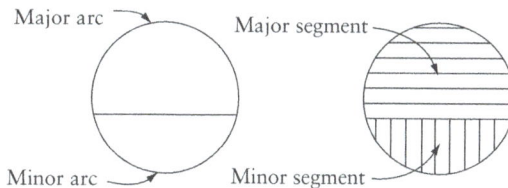

A **chord** of a circle is a straight line joining any two points on its circumference, and a chord which passes through the centre of the circle is called a **diameter**.

    A chord which is not a diameter divides the circumference into two arcs of different sizes, a **major arc** and a **minor arc** (see left-hand figure above). In addition, the chord divides the circle into two segments of different sizes, a **major segment** and a **minor segment** (see right-hand figure above). We will refer to these two arcs (respectively, segments) as being complements of each other.

In the figure on the left below, $P$, $Q$, and $R$ are points on the circumference of a circle.

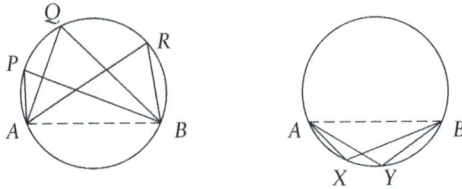

The angles $A\hat{P}B$, $A\hat{Q}B$ and $A\hat{R}B$ are said to be **subtended** at points $P, Q, R$ on the circumference by the chord $AB$, or by the minor arc $AB$. We say that the three angles are **all in the same major segment** $APQRB$.

Similarly, in the right-hand figure above, $A\hat{X}B$ and $A\hat{Y}B$ are angles subtended by the chord $AB$, or by the major arc $AB$, in the minor segment $AXYB$.

(1) *The angle which an arc of a circle subtends at the centre is twice that which it subtends at any point on the complementary arc of the circumference.*

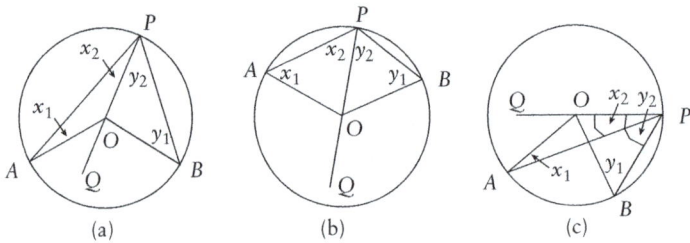

(a)                    (b)                    (c)

**Proof:** The three figures above represent the three possible cases, with the arc $AB$ subtending angle $A\hat{O}B$ at the centre and $A\hat{P}B$ at the circumference. In each figure, draw the line $PO$ and extend to any point $Q$.

$$OA = OP \quad \text{(radii),}$$

$$\text{therefore} \quad x_1 = x_2 \quad \text{(base angles of isosceles triangle),}$$

$$\text{so} \quad A\hat{O}Q = 2x_2.$$

In a similar way, $B\hat{O}Q = 2y_2$.
So, in figures (a) and (b),

$$A\hat{O}B = A\hat{O}Q + B\hat{O}Q$$
$$= 2(x_2 + y_2)$$
$$= 2 \times A\hat{P}B.$$

And, in figure (c),

$$A\hat{O}B = B\hat{O}Q - A\hat{O}Q$$
$$= 2(y_2 - x_2)$$
$$= 2 \times A\hat{P}B.$$

Hence, in every case, $A\hat{O}B = 2 \times A\hat{P}B.$    ◇

(2) *Angles in the same segment of a circle are equal.*

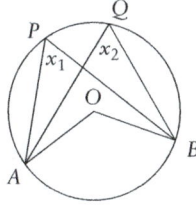

**Proof:**  Given any points $P$, $Q$ on the major arc of the circle $APQB$, join $A$ and $B$ to $O$, the centre of the circle. With the lettering in the figure above,

$$A\hat{O}B = 2x_1 \quad \text{(previous result (1))}$$
$$\text{and} \ \ A\hat{O}B = 2x_2 \quad \text{(previous result (1)),}$$
$$\text{therefore} \ \ x_1 = x_2.$$

◇

(3) *The angle in a semicircle is a right angle.*

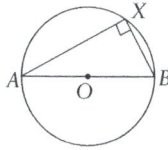

**Proof:**  In the figure above, $AB$ is a diameter of the circle, centre $O$, and $X$ is any point on the circumference of the circle.

$$A\hat{O}B = 2A\hat{X}B \quad \text{(result (1) above);}$$
$$\text{but} \ \ A\hat{O}B = 180°,$$
$$\text{therefore} \ \ A\hat{X}B = 90°.$$

◇

Before going on it is worth exploring an intuitive consequence of result (1) above. If we start with any triangle $ABC$, we easily convince ourselves that there will be a unique point $O$ on the perpendicular bisector of $AB$

which subtends an angle $\angle AOB = 2\hat{C}$. If we then draw the circle, centre $O$ and radius $OA = OB$, then it will pass through $A$ and $B$, being on their perpendicular bisector, and it will also pass through $C$. For if the line $AC$ cuts the circle at $C'$, then $\angle AC'B = 2\angle AOB = \angle ACB$, giving $C = C'$. Thus every triangle $ABC$ has a unique **circumcircle** – that is, any three points which are not all in the same straight line determine a unique circle passing through all three of them. We will demonstrate this more carefully later in this section.

Similar reasoning to that in the previous paragraph shows us that the converses of results (1)–(3) are true:

(1′) *If a point P makes $\angle APB$ equal to half the angle subtended by the arc AB of a circle at its centre, then P lies on the other arc of the circle.*

(2′) *If two points P, Q make $\angle APB = \angle AQB$, then all four points lie on the same circle.* (We study such **cyclic quadrilaterals** in the next subsection.)

(3′) *If triangle ABC has a right angle at C then AB is a diameter of the circumcircle ABC.*

## Cyclic quadrilaterals

If it is true that every triangle has a unique circumcircle, we can obviously choose a fourth point not on this circle, thus demonstrating that not every quadrilateral will have all four points lying on the circumference of a circle. A **cyclic quadrilateral** is one for which the four vertices do lie on a circle. In the figure below, $ABCD$ and $PQRS$ are cyclic quadrilaterals. Opposite angles $A, C$ of a cyclic quadrilateral $ABCD$ lie in opposite segments determined by the chord $BD$.

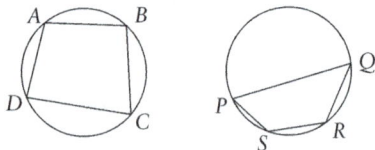

(4) *The opposite angles of a cyclic quadrilateral are supplementary.*

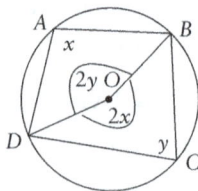

**Proof:**   With the lettering of the diagram above,

$$B\hat{O}D = 2y \quad \text{(result (2))},$$

and similarly  reflex $B\hat{O}D = 2x$;

therefore $2x + 2y = 360°$ (angles at a point),

therefore $x + y = 180°$. ◇

(5) *An exterior angle of a cyclic quadrilateral is equal to the interior opposite angle.*

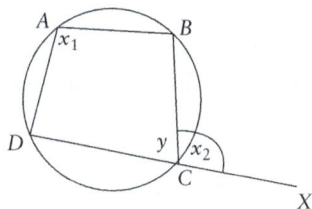

**Proof:** In the figure above,

$$x_1 + y = 180° \text{ (result (4))},$$

$$\text{and} \quad x_2 + y = 180° \text{ (angles on a straight line)},$$

$$\text{therefore} \quad x_1 = x_2. \qquad ◇$$

**Tangents to a circle**

A tangent to a circle is a line drawn to touch the circle – intersecting it at precisely one point. A tangent is perpendicular to the radius drawn to its point of contact. It follows from this that the perpendicular to a tangent at its point of contact passes through the centre of the circle.

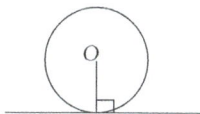

(6) *The tangents to a circle from an exterior point are equal.*

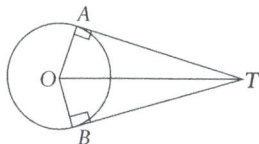

**Proof:** In the triangles $OAT$ and $OBT$, $\hat{A} = \hat{B} = 90°$, because each radius is (by definition of tangent) perpendicular to the tangent at point of contact. Also, $OA$ and $OB$ are both radii of the circle, so are equal, and $OT$ is a common side. Hence the two triangles $OAT$ and $OBT$ are congruent, from which it follows that $TA = TB$. ◇

Notice that this congruence means that the angles made by the two tangents with the line $OT$ are also equal. This leads naturally to the next idea.

## Incircle of a triangle

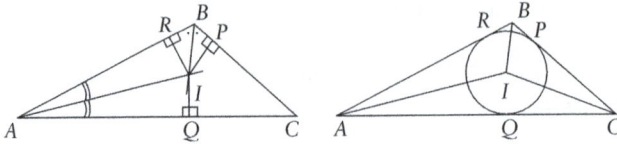

A circle which has all three sides of a triangle as tangents is called the **incircle** of the triangle. There is a unique incircle for any triangle. This is proved in a way similar to the previous result (6). Let $ABC$ be the triangle, and let the angle bisectors of $\hat{A}$ and $\hat{B}$ meet at a point $I$. Now construct the perpendiculars $IP, IQ, IR$ from $I$ to the three sides of the triangle as shown in the diagram. It is easy to see that triangles $AIR$ and $AIQ$ are congruent (right angle, angle and side) and that triangles $BIR$ and $BIP$ are congruent. This implies that the three perpendiculars are equal, and therefore the circle with $I$ as centre and $IP = IQ = IR$ as radius is the incircle of the triangle.

## Contact of circles

Two circles are said to **touch** each other at the point $T$ if they are both tangent to the same straight line $PT$ at the point $T$.

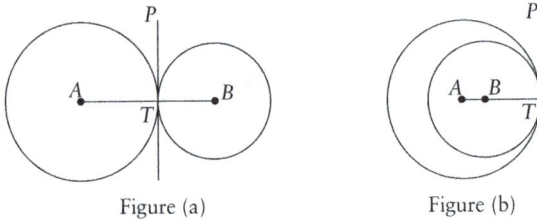

Figure (a)                          Figure (b)

In Figure (a) the two circles touch each other **externally**, and in Figure (b) they touch **internally**.

The lines $ATB$ (in Figure (a)) and $ABT$ (in Figure (b)) are straight lines, because the two radii $AT, BT$, are perpendicular to the same line $PT$. The straight line joining $A$ and $B$ in each of the figures above is called the **line of centres**. Hence, for two circles that touch each other (internally or externally), **the point of contact lies on the line of centres**. The distance between their centres is the **sum** of their radii if the circles touch externally, and the **difference** of their radii if they touch internally.

## Alternate segments

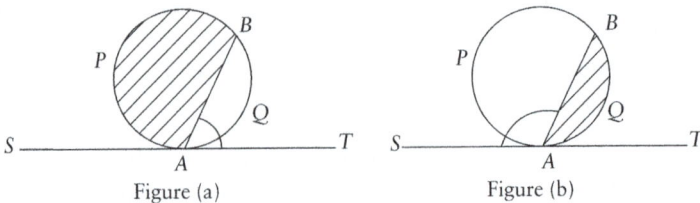

Figure (a)                          Figure (b)

In each figure above, the line *SAT* is tangent to the circle at *A*, and the chord *AB* divides the circle into two segments *APB* and *AQB*. In Figure (a), the segment *APB* is the **alternate segment** to angle $T\hat{A}B$, that is, it is on the other side of *AB* from $T\hat{A}B$. Similarly, in Figure (b), the segment *AQB* is the **alternate segment** to angle $S\hat{A}B$.

(7) *If a straight line is tangent to a circle, and from the point of contact a chord is drawn, each angle which the chord makes with the tangent is equal to the angle in the alternate segment.*

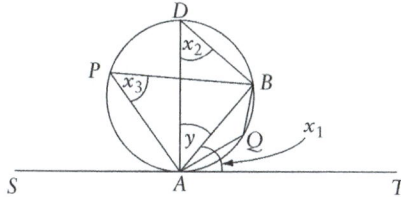

**Proof:**  Let *P* be any point on the major arc *AB*, and let *AD* be the diameter drawn from the point of contact. Let the angles $x_1$, $x_2$, $x_3$, *y* be as shown in the figure above. We have $x_1 + y = 90°$, since the tangent is perpendicular to the radius at point of contact; also, $x_2 + y = 90°$, since *AD* is a diameter (see result (3)). Hence $x_1 = x_2$. But $x_2 = x_3$, as angles in the same segment (result (2)). Therefore, $x_1 = x_3$, that is:

$$T\hat{A}B = A\hat{P}B.$$

Also,

$$S\hat{A}B = 180° - x_1$$
$$= 180° - x_3 \quad (x_1 = x_3 \text{ proved above})$$
$$= A\hat{Q}B \quad \text{(opposite angles of cyclic quad. – result (4))}.$$

◇

**Circumcircle of a triangle**

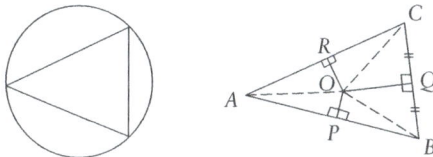

A circle which passes through the three vertices of a triangle is called the **circumcircle** of the triangle, or sometimes the **escribed** circle. There is a unique circumcircle for any triangle. Its construction and uniqueness follow from the following result:

(8) *The perpendicular bisector of a chord of a circle passes through its centre.*

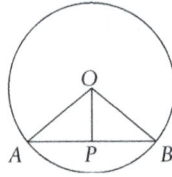

**Proof:**   The line from centre $O$ to midpoint $P$ of chord $AB$ is the perpendicular bisector, because the two triangles $AOP$ and $BOP$ in the figure above are congruent, since they have equal sides, hence $A\hat{P}O = B\hat{P}O = 90°$.   ◇

The result (8) can be re-expressed thus: the centre of any circle passing through two given points will lie on the perpendicular bisector of the line joining them. In a triangle $ABC$, let the perpendicular bisectors of two of the three sides be drawn, say $RO$ and $QO$ meeting at the point $O$, as in the right hand figure above. The triangles $A\hat{O}R$ and $C\hat{O}R$ are clearly congruent (two sides and included right angle are equal) hence $AO = CO$. Similarly, the triangles $C\hat{O}Q$ and $B\hat{O}Q$ are congruent, hence $CO = BO$. It is now clear that the circle centre $O$ and radius $AO$ will pass through all three vertices $A, B, C$ and is the circumcircle of the given triangle. It then follows from result (8) that the three perpendicular bisectors of the sides of the triangle all meet at the point $O$, which is called the **circumcentre** of the triangle. It is now possible to relate the radius of this circumcircle to the sine rule we gave earlier on page 11:

(9) *Extended sine rule:*

$$\frac{a}{\sin A} = \frac{b}{\sin B} = \frac{c}{\sin C} = 2R$$

*where $a, b, c$ are the sides of a triangle, $A, B, C$ are the corresponding points, and $R$ is the radius of the circumcircle.*

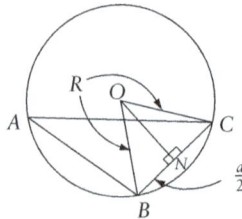

**Proof:**   Consider the triangle $ABC$ above, with its circumcircle centre $O$. Let $ON$ be the perpendicular bisector of $BC$, and let $R = OB = OC$ be the radius of the circumcircle. Now $B\hat{O}C = 2 \times \hat{A}$, being angles subtended by $BC$ at centre and circumference (see result (1)). Also, in the isosceles triangle $BOC$, $ON$ is a line of symmetry, so that $B\hat{O}N = N\hat{O}C = \hat{A}$. In triangle $BON$, we can now see that

$$\frac{a/2}{R} = \sin A,$$

$$\text{therefore} \quad \frac{a}{2R} = \sin A,$$

$$\text{so that} \quad 2R = \frac{a}{\sin A}.$$

◇

Now, before proceeding to the next section, here are the solutions to the appetizer problems with which we began this section.

**Solution of Appetizer Problem 1:**  The relationships between the various line segments are best expressed algebraically. Let the radius of the circle be $r$, and the side of the square be $x$. The dimensions of the various parts are then as labelled in the diagram, where $F$ is the centre of the circle.

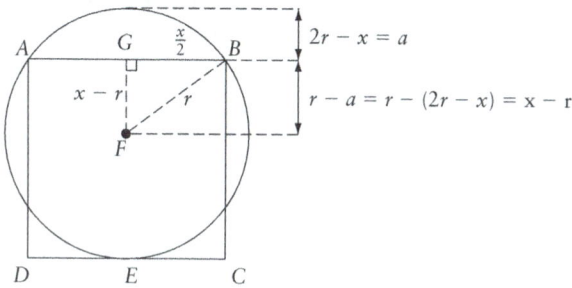

In the right-angled triangle $BFG$ above

$$(x - r)^2 + \left(\frac{x}{2}\right)^2 = r^2 \text{ (Pythagoras' theorem)},$$

$$\text{therefore} \quad \frac{5x^2}{4} - 2xr = 0,$$

$$\text{hence} \quad x = \frac{8r}{5}.$$

Therefore area of circle : area of square $= \pi r^2 : x^2$

$$= \pi r^2 : \frac{64r^2}{25}$$

$$= 25\pi : 64. \quad \text{Hence (A).}$$

**Solution of Appetizer Problem 2:**  Draw a good diagram first! (Ours is the one on the left.) Denote by $a$, $b$, $c$ respectively the pairs of (equal) lengths of the tangent lines. Because $3^2 + 4^2 = 5^2$, the converse of Pythagoras' theorem tells us we have a right-angled triangle. Draw the lines from the incentre to the points of tangency. We know that these meet at right angles, so that a square of side $c$ is formed at the right angle, and hence $r = c$ is the required radius. Solving the equations $a+b = 5, b+c = 4, a+c = 3$ gives $a = 2, b = 3, c = 1$. Hence (A).

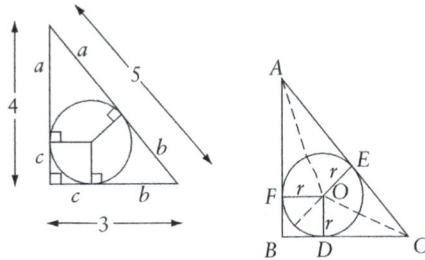

Alternatively we could use the diagram on the right, drawing the lines $OA, OB, OC$ from the incentre $O$ to the points $A, B, C$, and the perpendiculars $OD, OE, OF$ from the incentre to the points $D, E, F$ of tangency. Let the common length of these perpendiculars (the unknown radius) be $r$. Area of triangle $ABC$ = area of triangle $BOC$+ area of triangle $AOC$+ area of triangle $AOB$. The lines $OD, OE, OF$ are the altitudes of these three triangles, hence

$$\frac{1}{2} \times 3 \times 4 = \frac{4r}{2} + \frac{3r}{2} + \frac{5r}{2},$$

therefore    $6r = 6,$    giving    $r = 1$, as before.

**Solution of Appetizer Problem 3:**    Draw a diagram for each of inscribed and escribed circles. Since we are given $a^2 + b^2 = c^2$, the triangle $ABC$ shown in each of the diagrams is right-angled at $C$ by the converse of Pythagoras' theorem.

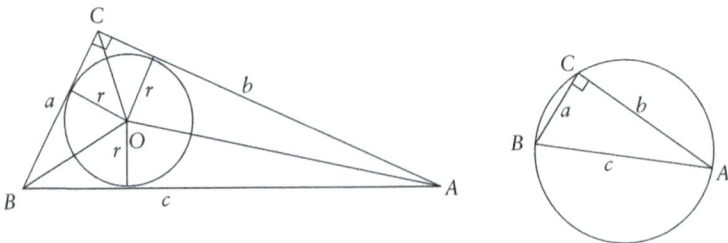

Consider the left-hand diagram, in which the centre of the inscribed circle is $O$ and the perpendiculars have been drawn from $O$ to the points of tangency. The area of the triangle is given by $T = \frac{ab}{2}$. To find the radius, r, of the inscribed circle, follow the second method of the previous problem.

Observe that the triangle is made up of three smaller triangles:

$$\text{area } \triangle ABC = \text{area } \triangle AOB + \text{area } \triangle AOC + \text{area } \triangle BOC$$

$$= \frac{1}{2}cr + \frac{1}{2}br + \frac{1}{2}ar$$

$$= \frac{r}{2}(a+b+c) = T = \frac{ab}{2},$$

$$\text{giving} \quad r = \frac{ab}{a+b+c}.$$

Now, to find the radius $R$ of the escribed circle, use the right-hand diagram. We know that $B\hat{C}A = 90°$, hence $BA$ is a diameter of circle $BAC$, so that $R = \frac{c}{2}$. Hence

$$\frac{r}{R} = \frac{ab}{a+b+c} \times \frac{2}{c}$$

$$= \frac{2ab}{c(a+b+c)}. \quad \text{Hence (A).}$$

## 1.2   Advanced geometry of the triangle

Here, again, as a tempting appetizer is a typical problem of the kind you should be able to solve when you have worked through this section. You are invited to try it now if you wish. You may find it hard, but by the end of this section it should look much easier to you. The solution is given at the end of the section.

**Appetizer Problem:**   *In the figure below, BX : XC = 2 : 3 and CY : YA = 1 : 2. If the area of the triangle COY is a and that of COX is 3b what is the area of the quadrilateral OXBZ?*

(A) $\frac{9b}{2}$  (B) $2a + 3b$  (C) $\frac{5a}{2} + 2b$  (D) $\frac{3a}{2} + 3b$  (E) $a + 3b$

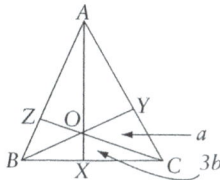

The idea of similar figures (similar triangles being a particular example) is a crucial mathematical concept which, as we shall see shortly, is an 'open sesame', or golden key, to many useful results.

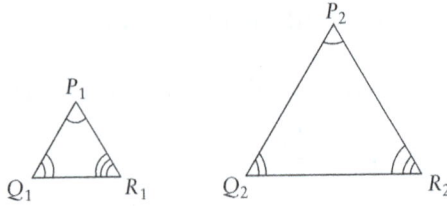

If, as in the figure, two triangles $P_1Q_1R_1$, $P_2Q_2R_2$ have corresponding angles equal, while the sides may have different sizes, they are called **similar** triangles. It is a basic fact (or axiom) of Euclidean geometry that if triangles $P_1Q_1R_1$, $P_2Q_2R_2$ are similar, then corresponding sides are 'proportional'; that is, they are in a fixed ratio. Thus the only difference between the two triangles is in scale. It follows that

$$\frac{P_1Q_1}{P_2Q_2} = \frac{Q_1R_1}{Q_2R_2} = \frac{P_1R_1}{P_2R_2}. \tag{1.8}$$

(1.8) is called a **ratio statement** and expresses the fact that in triangles $P_1Q_1R_1$ and $P_2Q_2R_2$, the ratio of the corresponding sides is fixed. If this ratio is 1, the triangles are of course congruent.

The converse also applies: that is, if for a pair of triangles it is given that the ratio of the corresponding sides is fixed, then it follows that the triangles have corresponding angles equal, and so they are similar. There is another situation that allows us to deduce similarity of triangles: when just two sides are proportional and the included angles are equal:

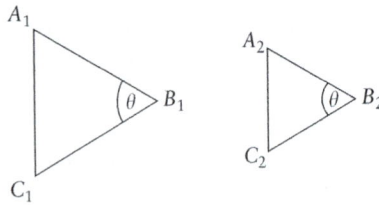

If, for the triangles $A_1B_1C_1$ and $A_2B_2C_2$, it is given that

$$\frac{A_1B_1}{B_1C_1} = \frac{A_2B_2}{B_2C_2}, \tag{1.9}$$

and that

$$A_1\hat{B}_1C_1 = A_2\hat{B}_2C_2 = \theta \quad \text{(the included angle)} \tag{1.10}$$

then the two triangles $A_1B_1C_1$ and $A_2B_2C_2$ are similar.

Note that we could write the requirement (1.9) equivalently as

$$\frac{A_1B_1}{A_2B_2} = \frac{B_1C_1}{B_2C_2}.$$

Simple yet very useful results arise from these ideas:

**Theorem 1**    *The line joining the midpoints of two sides of a triangle is parallel to the third side and equal to half of it, that is:*
*if, in triangle ABC below, AP = PB and AQ = QC, then PQ ∥ BC and* $\frac{PQ}{BC} = \frac{1}{2}$.

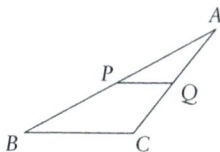

**Proof:**   In triangles $APQ$ and $ABC$ the angle $P\hat{A}Q$ is common. Also, $\frac{AP}{AB} = \frac{AQ}{AC} = \frac{1}{2}$ since $P$ and $Q$ are given to be the midpoints of $AB$ and $AC$ respectively. Hence $APQ$ and $ABC$ are similar triangles. Now the ratio statement for this pair gives:

$$\frac{PQ}{BC} = \frac{AP}{AB} = \frac{AQ}{AC} = \frac{1}{2}, \tag{1.11}$$

which proves one part of the theorem. For the other part, $PQ \parallel BC$ (that is, $PQ$ is parallel to $BC$) follows from the fact that $APQ$ and $ABC$, being similar, have corresponding angles equal, so that $A\hat{P}Q = A\hat{B}C$.    ◇

The result in Theorem 1 is frequently the basis of Olympiad problems. It is a special case of the more general result (referring to the same diagram) which the reader is challenged to prove:

**Theorem 2**    *Let P, Q be any points on the respective sides AB, AC of triangle ABC. Then PQ ∥ BC if and only if APQ and ABC are similar triangles. (This means that from each we can deduce the other; i.e. they are logically equivalent.)*

Another important result follows from the idea of similar triangles. We call a collection of lines **concurrent** if they all pass through the same point.

**Theorem 3**    *The three medians (lines joining vertices to the midpoints of the opposite sides) of a triangle are concurrent, and trisect each other at the common point of intersection. That is:*
*if, in triangle PQR below, P′, Q′ and R′ are the midpoints of QR, PR and PQ respectively, then there is a common point of intersection of PP′, QQ′ and RR′, which we call G, and*

$$PG : P'G = QG : Q'G = RG : R'G = 2 : 1. \tag{1.12}$$

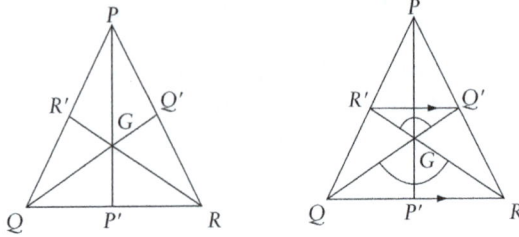

**Proof:**   In triangle $PQR$, since $R'$ and $Q'$ are the midpoints of $PQ$ and $PR$ respectively, Theorem 1 gives $R'Q' \parallel RQ$ and $RQ = 2R'Q'$. Now (referring to the right-hand diagram) in triangles $GR'Q'$ and $GRQ$,

$$R\hat{G}Q = R'\hat{G}Q' \quad \text{(vertically opposite angles)}$$

and

$$Q\hat{R}G = Q'\hat{R}'G \quad \text{(alternate angles since } R'Q' \parallel RQ).$$

Hence $GR'Q'$ and $GRQ$ are similar, so that the ratio statement gives

$$\frac{GR'}{GR} = \frac{R'Q'}{RQ} = \frac{GQ'}{GQ} \quad \text{hence} \quad \frac{GR'}{GR} = \frac{1}{2},$$

$$GR' : GR = 1 : 2 \text{ or equivalently, } GR : GR' = 2 : 1.$$

In a similar manner, it can be shown that $PG : P'G = 2 : 1$ and $QG : Q'G = 2 : 1$ so that each pair of medians intersects at the same point, and (1.12) holds. It is easy to show that the converse is also true: if $G$ is a point dividing one median in ratio $2 : 1$ then the lines joining the other vertices to $G$ will be medians.                    ◇

**Example:**   Triangle $ABC$ is an equilateral triangle with perimeter 3 cm. Five points are marked in the area bounded by the line $AB, BC$ and $AC$ (points may also be marked on any of the boundary lines). The probability of finding two points within 0.5 cm of each other is

(A) 1   (B) 0.5   (C) 0.2   (D) 0.3   (E) cannot be determined

**Solution:**

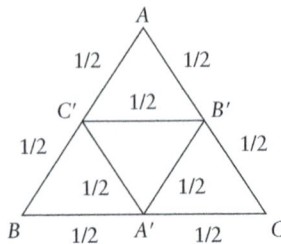

The triangle is shown above with medians $AA'$, $BB'$ and $CC'$. Using Theorem 1, we see that

$$B'C' = A'C' = A'B' = \frac{1}{2}.$$

Now we have four smaller congruent equilateral triangles making up the larger triangle as shown above.

We are now going to use the **pigeon-hole principle** (see Toolchest 8 for more): if $k+1$ letters are posted into $k$ pigeon-holes, then at least two letters will share the same hole. An easy example: among three normal human beings at least two will be of the same sex. By the pigeon-hole principle, we see that, in our problem, there will be one triangle containing more than one point (inside or on the boundary).

The furthest two such points can get from each other is 0.5 cm, so that we are guaranteed of a pair such that the distance between the points is at most 0.5 cm. The required probability is therefore 1, hence (A).

**Example:**   $ABC$ is a triangle with $AC = 9$, $BC = 40$ and $AB = 41$. The radius of the circle passing through the midpoints of $AC$ and $BC$ and the point $C$ is

(A) 10   (B) 9   (C) 8.2   (D) 10.25   (E) 12.5

**Solution:**   Observe that $40^2 + 9^2 = 41^2$ so that $ABC$ is right angled at $C$, using the converse of Pythagoras' theorem.

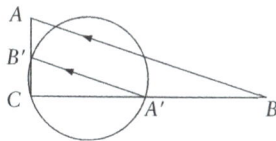

Applying Theorem 1 we have   $A'B' = \dfrac{41}{2}$.

But since $B'\hat{C}A' = 90°$, $A'B'$ is the diameter of the required circle, whose radius is therefore $\frac{1}{2}\left(\frac{41}{2}\right) = 10.25$, so that the correct answer is (D).

**Theorem 4   (The common altitude theorem)** *If two triangles $ABC$, $ADE$ have a common altitude $AF$, then*

$$\textit{Area } \triangle ABC : \textit{Area } \triangle ADE = BC : DE.$$

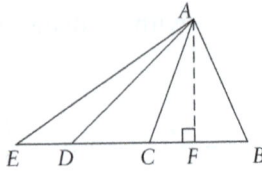

That is, the areas are in the same proportion as the bases. The proof of this is easy, following from the definition of area of a triangle as 1/2 base × altitude.

**Theorem 5    (The common base theorem)** *Two triangles ABC, A'BC have a common base BC. If the line AA' joining their vertices meets the base BC at P, then*

$$Area \triangle ABC : Area \triangle A'BC = AP : A'P.$$

**Proof:** We give proof only for the case when $A'$ is inside triangle $ABC$. The alternative – that $A'$ is on or outside the triangle – breaks up into a number of different possibilities, but the proofs are very similar, and are left as an exercise for the reader.

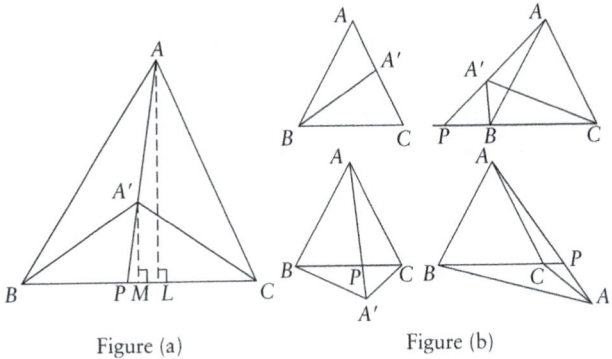

Case (i): A' inside △ABC          Case (ii): A' outside △ABC

Figure (a)                              Figure (b)

In Figure (a) above

$$\frac{\text{Area of triangle } ABC}{\text{Area of triangle } A'BC} = \frac{\frac{1}{2} \cdot BC \cdot AL}{\frac{1}{2} \cdot BC \cdot A'M}$$

$$= \frac{AL}{A'M}.$$

But triangles $ALP$ and $A'MP$ are similar (do you see why?) so that

$$\frac{AL}{A'M} = \frac{LP}{MP} = \frac{AP}{A'P}, \quad \text{and the result follows.}$$

◇

**Theorem 6**  (**The angle bisector theorem of a triangle**) *If, in triangle ABC, D lies on AC and BD bisects angle AB̂C, then*

$$\frac{AB}{BC} = \frac{AD}{DC}.$$

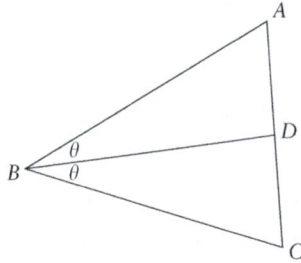

**Proof:**  Using Theorem 5, with common base $BD$,

$$\frac{\text{Area of triangle } ABD}{\text{Area of triangle } DBC} = \frac{AD}{DC}. \tag{1.13}$$

Also,

$$\frac{\text{Area of triangle } ABD}{\text{Area of triangle } DBC} = \frac{\frac{1}{2} \cdot AB \cdot BD \sin\theta}{\frac{1}{2} \cdot BC \cdot BD \sin\theta} = \frac{AB}{BC} \tag{1.14}$$

where $\theta = A\hat{B}D = D\hat{B}C$.

From (1.13) and (1.14), the result then follows.                    ◇

**Example:**  In the diagram below $ABC$ is a right angled triangle and $BD$ an angle bisector. If $AB = 3$ cm and the area of triangle $ABD$ is 9 cm², what is the length $DC$?

(A) 3 cm   (B) 2 cm   (C) 6 cm   (D) 1 cm   (E) Cannot be determined from given information

**Solution:**

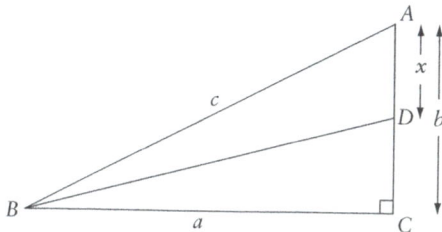

Let the area of $ABD$ be $A$. Then $A = \frac{ax}{2}$, hence $a^2x^2 = 4A^2$.

Now, using the angle bisector theorem:

$$\frac{AB}{BC} = \frac{AD}{DC},$$

$$\text{hence} \quad \frac{AB^2}{BC^2} = \frac{AD^2}{DC^2},$$

$$\text{therefore} \quad \frac{c^2}{a^2} = \frac{x^2}{DC^2},$$

$$\text{and so} \quad DC = \frac{2A}{c} = \frac{2 \times 9}{3} = 6 \text{ cm}. \quad \text{Hence (C).}$$

Our next main result, the triangle inequality, is one of those curious creatures in the mathematical zoo that seems ridiculously obvious but isn't so easy to prove without smuggling in unproved assumptions (like the corollary below). But it is worth displaying a proof here, because the exercise of 'forgetting the obvious' and reasoning logically from what has already been proved or postulated is good training for problem solving. We will approach the triangle inequality by proving a preliminary lemma, and we will base our proof on the very first formal result in this Toolchest: the angle sum of a triangle is two right angles.

**Lemma 1**   In a right-angled triangle, the hypotenuse (side opposite the right angle) is greater than either of the other two sides.

**Proof:**

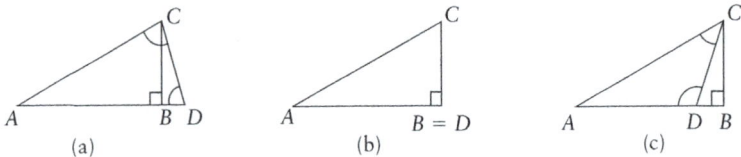

(a)          (b)          (c)

Let the triangle be $ABC$ with right angle at $B$ and let $D$ be the point on one of the sides, say $AB$ (extended if necessary), such that $AD = AC$. There are precisely three possibilities, shown in the three diagrams. Either (a) $AD > AB$, which is what we want to prove; or (b) $AD = AB$, with $B = D$; or (c) $AD < AB$. In case (b), the triangle would be isosceles, hence have two base angles each right angles, so the sum of the three angles would exceed two right angles, contradicting a basic fact about triangles. In case (c) we would have $ADC$ isosceles, hence

$$A\hat{C}D = A\hat{D}C = A\hat{B}C + B\hat{C}D > A\hat{B}C = 90°,$$

because $A\hat{D}C$ is an exterior angle of triangle $BCD$. As before, this implies the false proposition that the triangle $ACD$ has angle sum greater than two right angles. We are left with only (a).                    ◇

**Theorem 7**   (**Triangle Inequality**) *The sum of the lengths of any two sides of a triangle is greater than the length of the third side.*

**Proof:**

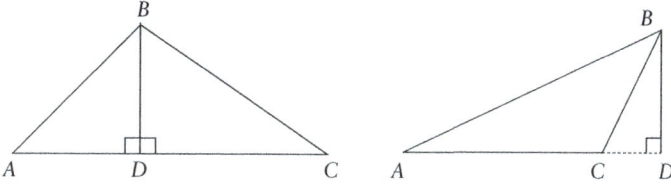

Let the triangle be $ABC$ and construct the perpendicular from $B$ to $AC$. Either $B$ lies in between $A$ and $C$, as in diagram (a) or it does not, as in diagram (b). Now, using the lemma,

$$\text{Case (a):}\quad AB + BC > AD + DC = AC.$$
$$\text{Case (b):}\quad AB + BC > AD + CD > AC.$$

Similarly, $AB + AC > BC$ and $AC + BC > AB$.          ◇

**Corollary:**   In any triangle, the larger of two sides is opposite the larger of the corresponding angles.

**Proof:**

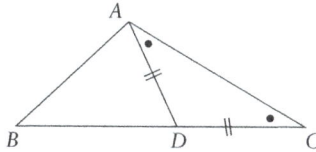

In the figure, we assume $\hat{C} < \hat{A}$. Construct $AD$ to meet side $BC$ of the triangle at $D$, such that $D\hat{A}C = D\hat{C}A$, using the assumption. Now triangle $ACD$ is isosceles so that $AD = DC$. Hence, applying the triangle inequality in triangle $ABD$:

$$AB < BD + AD$$
$$= BD + DC = BC$$
$$\text{therefore}\quad AB < BC.$$

◇

*Hello firebrigade? Yes, my cat is sitting on such a high pole that, according to the triangle inequality, my ladder is not long enough to form the hypotenuse to a suitable right-angled triangle. I desperately need a ladder long enough to allow the definition of a suitable triangle.*

**Example:**

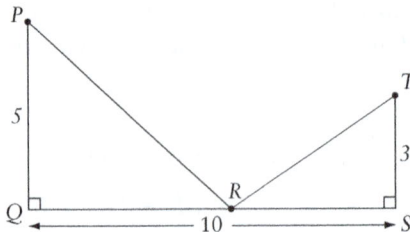

A person wishes to move from point $P$ to point $T$ via a point $R$ on line $QS$. The lines $PQ$ and $TS$ are perpendicular to $QS$, and $PR$, $RT$ are straight lines. We are given that $PQ = 5$ km, $TS = 3$ km and $QS = 10$ km. What is the distance $QR$ that makes the total distance travelled ($PR + RT$) minimum?

(A) 6.25 km    (B) 9.25 km    (C) 3.25 km    (D) 8 km    (E) 7 km

**Solution:**

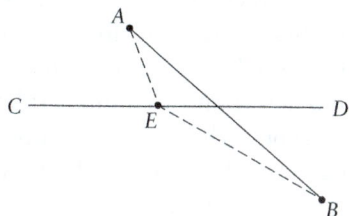

If you have two points $A$ and $B$ on different sides of a line $CD$, then the shortest distance between the two points is $AB$, the line that joins them.

This is equivalent to saying that the sum of two sides of a triangle $(AE + BE)$ is greater than the third, since any other way of moving from $A$ to $B$ forms a triangle, and this is just the triangle inequality (Theorem 7).

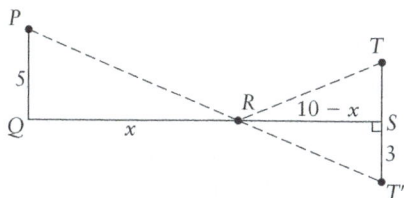

Now reflect $TS$ through $QS$ to obtain $T'$. Join $PT'$, the point in which $PT'$ intersects $QS$ is the point on $QS$ to which the person should walk. This can easily be seen from the fact that $PR + RT = PR + RT' = PT'$. Note that we are using congruence of triangles $RTS$ and $RT'S$.

Now triangles $QPR$ and $ST'R$ are similar so that

$$\frac{x}{5} = \frac{10 - x}{3}$$

therefore    $3x = 5(10 - x),$

so    $x = 6.25.$   Hence (A).

From Euclid (about 300 BCE) until the fourth century after Christ, the centre of operations of the Greek mathematicians was no longer Athens but Alexandria on the North coast of Africa, named after Alexander the Great. A great Library and Museum (meaning 'Temple of the Muses') was built there, and scholars would come from far and wide to meet and study. After Alexander's death in 323 BCE his Macedonian general Ptolemy I Soter ruled Egypt, and he and his successors went to great lengths to attract the best scholars and to obtain, by fair means or foul, the best manuscripts from throughout the Greek world. It was a kind of early University, where the eminent Fellows received good pay and exemption from taxes. One of the chief librarians was Eratosthenes (276–194 BCE), a friend and correspondent of the great Archimedes (c.287–212 BCE). Eratosthenes worked out a remarkably good estimate for the radius of the Earth, and is also famous for 'Eratosthenes' sieve' for prime numbers (see Toolchest 4, Section 1). Other great mathematicians who studied and taught there were: Apollonius (250–175 BCE), who wrote the first systematic and comprehensive work on the *Conics*; Menelaus (c.100 CE), who wrote the *Spherica* – the earliest known work on spherical trigonometry; and Claudius Ptolemy (c.100–178 CE, no relation of the kings of Egypt), who wrote an important *Geography* containing the earliest discussion of longitude, latitude, and projections for map-making. Ptolemy wrote a very influential *Mathematical Collection* in 13 books, containing a complete mathematical description (one of the first **mathematical models**) of the Greek model for the Earth-centred system of Sun, Moon and planets, including all the plane and spherical trigonometry necessary. This work was respectfully called by later Islamic scientists *al-magisti*, meaning 'the greatest', and the name Ptolemy's *Almagest* has stuck! We shall encounter 'Menelaus' theorem' and 'Ptolemy's theorem' later in this Toolchest. In other Toolchests will appear the name of Diophantus, who (followed by two more mathematicians: Theon, and his daughter Hypatia) lived nearer the end of the Alexandrian period, when political and social instability would bring the total destruction of the Library and the consequent rise of Baghdad as the new intellectual centre of the region. Another important mathematician, who flourished in Alexandria in the latter part of the first century CE, was Heron. He wrote many works, included the *Dioptra*, about applications of similar triangles to determining heights and distances, and digging tunnels through mountains by starting at both ends! His *Catoptrica* has applications of geometry to optics, his *Mechanics* includes the parallelogram of velocities, and his *Metrica* is a handbook of practical mensuration – calculating areas and volumes. In this work appears the following famous result:

**Theorem 8  (Heron's formula)** *For a triangle ABC, with the usual notation,*

$$\text{area of triangle } ABC = \sqrt{s(s-a)(s-b)(s-c)}$$

*where s, the semiperimeter, is given by* $s = \dfrac{a+b+c}{2}.$

**Proof:**

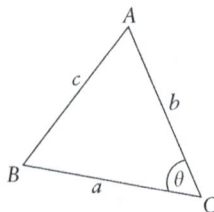

Let the symbol $\Delta$ denote the area of triangle $ABC$. Then

$$\Delta = \frac{1}{2}ab\sin\theta$$

therefore   $\dfrac{2\Delta}{ab} = \sin\theta,$

so that   $\sin^2\theta = \dfrac{4\Delta^2}{a^2b^2}.$

Using the result that $\cos^2 A + \sin^2 A = 1$ for any angle $A$, we have

$$\cos^2\theta = 1 - \frac{4\Delta^2}{a^2b^2}. \tag{1.15}$$

Using the cosine rule $c^2 = a^2 + b^2 - 2ab\cos\theta$, we have

$$\frac{a^2 + b^2 - c^2}{2ab} = \cos\theta,$$

therefore   $\left(\dfrac{a^2 + b^2 - c^2}{2ab}\right)^2 = \cos^2\theta = 1 - \dfrac{4\Delta^2}{a^2b^2}$   by (1.15),

hence   $(a^2 + b^2 - c^2)^2 = 4a^2b^2 - 16\Delta^2.$

Hence   $(4\Delta)^2 = (2ab)^2 - (a^2 + b^2 - c^2)^2$
$$= (2ab + a^2 + b^2 - c^2)(2ab + c^2 - a^2 - b^2)$$
$$= ((a+b)^2 - c^2)(c^2 - (a^2 + b^2 - 2ab))$$
$$= (a+b+c)(a+b-c)(c^2 - (a-b)^2)$$
$$= (a+b+c)(a+b-c)(c+a-b)(c+b-a) \tag{1.16}$$

and, since $s = \dfrac{a+b+c}{2}$, we have $2s = a+b+c$, hence

$2s - 2a = a+b+c - 2a = b+c - a,$ and similarly,

$2s - 2b = a+c - b$

$2s - 2c = a+b - c.$

Therefore (1.16) can be written as

$$4^2\Delta^2 = (2s)(2s - 2c)(2s - 2b)(2s - 2a),$$

therefore   $2^4\Delta^2 = 2^4 s(s-c)(s-b)(s-a),$

hence   $\Delta^2 = s(s-a)(s-b)(s-c),$

so that   $\Delta = \sqrt{s(s-a)(s-b)(s-c)},$   as required.

◇

**Example:**  A circle is drawn through the vertices $P, Q$ and $R$ of a triangle $PQR$ with $PQ = 4\,\text{cm}, QR = 6\,\text{cm}$ and $PR = 8\,\text{cm}$ as shown below. What is the radius of the circle?

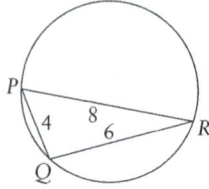

(A) $\frac{16}{15}\sqrt{15}\,\text{cm}$   (B) $\frac{1}{2}\sqrt{16}\,\text{cm}$   (C) $10\,\text{cm}$   (D) $4\sqrt{6}\,\text{cm}$   (E) $18\,\text{cm}$

**Solution:**   Using the extended sine rule

$$\frac{p}{\sin P} = \frac{q}{\sin Q} = \frac{r}{\sin R} = 2M$$

where $M$ is the radius of the circumcircle (we choose not to use the standard $R$ here in order to avoid confusion with the point $R$). Also, $\frac{1}{2}pq \sin R = \Delta$, where $\Delta$ is the area of triangle $PQR$, hence

$$\sin R = \frac{2\Delta}{pq} \quad \text{and} \quad \frac{r}{\sin R} = 2M, \quad \text{giving} \quad \frac{r}{\frac{2\Delta}{pq}} = 2M \quad \text{hence} \quad M = \frac{pqr}{4\Delta}.$$

Next, putting $s = \dfrac{4 + 6 + 8}{2} = 9$ in Heron's formula,

$$\Delta = \sqrt{9(9-4)(9-6)(9-8)} = 3\sqrt{15}.$$

Therefore   $M = \dfrac{pqr}{4\Delta} = \dfrac{8 \times 6 \times 4}{4 \times 3\sqrt{15}} = \dfrac{16}{15}\sqrt{15}.$   Hence (A).

Can we extend this result to quadrilaterals? A formula which is similar to Heron's formula was found, for the special case of a cyclic quadrilateral, by Indian mathematician Brahmagupta, born in 598 CE in northwestern India. This yields a very quick way of evaluating the area of a cyclic quadrilateral and has advantages similar to those of Heron's formula.

**Theorem 9**   (Brahmagupta's formula, 7th century CE)

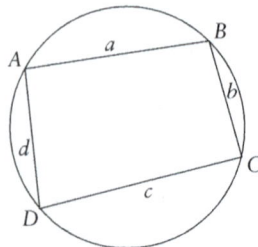

*In any cyclic quadrilateral ABCD, with sides a, b, c, d, the area is given by:*

$$\sqrt{(s-a)(s-b)(s-c)(s-d)}, \quad where \ \ s = \frac{a+b+c+d}{2}.$$

**Proof:**

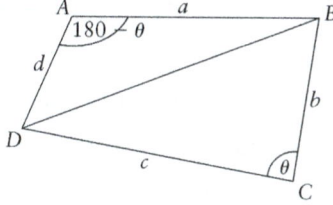

In the quadrilateral $ABCD$ above, let angle $B\hat{C}D = \theta$. Then

angle $D\hat{A}B = (180 - \theta)$, since $ABCD$ is cyclic, therefore

area $\triangle ABD = \frac{1}{2}ad\sin(180-\theta) = \frac{1}{2}ad\sin\theta = \Delta_1$, say, and

area $\triangle BCD = \frac{1}{2}bc\sin\theta = \Delta_2$, say.

Now,

$$area\ ABCD = \Delta_3 = \Delta_1 + \Delta_2 = \frac{1}{2}\sin\theta(ad+bc). \quad (1.17)$$

By the cosine rule, applied to triangle $ABD$,

$$BD^2 = a^2 + d^2 - 2ad\cos(180-\theta) = a^2 + d^2 + 2ad\cos\theta,$$

and, using triangle $BDC$, we also have: $BD^2 = b^2 + c^2 - 2bc\cos\theta$. Hence

$$b^2 + c^2 - 2bc\cos\theta = a^2 + d^2 + 2ad\cos\theta$$

therefore $\quad \dfrac{b^2 + c^2 - a^2 - d^2}{2} = (ad+bc)\cos\theta.$

Squaring gives

$$\frac{(b^2 + c^2 - a^2 - d^2)^2}{4} = (ad+bc)^2\cos^2\theta. \quad (1.18)$$

Squaring (1.17) gives

$$4\Delta_3^2 = (ad+bc)^2\sin^2\theta. \quad (1.19)$$

Since $\cos^2\theta + \sin^2\theta = 1$, adding (1.18) and (1.19) gives

$$\frac{(b^2 + c^2 - a^2 - d^2)^2}{4} + 4\Delta_3^2 = (ad+bc)^2.$$

$$\text{Therefore } 4\Delta_3^2 = (ad+bc)^2 - \frac{(b^2+c^2-a^2-d^2)^2}{4},$$

$$\begin{aligned} \text{hence } 16\Delta_3^2 &= (2ad+2bc)^2 - (b^2+c^2-a^2-d^2)^2 \\ &= (2ad+2bc-c^2-b^2+a^2+d^2)^2(2ad+2bc \\ &\quad + b^2+c^2-a^2-d^2) \\ &= ((a+d)^2-(b-c)^2)((b+c)^2-(a-d)^2) \\ &= (a+d-b+c)(a+d+b-c)(b+c+d-a)(b+c+a-d). \end{aligned}$$

$$(1.20)$$

Now $\dfrac{a+b+c+d}{2} = s$ yields $a+b+c+d = 2s$,

$$\begin{aligned} \text{so that } a+d-b+c &= 2s-2b \\ a+d+b-c &= 2s-2c \\ b+c+d-a &= 2s-2a \\ b+c+a-d &= 2s-2d. \end{aligned}$$

Hence the right-hand side of equation (1.20) may be written as $(2s-2b)$ $(2s-2c)(2s-2a)(2s-2d)$, so we have, finally,

$$16\Delta_3^2 = 2^4(s-a)(s-b)(s-c)(s-d),$$

$$\text{hence} \quad \Delta_3 = \sqrt{(s-a)(s-b)(s-c)(s-d)}.$$

$\diamond$

Brahmagupta (together with another Indian mathematician Aryabhata who lived about a century earlier) contributed to developing extensive sine tables. It is interesting to reflect upon one of the stories concerning their influence on the derivation of our modern word 'sine': The Sanskrit word *jya-ardha* means chord-half, and was often shortened to *jya* or *jiva*. When the Arabs translated the Hindu works into Arabic they simply called it *jiba*, which in Arabic is written without vowels, as *jb*, so confusing later writers into thinking the word was *jaib*, meaning bosom or breast. When Arabic trigonometry was then translated into Latin around the 12th century, the translators used the Latin word for bosom: *sinus*, which eventually became our *sine*! This story may be partly guesswork, but it serves to illustrate the fact that modern mathematics has descended to us through the contributions of many cultures over long periods of time!

**Theorem 10** (**Ceva's theorem**) *If three straight lines through the vertices P, Q and R of a triangle PQR are concurrent at G and meet the opposite sides of the triangle at M, N and L respectively, then:*

$$\frac{PN}{NR} \times \frac{RM}{MQ} \times \frac{QL}{LP} = 1.$$

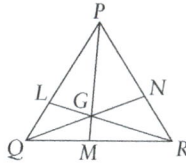

**Proof:**   In the figure, consider triangles $PGQ$ and $QGR$. These have a common base $QG$ and so, by the common base theorem:

$$\frac{\text{Area } PGQ}{\text{Area } QGR} = \frac{PN}{NR}.$$

Similarly:  $\dfrac{RM}{QM} = \dfrac{\text{Area } PGR}{\text{Area } PGQ},$

$$\text{and}\quad \frac{QL}{PL} = \frac{\text{Area } QGR}{\text{Area } PGR},$$

$$\text{hence}\quad \frac{PN}{NR} \times \frac{RM}{QM} \times \frac{QL}{PL} = \frac{\text{Area } PGQ}{\text{Area } QGR} \times \frac{\text{Area } PGR}{\text{Area } PGQ} \times \frac{\text{Area } QGR}{\text{Area } PGR}$$

$$= 1.$$

◇

This theorem, together with Menelaus' theorem (11), was published by the Italian Giovanni Ceva in 1678. The converse of Ceva's theorem is also true but we shall not prove it here.

**Example:**   In an Olympiad Georgina was challenged to bisect a line using an ungraduated ruler, a compass and pencil only. This is how she proceeded:

**Step 1**
She used the compass, in the standard way, to draw a line parallel to the line $AB$, by constructing two right angles.

**Step 2**
She then drew straight lines $AM$ and $BM$ which cut $AB$'s parallel counterpart at $K$ and $L$ respectively

**Step 3**
She joined $AL$ and $KB$, these meet at $N$

**Step 4**
She joined $MN$ and this line produced met $AB$ in $F$. She then claimed that $F$ is the midpoint of $AB$.

Is this method valid? Justify your answer.

**Solution:**   In triangle $MAB$, using Ceva's theorem gives:

$$\frac{MK}{AK} \times \frac{AF}{FB} \times \frac{BL}{LM} = 1. \tag{1.21}$$

But since $KL \parallel AB$, triangles $MKL$ and $MAB$ are similar so that

$$\frac{MK}{MK + AK} = \frac{ML}{ML + LB}, \quad \text{hence} \quad \frac{MK}{AK} = \frac{ML}{LB}.$$

Therefore (1.21) becomes

$$1 = \frac{ML}{LB} \times \frac{AF}{FB} \times \frac{BL}{LM} = \frac{AF}{FB}$$

therefore $AF = FB$.

That is, $F$ is indeed the midpoint of $AB$, and Georgina's method is correct. Here is a second proof, using similar triangles only:

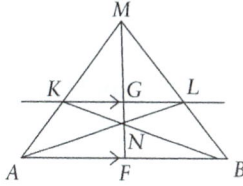

The triangles $AFN$ and $LGN$ clearly have equal angles, so are similar, and the same is true of triangles $BFN$ and $KGN$, hence:

$$\frac{AF}{LG} = \frac{NF}{NG} = \frac{FB}{GK},$$

therefore $\quad \dfrac{AF}{FB} = \dfrac{LG}{GK}.$ \hfill (1.22)

Also, because triangles $AFM$ and $KGM$ are similar, and triangles $BFM$ and $LGM$ are similar,

$$\frac{AF}{KG} = \frac{MF}{MG} = \frac{FB}{GL},$$

therefore $\quad \dfrac{AF}{FB} = \dfrac{KG}{GL}.$ \hfill (1.23)

From (1.22) and (1.23), we have

$$\frac{AF}{FB} = \left(\frac{AF}{FB}\right)^{-1}; \quad \text{hence} \quad \frac{AF}{FB} = 1, \quad \text{so } AF = FB.$$

The fact that the medians of any triangle are concurrent (we proved this as Theorem 3) is also easily proved using the converse of Ceva's theorem:

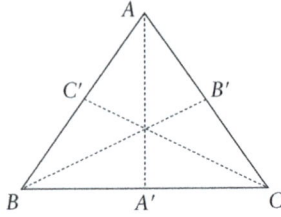

If, in triangle $ABC$, the lines $AA'$, $BB'$ and $CC'$ are the medians, as shown in the figure above (where you must temporarily disbelieve in the concurrence at $G$!) then

$$AC' = BC', \quad BA' = A'C \quad \text{and} \quad CB' = B'A, \quad \text{so that}$$

$$\frac{AC'}{BC'} \times \frac{BA'}{A'C} \times \frac{CB'}{AB'} = 1.$$

Hence, $AA'$, $BB'$ and $CC'$ are concurrent by the converse of Ceva's theorem.

The point $G$ of concurrence is called the **centroid** of triangle $ABC$. It is the geometrical analogue of the **centre of gravity** of a physical triangular lamina. We can also prove that the altitudes of a triangle are concurrent at a point called the **orthocentre**, and we have earlier proved that the perpendicular bisectors of the sides are concurrent at the **circumcentre** $O$ (page 20), while the angle bisectors are concurrent at a point $I$ called the **incentre** (page 18).

It is a fascinating fact that, for any given triangle, the centroid, the orthocentre and the circumcentre are collinear, and the line they form is called the **Euler line**, after its discoverer Leonhard Euler (1707–1783), perhaps the most prolific mathematician of all time. Another intriguing fact is that the centroid $G$ and the incentre $I$ are collinear with the centroid $B$ of the *boundary* of the triangle, which is itself the incentre of the triangle formed by the mid-points of the three sides. Moreover, $IB : IG = 3 : 2$. This seems to have been proved for the first time by Apostol and Mnatsakanian in *American Mathematical Monthly*, **111**, p. 10, in December 2004; it seems that the humble triangle will never cease to surprise us with more of its secrets!

After being forgotten for a while, the following theorem was rediscovered and published by Giovanni Ceva in 1678.

**Theorem 11   (Menelaus of Alexandria, 100 CE)**

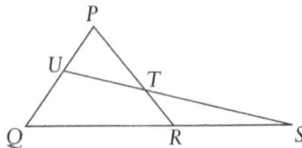

*Three points S, T, U on the sides of a triangle PQR shown above are collinear if and only if:*

$$\frac{QS}{RS} \times \frac{RT}{PT} \times \frac{PU}{QU} = 1. \qquad (1.24)$$

**Proof:**  'If and only if' means the implication goes both ways: collinearity implies the equation, and the equation implies collinearity. We will only prove it in one direction, starting with the premise that the points are collinear.

Let a perpendicular be dropped to $SU$ (produced if necessary) from each of the vertices $P, Q, R$ of triangle $PQR$ in the figure below.

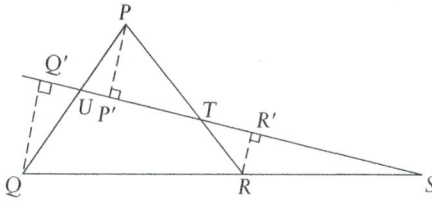

Referring to the figure, we have, by similar triangles:

$$\frac{QS}{RS} = \frac{QQ'}{RR'}, \quad \frac{RT}{PT} = \frac{RR'}{PP'}, \quad \frac{PU}{QU} = \frac{PP'}{QQ'}$$

so

$$\frac{QS}{RS} \times \frac{RT}{PT} \times \frac{PU}{QU} = \frac{QQ'}{RR'} \times \frac{RR'}{PP'} \times \frac{PP'}{QQ'} = 1.$$

◇

**Example:**

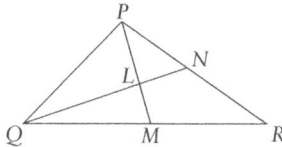

In the triangle shown in the figure, $QM : MR = 3 : 5$ and $RN : NP = 4 : 3$. If $QLN$ and $PLM$ are straight lines, what is the ratio $LN : QL$?

(A) $8 : 7$   (B) $7 : 8$   (C) $5 : 7$   (D) $7 : 5$   (E) $3 : 7$

**Solution:**  Applying Menelaus' theorem to the three points $P, L, M$, on the sides of the triangle $QNR$, we have:

$$\frac{RP}{PN} \times \frac{NL}{QL} \times \frac{QM}{MR} = 1.$$

Now $RN : NP = 4 : 3$ implies $RP : PN = (3 + 4) : 3 = 7 : 3$,   hence

$$\frac{7}{3} \times \frac{NL}{QL} \times \frac{3}{5} = 1, \quad \text{which implies} \quad \frac{NL}{QL} = \frac{5}{7}. \text{ Hence (C).}$$

Before going on to the next section, here is the solution to the appetizer problem you tasted at the beginning of this section.

**Solution of Appetizer Problem:**    Applying Ceva's theorem to triangle $ABC$ gives

$$\frac{BX}{XC} \times \frac{CY}{AY} \times \frac{AZ}{BZ} = 1,$$

$$\text{therefore} \quad \frac{2}{3} \times \frac{1}{2} \times \frac{AZ}{BZ} = 1,$$

$$\text{hence} \quad \frac{AZ}{BZ} = \frac{3}{1}. \tag{1.25}$$

We are given area $COX = 3b$. Using the common altitude theorem, we deduce that area $BOX = 2b$, since $BX : CX = 2 : 3$. Note that we could also use the common base theorem, since $OX$ is a common base for the two triangles, and $BC$ is the line joining their vertices. Similarly, since $CY : AY = 1 : 2$ and area $COY = a$ we deduce area $AOY = 2a$. The figure shows that

$$\text{area } YCB = \text{ area } BOX + \text{area } COX + \text{area } COY$$

$$= 2b + 3b + a$$

$$= 5b + a.$$

Now triangles $YCB$ and $YAB$ have common base $BY$ so once again the theorem gives

$$\frac{\text{area } YAB}{\text{area } YCB} = \frac{YA}{YC} = \frac{2}{1},$$

$$\text{therefore area } YAB = 2(5b + a) = 10b + 2a.$$

$$\text{Now} \quad \text{area } ABO = \text{area } YAB - \text{area } AOY$$

$$= (10b + 2a) - 2a = 10b,$$

$$\text{therefore} \quad \text{area } ZBO = \frac{1}{4} \times \text{ area } ABO \quad \text{(using (1.25))}$$

$$= \frac{1}{4} \times 10b.$$

Thus area $OXBZ = \text{area } ZBO + \text{area } BOX = \frac{10b}{4} + 2b = \frac{9b}{2}$. Hence (A).

## 1.3   Advanced circle geometry

*Hey, mom! I have done some measurements on those pies and
it seems the ratio of the circumference to the diameter is always the same.
Do you reckon I should call the ratio 'pie'?*

Here are two typical problems of the kind you should be able to solve when
you have worked through this section. You are invited to try them as soon
as you like. You may find them too hard for now, but by the end of this
section they will not look difficult to you. Their solutions are given at the
end of the section.

**Appetizer Problem 1:**   *In the diagram, PQ is a diameter and $\angle PTQ = \theta$.
What is the ratio of the areas: $\dfrac{Area \, \triangle SRT}{Area \, \triangle TQP}$?*

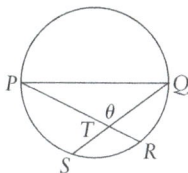

(A) $\cos^2 \theta$   (B) $\cos \theta$   (C) $\sin \theta$   (D) $\frac{1}{\cos^2 \theta}$   (E) $\sin^2 \theta$

**Appetizer Problem 2:**  *In the diagram below PQR is an equilateral triangle and arc PQSR is its circumcircle. If SR = 3 cm and QS = 2 cm, find PS.*

(A)  4 cm    (B)  9 cm    (C)  6 cm    (D)  5 cm    (E)  13 cm

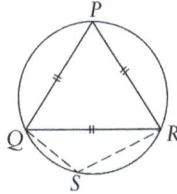

**Theorem 12    The principle of intersecting chords**

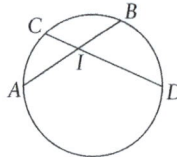

*If the chords AB and CD of the circle ABCD intersect at the point I, then:*

$$AI \times BI = CI \times DI.$$

**Proof:**

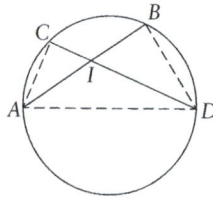

Join $AD, AC$ and $BD$ and observe that

$$A\hat{C}D = A\hat{B}D \quad \text{(subtended by same arc)}$$

$$\text{and} \quad C\hat{I}A = B\hat{I}D \quad \text{(vertically opposite)},$$

therefore triangles $CIA$ and $BID$ are similar, hence

$$\frac{CI}{BI} = \frac{IA}{ID},$$

therefore    $CI \times ID = BI \times IA,$  the required result.

◇

**Example:**

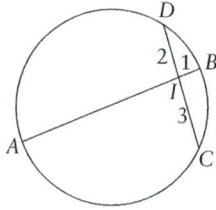

In the diagram above, $AB$ is a diameter of the circle $ACBD$ and $CD$ is a chord that intersects $AB$ at $I$ in such a way that $BI = 1$, $CI = 3$ and $DI = 2$. Find the radius of the circle.

(A) 2  (B) 3.5  (C) 3  (D) 5  (E) 2.5

**Solution:**  Let the radius be $r$, so that $AI = 2r - 1$.
 Using the intersecting chords principle we have

$$(2r - 1) \times 1 = 2 \times 3 = 6, \text{ so } r = 3.5. \text{ Hence (B).}$$

**Example:**  The chords $ED$ and $AB$ of the circle $AEBD$ meet at right angles at a point $F$ such that $EF = 6$, $AF = 2$ and $FD = 4$. Find the radius of the circle.

(A) 5  (B) 6  (C) $6\sqrt{2}$  (D) $5\sqrt{2}$  (E) $3\sqrt{3}$

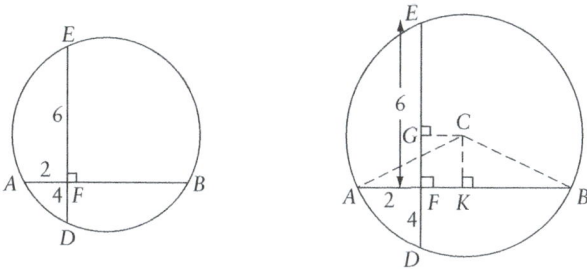

**Solution:**  Let $C$ be the centre of the circle. By the principle of intersecting chords $FB = 12$. Let $K$ and $G$ be points on $AB$ and $DE$, respectively, such that $AB \perp CK$ and $ED \perp GC$.

$$ED = 6 + 4 = 10, \text{ so } GD = GE = \frac{10}{2} = 5.$$

Hence

$$CK = GF = GD - FD = 5 - 4 = 1.$$
$$\text{and } AB = AF + FB = 2 + 12 = 14, \text{ so } AK = KB = \frac{14}{2} = 7.$$

Letting $r$ be the required radius, we have, in $\triangle ACK$:

$$r^2 = AK^2 + CK^2 = AC^2, \quad \text{by Pythagoras' theorem,}$$
$$= 7^2 + 1^2 = 50,$$

therefore $\quad r = 5\sqrt{2}$.  Hence (D).

**Theorem 13**    **(Ptolemy's theorem)** *In a cyclic quadrilateral PQRS,*

$$PQ \cdot SR + PS \cdot QR = PR \cdot SQ,$$

*that is, the sum of the products of opposite sides is equal to the product of the diagonals.*

**Proof:**

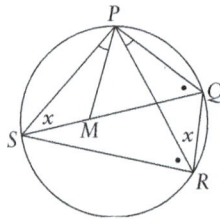

Draw the two diagonals $SQ$ and $PR$, and choose $M$, a point on $QS$, such that $S\hat{P}M = Q\hat{P}R$.

Focus on triangles $PMS$ and $PQR$, and observe that in the two triangles:

$$P\hat{S}M = P\hat{R}Q \quad \text{(angles subtended by same arc } PQ),$$
$$S\hat{P}M = Q\hat{P}R \quad \text{(by construction of } M),$$

so the triangles $PMS$ and $PRQ$ are similar. Hence

$$\frac{PS}{PR} = \frac{SM}{RQ}, \quad \text{so } PR \times SM = PS \times RQ. \tag{1.26}$$

Observe also that in triangles $PMQ$ and $PSR$:

$$M\hat{Q}P = S\hat{R}P \quad \text{(angles subtended by same arc } SP),$$
$$M\hat{P}Q = S\hat{P}R \quad \text{(equal angles plus same angle } M\hat{P}R),$$

so the triangles $PMQ$ and $PSR$ are similar. Hence

$$\frac{PQ}{PR} = \frac{QM}{RS}, \quad \text{so } PR \times QM = PQ \times RS. \tag{1.27}$$

Adding (1.26) and (1.27) gives

$$PR \cdot SM + PR \cdot QM = PS \cdot RQ + PQ \cdot RS,$$

therefore $\quad PR \cdot (SM + QM) = PS \cdot RQ + PQ \cdot RS.$

But $\quad SM + QM = QS$, hence the required result.

$\diamond$

As you might have guessed, Ptolemy's theorem is an 'open sesame' to many other important results. For example:

1. We can show that if $PQR$ is an equilateral triangle and $S$ lies on the arc $QR$ of the circumcircle of triangle $PQR$, then, $SR + QS = PS$.

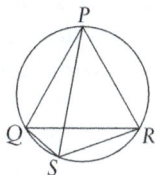

2. We can prove that $\sin(A - B) = \sin A \cos B - \cos A \sin B$ using Ptolemy's theorem.

**Example:**

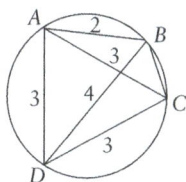

In the diagram, $ABCD$ is a cyclic quadrilateral with $AC = AD = CD = 3$, $AB = 2$ and $BD = 4$. Find the perimeter of the quadrilateral $ABCD$.

(A) 9   (B) 10   (C) 6   (D) 8   (E) 11

**Solution:**   Using Ptolemy's theorem

$$AD \times BC + AB \times DC = AC \times BD,$$
$$\text{therefore} \quad 3 \cdot BC + 3 \cdot AB = 3 \cdot BD$$
$$\text{hence} \quad BC + AB = BD = 4,$$
$$\text{and so} \quad BC + AB + CD + DA = 4 + 3 + 3 = 10. \quad \text{Hence (B).}$$

Hippocrates of Chios (mid-fifth century BCE) was mentioned in the introduction to this Toolchest. His notable achievement was to 'square a curvilinear area' for the first time in history; that is, to find a square or triangle whose area is equal to that of a given area bounded by circular arcs. This was referred to as a 'quadrature' of the given area. Hippocrates' triumph must have given him and his contemporaries renewed hope of solving that greatest of all problems of pre-modern mathematics: to 'square the circle' itself. (The problem is to find an exact compass and ruler construction, so that areas are proven equal by deduction from Euclidean axioms.) But nobody ever succeeded in doing so, and the problem was finally laid to rest as late as 1882 when Ferdinand Lindemann succeeded in proving that the number $\pi$ is worse than irrational, it is 'transcendental' – not the root of any polynomial equation with integral coefficients, hence not constructible. Here are three of Hippocrates' famous quadratures of 'lunes' or moon-shaped figures:

**Theorem 14**   (**Hippocrates' theorem**) *Referring to Figure (a), let triangle ABC be right angled at B. Draw the semicircles on diameters AB, BC and AC. (The latter passes through B – do you see why?) Then the sum of the areas of the two shaded crescents (or lunes) equals the area of the triangle ABC.*

*The other two Figures (b) and (c) represent other situations where Hippocrates showed that the shaded areas are equal.*

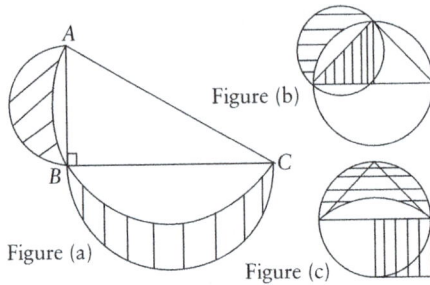

Figure (b)

Figure (a)

Figure (c)

**Proof:**   The proofs of (b) and (c) are similar to the proof of (a) which we now give in modern form. (Hippocrates would not have used the irrational number $\pi$ explicitly, but would have based his reasoning on the geometrical fact that 'areas of circles are as the squares of their radii'.)

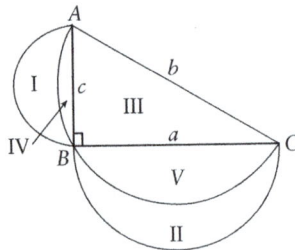

With the usual notation, we have

$$a^2 + c^2 = b^2, \quad \text{(by Pythagoras' theorem)},$$

therefore   $\pi a^2 + \pi c^2 = \pi b^2$,

and hence   $\dfrac{\pi a^2}{8} + \dfrac{\pi c^2}{8} = \dfrac{\pi b^2}{8}.$

That is, the area of semicircle with diameter $BC$ plus area of semicircle with diameter $AB$ equals area of semicircle with diameter $AC$.

The conclusion now follows by 'removing the common area'. To see this, write the equation in terms of the Roman symbols I through to V that correspond to the five different regions shown in the figure above:

$$(I + IV) + (II + V) = (III + IV + V),$$
$$\text{therefore} \quad I + II = III.$$

◇

**Example:** In the diagram, $ABC$ is a triangle right angled at $C$, and three semicircles are drawn on the straight lines $AB$, $BC$ and $CA$ as the respective diameters. The shaded areas between the semicircles are called the crescents of the triangle $ABC$. If the sides $a, b, c$ of the triangle are each an integral number of units, and the sum of the shaded areas is to be a whole number of square units, how many distinct such triangles are there with perimeter at most 30 cm?

(A) None   (B) 1   (C) 3   (D) 4   (E) Infinitely many

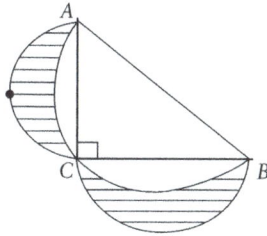

**Solution:** Let the three sides be $a, b, c$, where $c$ is the hypotenuse. By Hippocrates' theorem, the sum of the crescents is equal to the area $\frac{1}{2}ab$ of the triangle itself. Hence the required number of triangles equals the number of Pythagorean triples $(a, b, c)$ such that $a + b + c \le 30$ and $\frac{1}{2}ab$ is a whole number.

For each primitive Pythagorean triple $(a, b, c)$ there exist positive integers (see remarks below) $m$, $n$ with $m > n$, such that: $a = m^2 - n^2$, $b = 2mn$, $c = m^2 + n^2$, where we order the $a, b$ in such a way that the $b$ is even (precisely one of them must be even, see remarks below.) Now we want the area to be a natural number, that is, in symbols:

$$ABC = \frac{1}{2}ab \in \mathbb{N}$$

that is,   $\frac{1}{2}(m^2 - n^2)(2mn) = mn(m^2 - n^2) \in \mathbb{N}.$   (1.28)

$$\text{Also, we want} \quad a + b + c \le 30,$$
$$\text{that is,} \quad m^2 - n^2 + 2mn + m^2 + n^2 \le 30,$$
$$\text{hence} \quad m(m + n) \le 15. \quad (1.29)$$

Clearly, all $(m, n)$ will satisfy (1.28), and $(m, n) = (3, 2)$ satisfies (1.29), giving equality. In fact, it's quite easy to show that $(3, 2)$ is the only solution of the Diophantine equation $m(m+n) = 15$, and that there are only two other ordered pairs $(m, n)$ with $m > n$ satisfying the inequality (1.29), namely, $(2, 1)$ and $(3, 1)$. This gives us two primitive triples $(5, 12, 13)$, $(3, 4, 5)$, and the third is a multiple of $(3, 4, 5)$. No other multiples satisfy the inequality. Thus we have precisely three triangles with required properties, those given by $(5, 12, 13)$, $(3, 4, 5)$, $(6, 8, 10)$. Hence (C).

**Remark:**  A **Pythagorean triple** $(a, b, c)$ is a triple of positive integers like $(3, 4, 5)$ such that $a^2 + b^2 = c^2$; in that case a multiple $(ka, kb, kc)$ will also be a Pythagorean triple for any positive integer $k$. If $a, b, c$ are relatively prime (no common factor) then we call $(a, b, c)$ a **primitive** (or **reduced**) Pythagorean triple. In this solution we used the facts that every possible integral solution for the sides of a right angled triangle is a multiple of some primitive Pythagorean triple, and every primitive Pythagorean triple is given by the formula: $(m^2 - n^2, 2mn, m^2 + n^2)$, where $m, n$ are positive integers with $m > n$ and can be taken to be of opposite parity. By simply checking the algebra, it is easy to see that any triple given by this formula will be Pythagorean. Clearly the hypotenuse is the third since it is largest. It is harder to prove that we can get *every* Pythagorean triple this way, but even the ancient Babylonians seemed to have knowledge of this fact. It is very unlikely that they had a proof, but Diophantus of Alexandria gave a proof in the third century CE. See Toolchest 9, Miscellaneous Problem 78.

**Example:**  In the figure, each of the areas $A$, $B$, $C$ $D$ and $E$ (in cm$^2$) is an integer and the total area of the figure (in cm$^2$) is 10. What is the value of $E$ when $A + B$ is minimum?

(A) 8  (B) 2  (C) 6  (D) 9  (E) 4

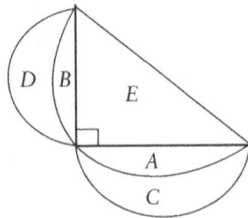

**Solution:**  Let $A + B = x$ (in cm$^2$) and $C + D = y$ (in cm$^2$). By Hippocrates' theorem we have $E = C + D = y$, hence, from what we are given,

$$10 = A + B + C + D + E = x + 2y, \quad \text{so} \quad x = 10 - 2y.$$

Since $A, B, C, D$ are positive integers, we know $x, y \geq 2$, so the minimum value of $x$ is 2, giving $y = 4$.
   Therefore, $E = y = 4$.  Hence (E).

Finally, here are solutions to the two problems we posed at the start of this section:

**Solution to Appetizer Problem 1:**   By the principle of intersecting chords we have

$$PT \cdot TR = QT \cdot TS, \text{ hence } \frac{TR}{QT} = \frac{TS}{PT}. \qquad (1.30)$$

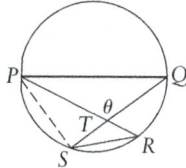

The required ratio becomes

$$\frac{\frac{1}{2}TS \cdot TR \sin\theta}{\frac{1}{2}PT \cdot QT \sin\theta} = \frac{TS \cdot TR}{PT \cdot QT} = \frac{TR}{QT} \cdot \frac{TS}{TP} = \frac{TS^2}{PT^2} \quad \text{(using (1.30))}.$$

Further, $\angle PTS = 180° - \theta$, and since $PQ$ is a diameter $\angle PSQ = 90°$, so we can use the right-angled triangle $\triangle PST$ to obtain $\cos(180° - \theta) = \frac{TS}{PT}$, and hence $\cos\theta = -\frac{TS}{PT}$.

The required ratio is therefore $\left(\frac{TS}{PT}\right)^2 = \cos^2\theta$.  Hence (A).

**Solution to Appetizer Problem 2:**

*Method 1:* Since $P, Q, S$ and $R$ are concyclic points, we use Ptolemy's theorem to obtain $PQ \cdot RS + PR \cdot QS = PS \cdot QR$. Now $PQ = PR = QR$, so $RS + QS = PS$, after simplifying. Therefore, $PS = 2 + 3 = 5$ cm. This is a very neat solution!

*Method 2:* This method, by its length and relative messiness, will help to make you appreciate Ptolemy's theorem, because using Ptolemy in that quick and elegant first method simply leapt over all the arduous steps (including no less than three applications of the cosine rule) we are about to take in Method 2!

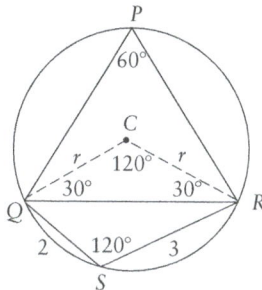

Let $C$ be the circle's centre and let its radius be $r$. Then $\angle QCR = 120°$ (angle at centre twice that at circumference). Therefore $\angle QSR = \frac{1}{2}(360 - 120) = 120°$. Using the cosine rule in triangle $QSR$ gives $QR^2 = 4 + 9 - 2 \cdot 3 \cdot 2 \cos 120° = 19$. Therefore $QR = PQ = PR = \sqrt{19}$ and now $\angle PQS = 60° + \angle RQS$. Therefore $\cos PQS = \cos 60° \cos RQS - \sin 60° \sin RQS$. Now, using the cosine rule in a different way in that same triangle $QSR$, gives $9 = 19 + 4 - 2\sqrt{19} \cdot 2 \cos R\hat{Q}S$. Hence $\cos R\hat{Q}S = \frac{7}{2\sqrt{19}}$. Since $\cos^2 R\hat{Q}S + \sin^2 R\hat{Q}S = 1$, we have $\sin R\hat{Q}S = \frac{3\sqrt{3}}{2\sqrt{19}}$. Therefore $\cos P\hat{Q}S = \frac{1}{2} \cdot \frac{7}{2\sqrt{19}} - \frac{\sqrt{3}}{\sqrt{2}} \cdot \frac{3\sqrt{3}}{2\sqrt{19}} = -\frac{1}{2\sqrt{19}}$. Finally, using the cosine rule in triangle $PQS$ gives $PS^2 = 19 + 4 - 2\sqrt{19} \cdot 2 \times -\frac{1}{2\sqrt{19}} = 23 + 2 = 25$. Hence $PS = 5$ cm as before.

## 1.4  Problems

**Problem 1:**  *A cardboard piece was cut from a regular polygon, and thrown away. Tatenda picked up this discarded piece, and what he discovered is shown below. How many sides did the polygon have?*

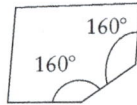

(A)  18   (B)  20   (C)  19   (D)  16   (E)  15

**Problem 2:**  *In the diagram, O is the center of the circle. If $\angle BAC = 40°$, then $\angle BCD$ equals:*

(A)  40°   (B)  60°   (C)  10°   (D)  50°   (E)  45°

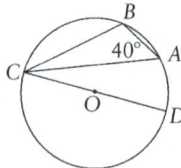

**Problem 3:**  *If a three metre stake casts a shadow 7 metres long, then the height of a tree which casts a shadow 63 metres long is*

(A)  18   (B)  21   (C)  245   (D)  27   (E)  30.5

**Problem 4:**  *In the figure, where the two arcs are quarter circles, the area of the shaded region is*

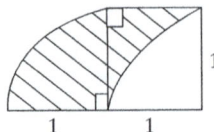

(A) $\pi$   (B) $\frac{\pi}{4}$   (C) $\frac{\pi}{4} + \frac{1}{2}$   (D) 1   (E) 4

**Problem 5:**   *Consider a large rectangle with sides of length a and b respectively. A small circle of radius r is drawn such that one of the vertices of the rectangle lies on the circumference of the circle and the circle cuts the rectangle at two other points X and Y. Find the length of XY.*

(A) $ab$   (B) $2r$   (C) $ab\sqrt{r}$   (D) $\frac{ab}{r}$   (E) $\sqrt{(a-r)^2 + (b-r)^2}$

**Problem 6:**   *The radius of a circle is increased by 1 cm. By how many cms is the circumference increased?*

(A) $2\pi$   (B) $\pi$   (C) 2   (D) 3   (E) 1

**Problem 7:**   *A piece of paper has the shape of the larger circular sector shown in the diagram, with dimensions and angles as shown. This paper is suitably folded to form the vertical cone shown to the right. Find the height of the cone.*

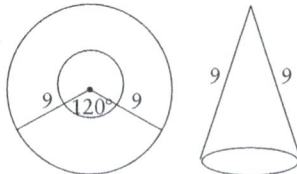

(A) 3   (B) $3\sqrt{5}$   (C) $3\sqrt{2}$   (D) $9\sqrt{2}$   (E) 9

**Problem 8:**   *In the diagram below, the line ST is tangent to the smaller of two concentric circles, and is 36 cm long. Find the area of the annulus (shaded region):*

(A) $324\pi$   (B) $81\pi$   (C) $1296\pi$   (D) $162\pi$
(E) Cannot be determined from given information

**Problem 9:**   *The figure in the diagram is formed by two overlapping circles. The circles have radii 1 and 3 respectively. If the area of the shaded region is $\frac{\pi}{2}$, then the total area of the figure is*

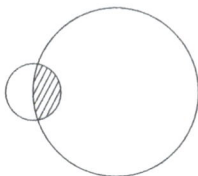

(A) $10\pi$   (B) $\frac{19\pi}{2}$   (C) $8\pi$   (D) $\frac{7\pi}{2}$   (E) $9\pi$

**Problem 10:**   *In the figure, UVW is an equilateral triangle, UV is parallel to XY and to FG, and ∠XJV = 70°. Find ∠FHG*

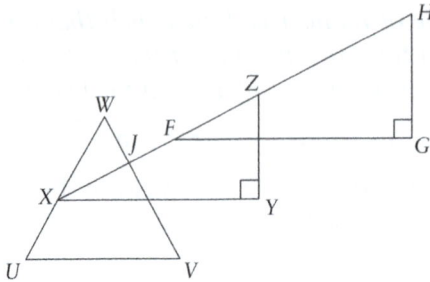

(A) 70°  (B) 20°  (C) 60°  (D) 40°  (E) 50°

**Problem 11:**   *The figure shows the square ABCD and the equilateral triangle EBC. The size of ∠AED is*

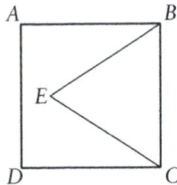

(A) 90°  (B) 120°  (C) 135°  (D) 150°  (E) None of these

**Problem 12:**   *The diagram below shows a circle with centre O and AB parallel to CD. Given the angles 75° and θ as shown, find the value of θ.*

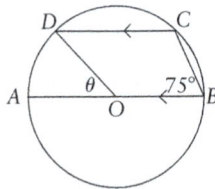

(A) 15°  (B) 45°  (C) 60°  (D) 75°  (E) 30°

**Problem 13:**   *The ratio of the radii of concentric circles is 1 : 3. If AC is a diameter of the larger circle, BC is a chord of the larger circle and is also tangent to the smaller circle, and AB = 12, then the radius of the larger circle is*

(A) 13  (B) 18  (C) 21  (D) 24  (E) 26

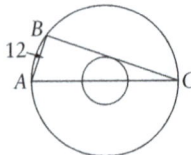

**Problem 14:**   *In the diagram below AC is the bisector of ∠DAB and BCD is a straight line. If we are given that CD = 6, BD = 10 and AD = 9, what is the length of AB?*

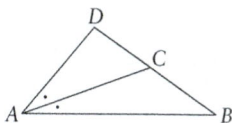

(A)  3  (B)  4  (C)  6  (D)  8  (E)  9

**Problem 15:**   *In the diagram, the value of α is*

(A)  36°   (B)  40°  (C)  20°  (D)  30°  (E)  85°

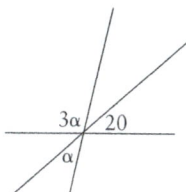

**Problem 16:**   *Two circles of equal radius r intersect at A and B as shown. Each circle passes through the centre of the other circle. Find the length of AB.*

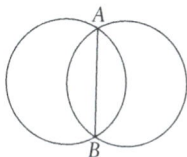

(A)  *r*  (B)  2*r*  (C)  *r*√2  (D)  *r*√3  (E)  $\frac{r\sqrt{3}}{2}$

**Problem 17:**   *In the figure ∠A = 50° and the circle with centre J touches BC, AB produced and AC produced, as shown. What is the angle ∠BJC?*

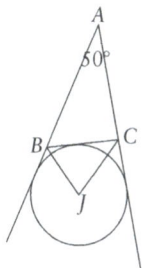

(A)  130°  (B)  65°  (C)  50°  (D)  60°  (E)  70°

**Problem 18:** *A rectangle is inscribed in a circle centre O as shown below, with AB = 5, BC = 12 and angle ∠BOA = θ. Find sin θ.*

(A) $\frac{120}{169}$  (B) $\frac{60}{169}$  (C) $\frac{5}{13}$  (D) $\frac{10}{13}$  (E) $\frac{12}{13}$

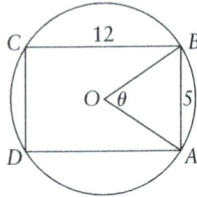

**Problem 19:** *Two identical circles intersect as shown. The area of the shaded region is equal to the sum of the areas of the two unshaded regions. Suppose the area of the shaded region is 24π. Find the circumference of each circle.*

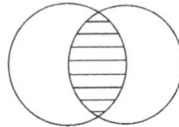

(A) $6\pi\sqrt{2}$  (B) $12\pi$  (C) $24\pi$  (D) $\pi\sqrt{24}$  (E) $30\pi$

**Problem 20:** *A triangle ABC has sides of length 5, 12, and 13 cms respectively. What is the radius of the circle that passes through the vertices of the triangle?*

(A) 30  (B) 15  (C) 6.5  (D) 8  (E) 7

**Problem 21:** *A rectangle 18 m long and 8 m wide is cut as shown below, and the pieces are rearranged to form a square. What is the perimeter of the square in metres?*

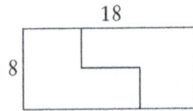

(A) 36  (B) 20  (C) 52  (D) 30  (E) 48

**Problem 22:** *A circle PQRS is inscribed in a quadrilateral ABCD so that it touches AB at S, BC at R, CD at Q and AD at P. If AB = 10 cm, CR = 4 cm and DQ = 3 cm, what is the perimeter of the quadrilateral ABCD in cm?*

(A) 34  (B) 30  (C) 27  (D) 32  (E) 33

**Problem 23:** *Consider a square-based pyramid. Suppose that the height of the pyramid is increased by 10% and the sides of its square base are reduced by 10%. How does the volume of the pyramid change?*

(A) 33.1% increase   (B) 1% decrease   (C) 8.9% increase
(D) 10.9% decrease   (E) 119% decrease

**Problem 24:** *Two circles of unit radius are drawn such that the centre of each is on the circumference of the other. Find the area of the region of intersection between the two circles.*

(A) $\frac{\pi}{3} + \frac{\sqrt{3}}{2}$   (B) $\frac{2\pi}{3} - \frac{\sqrt{3}}{2}$   (C) $\frac{\pi}{3} - \frac{\sqrt{3}}{2}$   (D) $\frac{\pi}{6} - \frac{\sqrt{3}}{4}$   (E) $\frac{\pi}{6} + \frac{\sqrt{3}}{2}$

**Problem 25:** *The figure shows a regular hexagon with area R. Find in terms of R, the area of the shaded portion.*

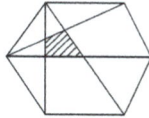

(A) $\frac{R}{3}$   (B) $\frac{R}{6}$   (C) $\frac{R}{12}$   (D) $\frac{R}{18}$   (E) $\frac{R}{24}$

**Problem 26:** *An equilateral triangle is drawn inside a unit square such that one of its vertices coincides with a vertex of the square. What is the maximum possible area of all such triangles?*

(A) $\frac{7\sqrt{3}}{4} + 3$   (B) $2\sqrt{3} + 3$   (C) $2\sqrt{3} - 3$   (D) $\frac{7\sqrt{3}}{4} - 3$   (E) $2 - \sqrt{3}$

**Problem 27:** *Three identical circles, each of radius a, are drawn tangent to each other and to a circle of radius r as shown. Express a in terms of r.*

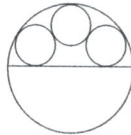

(A) $\frac{r}{4}$   (B) $\frac{r}{2}$   (C) $\frac{2r}{3}$   (D) $\frac{r}{3}$   (E) $\frac{r}{6}$

**Problem 28:** *A circular sheet of paper of radius 6 cm is cut into six equal sectors. Each sector is formed into the curved surface of a cone, with no overlap. What is the height of each cone?*

(A) $\sqrt{27}$   (B) $\sqrt{32}$   (C) $\sqrt{35}$   (D) 6   (E) 7

**Problem 29:** *The side AB of the rectangle ABCD is trisected at E and F so that AE = EF = FB. The lines DF and EC meet at the point O. Find the value of the ratio* $\dfrac{area \ of \ triangle \ OFC}{area \ of \ rectangle \ ABCD}.$

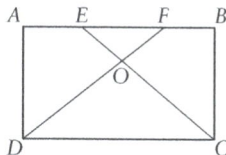

(A) $\frac{1}{4}$  (B) $\frac{1}{6}$  (C) $\frac{5}{23}$  (D) $\frac{4}{25}$  (E) $\frac{1}{8}$

**Problem 30:**   *Consider a circle with centre O as shown. The lines AB and COD are parallel, and so too are FD and EOB. Let angle $A\hat{B}E = \theta$. Express angle $E\hat{O}F$ in terms of $\theta$.*

(A) $90 - \theta$  (B) $\theta$  (C) $180 - \theta$  (D) $45 - \theta$  (E) $\frac{\theta}{2}$

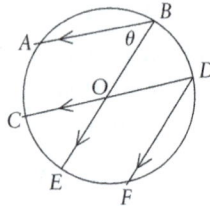

**Problem 31:**   *The figure shows a right-angled triangle ABC. The perpendicular from A meets BC at point D. Point E is on AC and is such that $BE = EC = CD = 1$. Find the length of AC.*

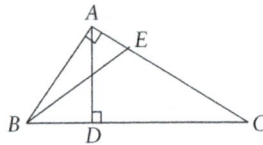

(A) $2^{\frac{1}{3}}$  (B) $3^{\frac{1}{3}}$  (C) $2^{\frac{1}{3}} - \frac{1}{6}$  (D) $\frac{4}{3}$  (E) $\frac{3}{2}$

**Problem 32:**   *In the figure below, AB is perpendicular to BC and $AB = c$ units long. AC and DB intersect at a point I and IG is perpendicular to BC. If $DC = b$ units and $BC = a$ units, what is IG?*

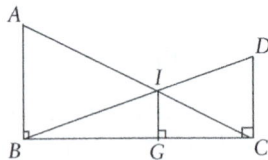

(A) $\frac{ab + ac + bc}{a}$  (B) $\frac{bc}{b+c}$  (C) $\frac{ac}{a+c}$  (D) $\frac{ab + ac + bc}{b}$  (E) $\frac{a^2 + b^2 + c^2}{a+b+c}$

**Problem 33:**   *In the figure below, ABCD is a cyclic quadrilateral with AB the diameter of the inscribing circle. If $AB = 36$ and $AD = BC = 12$, what is the length of CD?*

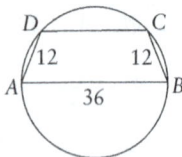

(A) 28  (B) 12  (C) 18  (D) 6  (E) 24

**Problem 34:** *In triangle XYZ, angle X = a, angle Y = 2a and angle Z = ra, where r ≥ 4. Let the sides of the triangle be x, y, z, opposite the angles X, Y, Z, respectively. The value of $y^2 - x^2$ is then:*

(A) $xz$    (B) $xy$    (C) $yz$    (D) $z^2$    (E) $rz$

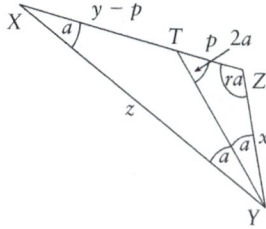

**Problem 35:** *A circle with centre P and radius 10 is tangent to the sides of an angle of 60°. A larger circle with centre Q is tangent to the sides of the angle, and also to the first circle. The radius of the larger circle is*

(A) $30\sqrt{3}$    (B) 20    (C) $20\sqrt{3}$    (D) 30
(E) Impossible to determine from given information

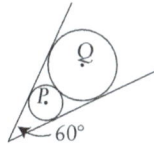

**Problem 36:** *The figure shows a square EFGH and a rectangle ABCD with meeting points P and Q as shown. If this square and rectangle have the same area, and AP = EP = DQ = FQ = 1, then BC is equal to*

(A) 2    (B) $2 + \sqrt{2}$    (C) 4    (D) $2 + \sqrt{7}$    (E) 5

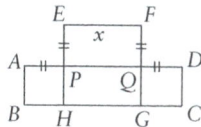

**Problem 37:** *In the diagram below, ABC is an equilateral triangle whose base AB is a diameter of the circle, with AB = 8 cm and AC, BC cutting the circle at P, Q, respectively, as shown. The shaded area (in square cm) is:*

(A) 64    (B) 8    (C) $4(\sqrt{3} - \frac{\pi}{3})$    (D) $8(\sqrt{3} - \frac{\pi}{3})$    (E) $4(\sqrt{3} - \frac{\pi}{2})$

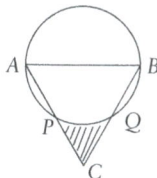

**Problem 38:** *In the diagram below, $A, B, C$ and $D$ are concyclic points, $DA = 3$ cm, $AC = (1 + 2\sqrt{6})$ cm and $BC = 2$ cm. If $BD$ is the circle's diameter, find the length of $BD$.*

(A) $5\sqrt{6}$ cm  (B) $6\sqrt{6}$ cm  (C) $6\sqrt{2}$ cm  (D) $6$ cm  (E) $5$ cm

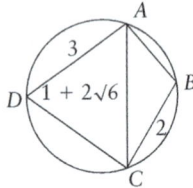

**Problem 39:** *The diameter $CD$ of a circle with centre $G$ is produced to a point $K$ so that $DK$ equals the radius of the circle. Another circle with diameter $GK$ is drawn. Find the area (in square cm) common to the two circles if $GK = 8$ cm.*

(A) $16$  (B) $\left(\frac{8\pi}{3} - 2\sqrt{3}\right)$  (C) $16\pi$  (D) $\left(\frac{8\pi}{3} - \sqrt{3}\right)$  (E) $2\left(\frac{16\pi}{3} - 4\sqrt{3}\right)$

**Problem 40:** *In the figure below, $A, B$ and $C$ are points on the circle, $CD$ is tangent to the circle at $C$, $BC$ is a diameter of the circle and $BD$ cuts the circle at $A$. If $AB = 5$ cm and $AD = 4$ cm, find $CD$.*

(A) $25$ cm  (B) $20$ cm  (C) $9$ cm  (D) $6$ cm  (E) $10$ cm

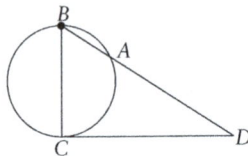

**Problem 41:** *In the figure below, $ABCD$ is a trapezium with $\angle ADC = \angle BAD = 90°$. The circle with centre $B$ and radius $AB$ cuts lines $BD$ and $BC$ at $M, N$, respectively, and $ABNM$ is a parallelogram. Find the shaded area $DMNC$ in square cm if the radius of the circle is $1$ cm.*

(A) $\left(\sqrt{3} - \frac{\pi}{6}\right)$  (B) $\left(\sqrt{3} - \frac{\pi}{12}\right)$  (C) $\left(\sqrt{2} - \frac{\pi}{6}\right)$  (D) $\frac{\pi}{6}$  (E) $\sqrt{6}$

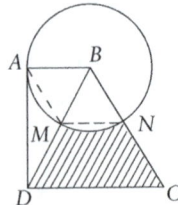

**Problem 42:** *In the figure below is a circle with centre $O$, and another circle with centre at the midpoint $B$ of the radius $OA$ and passing through $O$. A small circle is drawn as shown, tangent to both the larger circles, such*

*that its centre N is directly above B (that is, BN ⊥ AB). If the perimeter of the triangle NBA is 8 cm, find the value (in cm) of the radius of the small circle.*

(A) $\frac{2}{3}$   (B) 2   (C) 1   (D) $\sqrt{2}$   (E) $(\sqrt{2} - 1)$

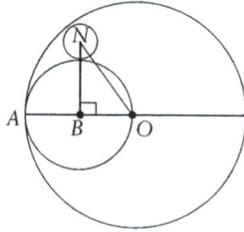

**Problem 43:**   *Triangle ABC has perimeter 3, and the sum of the squares of its sides is 5. Find the value of the sum of the three heights of the triangle if the radius of the circumcircle is 1.*

(A) 1   (B) 3   (C) 5   (D) 2   (E) 4

**Problem 44:**   *In the diagram below, ABCD is a square, side DC is produced to a point G so that CG = DC, and M is the midpoint of CG. The line AG meets BD, BC and BM at E, F and K, respectively. Find the ratio AE : EF : FK.*

(A) 1 : 2 : 1   (B) 2 : 1 : 1   (C) 2 : 1 : 2   (D) 3 : 2 : 1   (E) 2 : 3 : 1

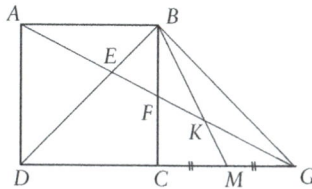

**Problem 45:**   *In the figure, ABED is a circle of radius 4 cm and ABC, DCE are right angled triangles with their right angles positioned as shown. If AB = 2 cm and C is the midpoint of BE, find AD in cm.*

(A) $(3\sqrt{2})$   (B) 3   (C) $\sqrt{2}$   (D) $(2\sqrt{6})$   (E) 8

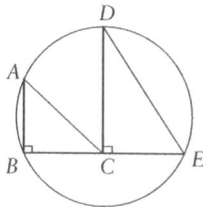

**Problem 46:**   *In the figure below, BG is a median of triangle ABC and the lines AM, AK and AL cut BC so that BM = MK = KL = LC. Let them*

*cut BG at D, E and F respectively. If the area of triangle AEB is $A_1$ and the area of triangle ABC is $A_2$, find the ratio $A_2 : A_1$.*

(A)  $3:1$    (B)  $1:3$    (C)  $3:2$    (D)  $2:3$    (E)  $4:3$

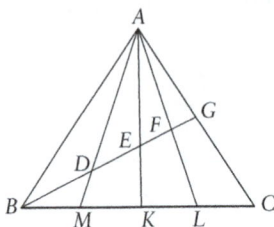

**Problem 47:**   *Three circles are tangent to each other and to the straight line. The radii of the circles with centres A, B and C are a, b and c, respectively. Then c, expressed in terms of a and b, equals:*

(A)  $\dfrac{ab}{(\sqrt{a}+\sqrt{b})^2}$    (B)  $\dfrac{\sqrt{ab}}{a+b}$    (C)  $\dfrac{ab}{(a+b)^2}$    (D)  $\dfrac{\sqrt{ab}}{\sqrt{a}+\sqrt{b}}$    (E)  $\dfrac{ab}{a+b}$

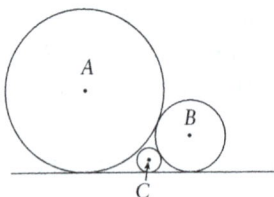

**Problem 48:**   *In the diagram, PQ and TS are perpendicular to QS, PQ = 12, TS = 8, QS = 20, and QR = x. If PRT is a right angle, then*

(A)  $x$ has two possible values whose difference is 4
(B)  $x$ has two possible values whose sum is 28
(C)  $x$ has only one value, and $x \geq 10$
(D)  $x$ has only one value, and $x < 10$
(E)  $x$ cannot be determined from the given information

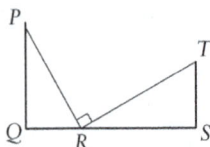

**Problem 49:**   *The lengths of the medians of a triangle are 15, 36 and 39. What is the perimeter of the triangle?*

(A)  $26 + 2\sqrt{601}$    (B)  $26 + 2\sqrt{61} + 2\sqrt{601}$    (C)  $26 + 4\sqrt{61} + 2\sqrt{601}$
(D)  $26 + 4\sqrt{61} + 4\sqrt{601}$    (E)  $26 + 2\sqrt{61} + 4\sqrt{601}$

**Problem 50:**   *Give a proof of Pythagoras' theorem (page 8) along the lines of Euclid's first deductive proof. Use the diagram on the left, and look for congruent triangles. Then think about areas.*

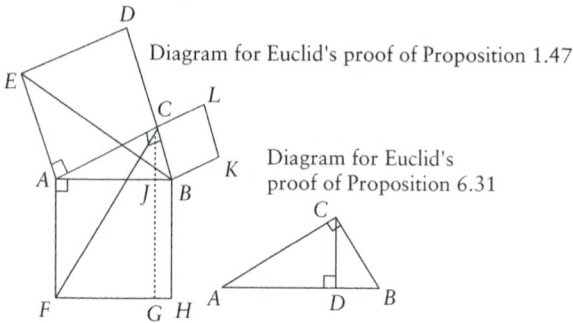

Diagram for Euclid's proof of Proposition 1.47

Diagram for Euclid's proof of Proposition 6.31

**Problem 51:**   *Give a proof of Pythagoras' theorem (page 8) along the lines of Euclid's second proof, using similarity of triangles. Use the diagram on the right, and think about ratios of sides.*

## 1.5   Solutions

**Solution 1:**   From the 'discovery', each exterior angle is $180 - 160 = 20°$, because all the exterior angles of a regular polygon are equal. Since the exterior angles of any polygon add up to $360°$, we have the number of sides as $\frac{360}{20} = 18$. Hence (A).

**Solution 2:**   Joining $BD$, we see that $C\hat{D}B = 40°$ (angles subtended by same chord) and $C\hat{B}D = 90°$ (angle in a semicircle). Therefore, in $\triangle CBD$, $B\hat{C}D = 90 - 40 = 50°$. Hence (D).

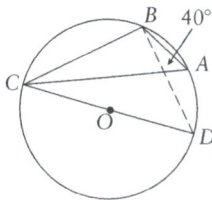

**Solution 3:**   Let the height of the tree be $x$. We then have two triangles, which are similar (each has a right angle and the angle of the Sun's elevation).

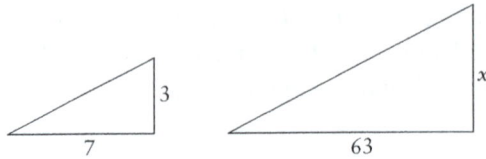

Thus $\frac{x}{3} = \frac{63}{7}$, so that $x = 27$. Hence (D).

**Solution 4:**   Denote the areas of the regions by $A$, $B$, $C$ as shown. Then $A = C$ and $B + C = 1$ so that $A + B = 1$. Hence (D).

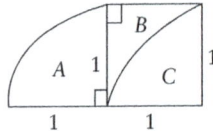

**Solution 5:**   Since $XY$ subtends an angle of $90°$, it must be a diameter and the result that $XY = 2r$ follows immediately. Hence (B).

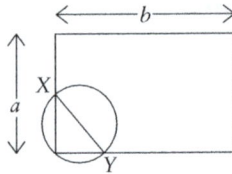

**Solution 6:**   Let the radius of the original circle be $R$ and the corresponding circumference $C_1$, and let the circumference of the modified circle be $C_2$.

$$C_1 = 2\pi R$$
$$C_2 = 2\pi(R + 1)$$

so that   $C_2 - C_1 = 2\pi(R + 1 - R) = 2\pi$.   Hence (A).

This problem is the basis of a 'shock' question: You have a communications cable that goes tightly right around the Earth (say around the equator), and you want to lengthen it so it can be raised up on pylons about 5 metres high all around. Approximately how much longer must your cable be? Most people will greatly overestimate the extra cable required – try it on your friends! (You don't need to know the radius of the Earth.)

**Solution 7:**   Observe that the circumference of the circular sector equals the circumference of the circular base of the cone. This allows us to find the radius $r$ of the cone, by setting

$$2\pi r = \frac{240}{360} \times 2\pi \times 9,$$

$$\text{therefore} \quad r = \frac{2 \times 9}{3} = 6.$$

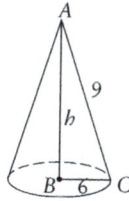

The triangle $ABC$ obtained from a suitable cross-section of the cone allows us to find the required height $h$:

$$h = \sqrt{9^2 - 6^2} \quad \text{(by Pythagoras)}$$
$$= \sqrt{(9+6)(9-6)}$$
$$= \sqrt{3^2 \cdot 5} = 3\sqrt{5}. \quad \text{Hence (B).}$$

**Solution 8:** Most people will immediately go for (E), as it is hard to believe there is sufficient information to determine the area! But the mathematics is unanswerable:

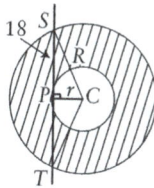

Let $R$ and $r$ (in cm) be the radii of the greater and smaller circle respectively. Let $C$ be the common centre of the circles and $CP$ be the perpendicular bisector of $ST$, which must meet $ST$ at the point $P$ of tangency (why?). Hence, in triangle $SPC$, Pythagoras' theorem gives $R^2 - r^2 = 18^2$. Let the area of the annulus be $A$, so that

$$A = \pi R^2 - \pi r^2$$
$$= \pi (R^2 - r^2)$$
$$= \pi (18)^2$$
$$= 324\pi. \quad \text{Hence (A).}$$

**Solution 9:** Let $A, B, C$ be the areas, as shown in the diagram.

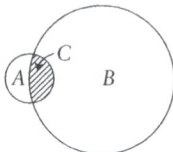

$$\text{Required area} = A + B + C$$

$$= (A + C) + (B + C) - C$$

$$= \text{area of small circle} + \text{area of large circle}$$

$$- \text{shaded area}$$

$$= \pi(1)^2 + \pi(3)^2 - \frac{\pi}{2}$$

$$= \frac{19\pi}{2}. \quad \text{Hence (B)}.$$

**Solution 10:** Because $XY \parallel UV$, we have $\angle XOJ = \angle UVW$. But the angles in the equilateral triangle $\triangle UVW$ are all $60°$, so $\angle XOJ = 60°$. Therefore, in $\triangle JOX$, $\angle OXJ = 180° - (70° + 60°) = 50°$.

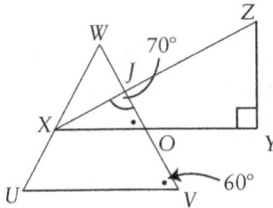

Referring now to the full diagram given previously, because $FG \parallel XY$, we have $\angle GFH = \angle YXH = \angle OXJ = 50°$, and hence $\angle FHG = 90° - 50° = 40°$, since $\hat{G} = 90°$. Hence (D).

**Solution 11:** Since the triangles $\triangle ABE$ and $\triangle ECD$ are isosceles (sides of a square are equal) and congruent (by symmetry),

$$\angle BAE = \angle BEA = \angle CDE = \angle CED = \frac{180° - 30°}{2} = 75°$$

So that $\angle AED = 360° - (60° + 75° + 75°) = 150°$. Hence (D).

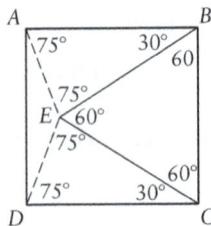

**Solution 12:** Draw line $AC$. In triangle $ABC$, $\angle ACB = 90°$ (angle in a semicircle), hence $\angle CAB = 90° - 75° = 15°$.

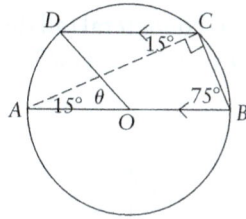

Since $AB$ and $CD$ are parallel, $\angle ACD = \angle CAB = 15°$, so that

$$\theta = \angle AOD = 2 \times \angle ACD \quad \text{(angle at centre} = 2 \times \text{angle at circumference)}$$
$$= 2 \times 15° = 30°. \quad \text{Hence (E)}.$$

**Solution 13:**   Draw the line $DE$ from the centre $D$ of the circles to the point $E$ of tangency, so that $\angle DEC$ is a right angle.

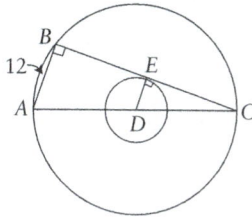

Since $\angle ABC$ is an angle in a semicircle it is also a right angle, and therefore triangles $DEC$ and $ABC$ are similar. Therefore

$$\frac{DE}{AB} = \frac{CD}{CA} = \frac{1}{2} \quad \text{(diameter} = 2 \times \text{radius)}$$

which gives $DE = \frac{1}{2} \cdot AB = \frac{1}{2} \cdot 12 = 6$. But $\dfrac{AD}{DE} = 3$, so $AD = 3 \times 6 = 18$. Hence (B).

**Solution 14:**   Label the angles as shown. In triangle $\triangle ADC$, $\dfrac{9}{\sin \alpha} = \dfrac{6}{\sin \theta}$, while in triangle $\triangle ABC$, $\dfrac{4}{\sin \theta} = \dfrac{AB}{\sin(180° - \alpha)} = \dfrac{AB}{\sin \alpha}$.

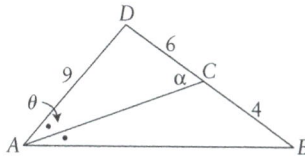

Hence $AB = 4 \cdot \dfrac{\sin \alpha}{\sin \theta} = 4 \cdot \dfrac{9}{6} = 6$. Hence (C). (Or, simply use the angle bisector theorem: Theorem 6 on page 29).

**Solution 15:**   The angle vertically opposite $\alpha$ is also $\alpha$, hence

$$3\alpha + \alpha + 20° = 180°, \quad \text{so } \alpha = 40°. \quad \text{Hence (B)}.$$

**Solution 16:** Triangle $APQ$ is equilateral (all the sides are radii of the equal circles) with height $AC = r\sin 60° = \frac{1}{2}r\sqrt{3}$ giving $AB = 2AC = r\sqrt{3}$. Hence (D).

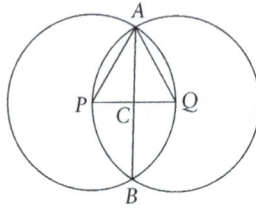

**Solution 17:** Draw the three radii to the points of tangency, and observe that these will be perpendicular to the respective tangents. Let the required angle be $x$, and label the other angles $y, z, b, c$ as shown, observing that

(1) $\angle JBK = \angle JBT = b$ because triangles $BKJ$ and $BTJ$ are congruent (they have side $BJ$ in common, $KJ = TJ$=radius of the circle $KTL$, and $BKJ = BTJ = 90°$), and similarly
(2) $\angle BCJ = \angle JCL = c$.

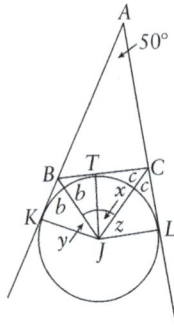

Now, in quadrilateral $AKJL$,

$$x + y + z = 360° - (90° + 90° + 50°) = 130°. \qquad (1.31)$$

Also:

$$b + y = 90° \text{ and } c + z = 90°,$$
$$\text{therefore } b + c + y + z = 180°. \qquad (1.32)$$
$$\text{But} \quad x = 130° - y - z, \text{ by (1.31)}, \qquad (1.33)$$
$$\text{and} \quad x = 180° - b - c, \text{ from } \triangle BCJ, \qquad (1.34)$$
$$\text{hence} \quad 2x = 310° - (z + y + b + c), \text{ adding (1.33)}$$
and (1.34),

$$= 310° - 180° = 130°, \quad \text{using (1.32)},$$

so that $\quad x = \dfrac{130}{2} = 65°.$  Hence (B).

**Solution 18:**   Using the converse of 'the angle in a semicircle is a right angle' theorem, we have that $AC$ and $BD$ are diameters, so pass through $O$.

Thus, diameter of circle $ABCD$ = length of diagonal = $\sqrt{(12^2 + 5^2)} = 13$.

Using the formula $\frac{1}{2}ab\sin\theta$ for the area of a triangle, we have:

$$\text{area } \triangle BOA = \frac{1}{2}\cdot\frac{13}{2}\cdot\frac{13}{2}\cdot\sin\theta = \frac{1}{2}\cdot 6\cdot 5, \quad \text{so} \quad \sin\theta = \frac{120}{169}. \text{ Hence (A).}$$

**Solution 19:**   We are given that $24\pi$ is twice the area of each unshaded region, hence each of these areas is $\frac{24\pi}{2} = 12\pi$. Now we have:

$$\text{area of each circle} = \text{area of shaded region}$$

$$+\text{area of one unshaded region}$$

$$= 24\pi + 12\pi = 36\pi$$

$$= \pi r^2, \quad \text{where } r \text{ is the radius of each circle,}$$

therefore   $r = 6$.

Thus the circumference is $2\pi r = 12\pi$. Hence (B).

**Solution 20:**   The key thing to notice here is that we have a *Pythagorean triple*:

$$5^2 + 12^2 = 13^2.$$

Pythagoras' theorem and its converse together state this equivalence: In any triangle $ABC$,

$$AB^2 + BC^2 = AC^2 \quad \Leftrightarrow \quad \angle ABC \text{ is a right angle.}$$

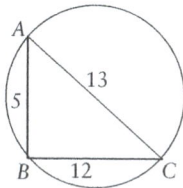

We deduce that $\hat{B}$ is a right angle, and then, using the converse of the theorem that the angle in a semicircle is a right angle, we conclude that $AC$ is the diameter of the circle passing through $A$, $B$, and $C$. Hence the required radius is $\frac{13}{2} = 6.5$ cm, making (C) the correct choice.

**Solution 21:**   The area is invariant, that is,

area of the square = area of rectangle = $18 \times 8 = 144 \, \text{m}^2$.

And, since there can be only one length of side for a square of given area,

$$\text{side of square} \; = \sqrt{144} = 12 \text{ m}$$

$$\text{therefore perimeter} \; = 12 \times 4 = 48 \text{ m. Hence (E).}$$

It's quite clear that the actual dissection of the rectangle is immaterial. This means that, while there are many ways in which a rectangle of a given finite area can be dissected and reassembled into a square, the square produced is unique.

**Solution 22:**   As tangents to the circle from a common point, $AP = AS$ and $BS = BR$, giving

$$AP + AS + BS + BR = 2(AS + BS)$$
$$= 2 \times 10 = 20.$$

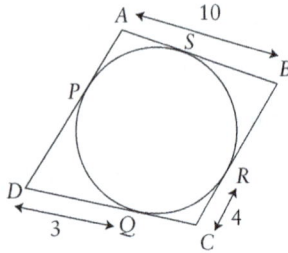

Similarly, $DP = PQ$ and $CQ = CR$, giving

$$DP + PQ + CQ + CR = 2(PQ) + 2(DP)$$
$$= 2 \times 3 + 2 \times 4 = 14.$$

Therefore the perimeter is $14 + 20 = 34$ cm.   Hence (A).

**Solution 23:**   Let the old pyramid of volume $V$ have base of side $s$ and height $h$, and let the new pyramid of volume $V_1$ have base of side $s_1$ and height $h_1$. Then

$$V_1 = \frac{1}{3}s_1^2 h_1 = \frac{1}{3}\left(\frac{9s}{10}\right)^2 \left(\frac{11h}{10}\right) = \frac{891}{1000}\left(\frac{1}{3}s^2 h\right) = \frac{891}{1000}V,$$

$$\text{so that} \quad V_1 = \frac{89.1}{100}V = 89.1\% \, V.$$

Therefore the percentage decrease is $100V - 89.1\,V = 10.9\,V$. Hence (D).

**Solution 24:**   The region of intersection is composed of two triangles each of area $A$ and four segments each of area $B$. We can easily see that each of the triangles is equilateral by observing that the sides are the radii of the two circles with equal radii.

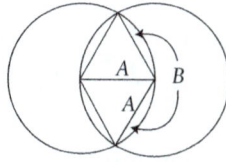

Since $A + B = $ *area of sector* $= \frac{r^2\theta}{2} = \frac{1}{6}1^2\pi = \frac{\pi}{6}$, and $A = $ *area of triangle* $= \frac{1}{2}ab\sin\theta = \frac{1}{2}1^2\sin 60° = \frac{\sqrt{3}}{4}$, we find the required area to be

$$2A + 4B = 4(A + B) - 2A$$

$$= \frac{2\pi}{3} - \frac{\sqrt{3}}{2}. \quad \text{Hence (B)}.$$

**Solution 25:** We use the symmetry of the regular hexagon, composed of six equilateral triangles of which the shorter diagonals of the hexagon form medians. As can be seen from the diagram below, with the third median of the equilateral triangle drawn, the shaded region is made up of two small triangles, of which six would make up the equilateral triangle. Thus the required area occupies

$$2 \times \frac{1}{6} \times \frac{1}{6} \times R = \frac{R}{18}. \quad \text{Hence (D)}.$$

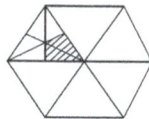

**Solution 26:** Let the shared vertex be $V$. We need to maximize the side of the triangle and keep the triangle's vertices in the square.

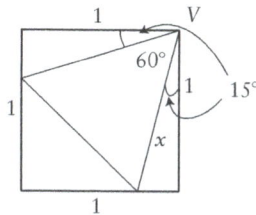

Clearly, for maximum area the two unknown vertices must lie symmetrically on two sides of the square. Now the two small angles on either side of the triangle's 60° angle at $V$ must be equal, and they sum to 30°, hence each is $15°$. Furthermore we know that $x \cos 15° = 1$. But

$$\cos 15° = \cos(45° - 30°) = \cos 45° \cos 30° + \sin 45° \sin 30°$$

$$= \frac{1}{\sqrt{2}} \cdot \frac{\sqrt{3}}{2} + \frac{1}{\sqrt{2}} \cdot \frac{1}{2} = \frac{1 + \sqrt{3}}{2\sqrt{2}},$$

hence $x = \dfrac{2\sqrt{2}}{1 + \sqrt{3}}$. Substituting into the formula for the area $A$ of the triangle, we get

$$A = \frac{1}{2} x^2 \sin 60°$$

$$= \frac{1}{2} \cdot \frac{8}{4 + 2\sqrt{3}} \cdot \frac{\sqrt{3}}{2}$$

$$= \frac{\sqrt{3}}{2 + \sqrt{3}}$$

$$= \frac{\sqrt{3}(2 - \sqrt{3})}{(2 + \sqrt{3})(2 - \sqrt{3})} = 2\sqrt{3} - 3. \text{ Hence (C).}$$

**Solution 27:**   From the diagram, we see that $A\hat{O}B = B\hat{O}C = C\hat{O}D = 60°$, so that, by symmetry, $\theta = Y\hat{O}D = 30°$, where $Y$ is the point of tangency of the circles.

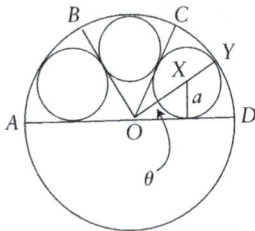

Let $X$ be the centre of the right-hand circle. Since $OX \sin \theta = a$ and $\sin 30° = \frac{1}{2}$, we have $OX = 2a$. Now:

$$r = OY = OX + XY$$

$$= 2a + a,$$

$$\text{therefore} \quad a = \frac{r}{3}. \text{ Hence (D).}$$

**Solution 28:**   The diagram shows one sector of radius 6. The circular arc length of this sector is $\frac{12\pi}{6} = 2\pi$. Referring now to the cone, we deduce that the circumference of its circular base is given by $2\pi r = 2\pi$ (from its given construction) and thus its radius $r$ must be 1.

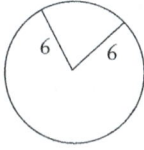

Now, by Pythagoras' theorem, the height of each cone is $\sqrt{6^2 - 1^2} = \sqrt{35}$. Hence (C).

**Solution 29:**   Referring to the diagram below, let $X$ be the midpoint of $EF$, and draw lines $OX$ and $CF$. Let $a = AE = EF = FB$, and $b = BC$.

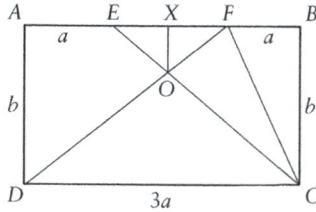

By symmetry, $XO$ is perpendicular to $AB$, hence triangles $EXO$ and $EBC$ are similar. Therefore

$$\frac{XO}{BC} = \frac{EX}{EB}, \text{ and so } XO = b\frac{EX}{EB}.$$

Since $EX = \frac{1}{2}EF$ and $EF = \frac{1}{2}EB$, we have

$$\frac{EX}{EB} = \frac{1}{4}, \text{ so that } XO = \frac{b}{4}.$$

Therefore

$$\text{area of triangle } EFO = 2 \times \frac{1}{2} \times \frac{a}{2} \times \frac{b}{4} = \frac{ab}{8};$$

$$\text{also, area of triangle } EFC = \frac{ab}{2}.$$

$$\text{Therefore area of triangle } OFC = \frac{ab}{2} - \frac{ab}{8} = \frac{3ab}{8}.$$

Finally, the required ratio is $\frac{3ab}{8} \times \frac{1}{3ab} = \frac{1}{8}$. Hence (E).

**Solution 30:**   Referring to the diagram below,

$$C\hat{O}F = O\hat{F}D + C\hat{D}F \quad \text{(ext.} = \text{sum of interior opp. angles).}$$

But   $C\hat{D}F = O\hat{F}D$,   (since $\triangle FOD$ is isosceles),

therefore   $C\hat{O}F = 2C\hat{D}F.$

Now,   $C\hat{D}F = C\hat{O}E$   (corresponding angles)

$= B\hat{O}D$   (vertically opposite angles)

$= A\hat{B}E = \theta$   (alternate angles).

Therefore  $E\hat{O}F = C\hat{O}F - C\hat{O}E = 2\theta - \theta = \theta.$  Hence (B).

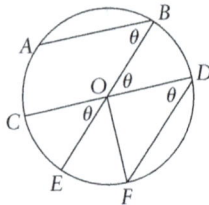

**Solution 31:**   It helps to get the diagram looking right according to the given information, by making a few trial diagrams first. Then label with the given information, and label the unknown $x = AC$. If we let $\theta = A\hat{C}D$ then, since $BE = EC = 1$ makes triangle $BEC$ isosceles, we have $E\hat{B}C = \theta$ and so the exterior angle $B\hat{E}A = \theta + \theta = 2\theta.$

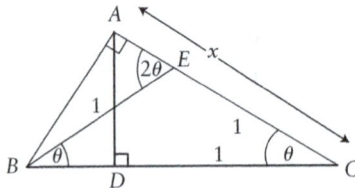

In triangle $ADC$ we see that $\cos\theta = \frac{1}{x}$, and in triangle $ABE$ we see that $\cos 2\theta = x - 1$, so, using the identity

$$\cos 2\theta = 2\cos^2\theta - 1,$$

we have   $x - 1 = \dfrac{2}{x^2} - 1,$

therefore   $x^3 = 2$,  so that $x = 2^{\frac{1}{3}}.$  Hence (A).

**Solution 32:**    Triangles $ABC$ and $IGC$ are similar. Therefore

$$\frac{AB}{IG} = \frac{BC}{GC} = \frac{AC}{IC},$$

$$\text{hence}\quad GC = \frac{IG \cdot a}{c}. \tag{1.35}$$

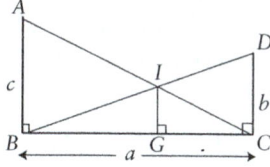

Triangles $DCB$ and $IGB$ are similar. Therefore

$$\frac{DC}{IG} = \frac{CB}{GB} = \frac{DB}{IB},$$

$$\text{hence}\quad GB = \frac{IG \cdot a}{b}. \tag{1.36}$$

Now $BC = GB + GC$, so that

$$a = IG\left(\frac{a}{b} + \frac{a}{c}\right) \quad \text{from (1.35) and (1.36) above,}$$

$$\text{therefore}\quad IG = \frac{abc}{ac + ab}$$

$$= \frac{bc}{b+c}. \quad \text{Hence (B).}$$

**Solution 33:**    It is best to draw in the diagonals $AC, BD$ on a separate diagram, but we will just imagine them in the diagram already given. Since $AB$ is a diameter, $A\hat{D}B$ and $A\hat{C}B$ are right angles. So, by Pythagoras' theorem:

$$AD^2 + BD^2 = AB^2 \tag{1.37}$$
$$AC^2 + BC^2 = AB^2. \tag{1.38}$$

But $BC = AD$ gives $BC^2 = AD^2$, so subtracting (1.38) from (1.37) gives us

$$BD^2 - AC^2 = AB^2 - AB^2 = 0,$$

$$\text{therefore}\quad BD = AC. \tag{1.39}$$

Now, by Ptolemy's theorem,

$$AD \cdot BC + DC \cdot BA = BD \cdot AC. \tag{1.40}$$

Using (1.39) in (1.40) and putting $BC = AD$ gives

$$AD^2 + DC \cdot AB = AC^2 = AB^2 - BC^2,$$
$$\text{therefore}\quad 12^2 + DC \cdot 36 = 36^2 - 12^2,$$
$$\text{so that}\quad DC = 28.\quad \text{Hence (A)}.$$

**Solution 34:** Draw the bisector $YT$ of angle $Y$, meeting $XZ$ at $T$, and let $ZT = p$ so that $XT = y - p$. Then triangle $XYT$ is isosceles, so we have $TY = y - p$ and we have exterior angle $ZTY = 2a°$. Therefore triangles $ZTY$ and $ZYX$ are similar, hence

$$\frac{ZT}{ZY} = \frac{TY}{YX} = \frac{ZY}{ZX},$$
$$\text{therefore}\quad \frac{ZY}{ZT} = \frac{YX}{TY} = \frac{ZX}{ZY},$$
$$\text{therefore}\quad \frac{x}{p} = \frac{z}{y-p} = \frac{y}{x},$$
$$\text{therefore}\quad (i)\ x^2 = py \quad \text{and} \quad (ii)\ y^2 - py = xz.$$

Subtracting (i) from (ii) gives

$$y^2 - x^2 = xz + py - py = xz.\quad \text{Hence (A)}.$$

**Solution 35:** Draw the perpendiculars $PR$, $QS$ from centres to points of tangency. Let the radius $QS$ of the larger circle be $r$.

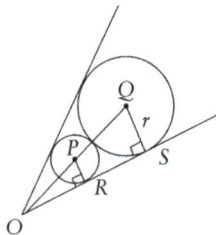

Both triangles $OPR$ and $OQS$ have sides in the ratio $2:1:\sqrt{3}$, since all triangles with angles $30°, 60°$ and $90°$ share this property. We are given that $PR = 10$, so that $OP = 20$ and hence $OQ = 20 + 10 + r = 30 + r$. Then, in triangle $OQS$, $\frac{r}{30+r} = \frac{1}{2}$, giving $r = 30$. Hence (D).

**Solution 36:** Suppose the square has side $x$, area $x^2$. Then $BC = x + 2$ and $AB = x - 1$. Thus area of rectangle is $(x - 1)(x + 2) = x^2$ (the two areas are given to be the same) giving unique solution $x = 2$ and thus $BC = 4$. Hence (C).

**Solution 37:** Let $M$ be the centre of the circle, and draw the two radii $MP$, $MQ$, so that $AM = MB = MP = MQ = 4$. Since triangle $ABC$ is given to be equilateral, we have $\angle PAM = 60°$, hence $\angle APM = 60°$ because

$AM = MP$, so that also $\angle AMP = 60°$ and triangle AMP is equilateral. Similarly $QMB = 60°$ and hence $PMQ = 60°$. Let the symbol '$\triangle$' stand for 'area in the figure bounded by the points ...'

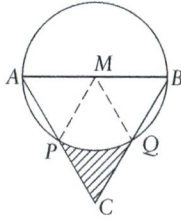

Then

$$\triangle ABC = \triangle AMP + \triangle BMQ + \triangle MPQ + \triangle PQC,$$

therefore $\triangle PQC = \triangle ABC - (\triangle AMP + \triangle BMQ + \triangle MPQ)$.

Now $\triangle AMP = \frac{1}{2} \cdot 4 \cdot 4 \cdot \sin 60° = 4\sqrt{3}$. Similarly, $\triangle BMQ = 4\sqrt{3}$, $\triangle MPQ = \frac{60}{360} \times \pi \times 4 \times 4 = \frac{8\pi}{3}$ and $\triangle ABC = \frac{1}{2} \times 8 \times 8 \times \sin 60 = 16\sqrt{3}$. Therefore

$$\triangle PQC = 16\sqrt{3} - 8\sqrt{3} - \frac{8\pi}{3}$$

$$= 8\left(\sqrt{3} - \frac{\pi}{3}\right) \text{ cm}^2. \text{ Hence (D)}.$$

**Solution 38:** Draw line $BD$ and let $BD = 2r$, $AB = h$ and $DC = x$. Let $\angle BDA = \theta$. Then $\angle ACB = \theta$ (angles subtended by same arc). Focus on triangle $ABC$.

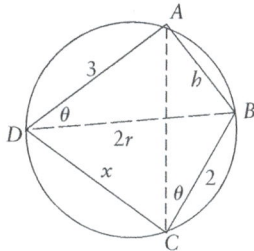

The cosine rule gives

$$h^2 = (1 + 2\sqrt{6})^2 + 2^2 - 4(1 + 2\sqrt{6})\cos\theta$$

$$= 29 + 4\sqrt{6} - 4(1 + 2\sqrt{6})\cos\theta$$

$$= 29 + 4\sqrt{6} - \frac{6}{r}(1 + 2\sqrt{6}), \tag{1.41}$$

because, in triangle $DAB$, $BD$ is the diameter so $\angle DAB = 90°$, giving $\cos\theta = \frac{3}{2r}$. Now, applying Pythagoras' theorem in triangle $DAB$,

$$BD^2 = AB^2 + DA^2,$$

$$\text{therefore}\quad 4r^2 = 29 + 4\sqrt{6} - \frac{6}{r}(1 + 2\sqrt{6}) + 9, \quad \text{(using (1.41))}$$

$$= 38 + 4\sqrt{6} - \frac{6}{r}(1 + 2\sqrt{6}),$$

$$\text{so that}\quad 4\sqrt{6}(r - 3) = 2(2r^3 - 19r + 3)$$

$$= 2(r - 3)(2r^2 + 6r - 1).$$

This shows that $r = 3$ cm is one solution, and so $BD = 2r = 6$ cm. Hence (D). (But there may be another solution too – check!)

**Solution 39:**  The diagram below illustrates all the given information.

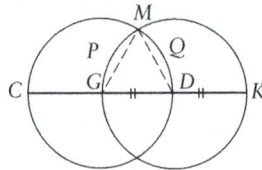

Let the second circle cut the first at $M$ as shown and let $GD = DK = r$. As radii of he respective circles, $GM = GD = r$, therefore triangle $GMD$ is equilateral. By symmetry, the required area $A$ is twice the sum of the area of sector $GMQD$ and the area of segment $GPM$. Hence

$$\frac{1}{2}A = \frac{60}{360} \times \pi r^2 + \left(\frac{60}{360} \times \pi r^2 - \frac{1}{2}r^2 \sin 60\right)$$

$$= \frac{\pi r^2}{3} - \frac{r^2\sqrt{3}}{4}.$$

But $GK = 2r = 8$, so $r = 4$, giving the required area as $2(\frac{16\pi}{3} - 4\sqrt{3})$ cm$^2$. Hence (E).

**Solution 40:**  It is a good idea to experiment with the diagram so that it better reflects the given line proportions. Then join $AC$ as shown below.

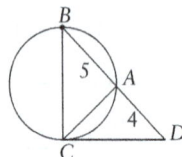

Then $\angle BAC = 90°$ (angle in a semicircle), and $\angle BCD = 90°$ (tangent $\perp$ to radius). Hence triangles $ABC$ and $CBD$ are equiangular and therefore similar. Thus we have:

$$\frac{AB}{CB} = \frac{BC}{BD} = \frac{AC}{CD}.\tag{1.42}$$

Now, using Pythagoras' theorem,

$$CD^2 = BD^2 - BC^2$$
$$= BD^2 - AB \cdot BD \quad \text{(because } BC^2 = AB \cdot BD \text{ from (1.42))}$$
$$= BD(BD - AB) = BD \cdot AD = 9 \times 4 = 36.$$

Therefore $CD = \sqrt{36} = 6$ cm. Hence (D).


**Solution 41:** Concentrate on parallelogram $ABNM$. We have $AB = BM = BN = r$, the radius of the circle. Observe that $MN = AB = r$ and $AM = BN = r$ (opp. sides of a parallelogram), so we actually have a rhombus. Therefore triangle $BMN$ and triangle $BAM$ are equilateral and all their angles are $60°$. In the right angled triangle $ABD$, then, $BD = \sec 60° \times AB = 2 \times 1 = 2$. Thus the point $M$ is the midpoint of $BD$.

Since $BAMN$ is a parallelogram, $AB \parallel MN$; but $AB \perp DC$, so $MN \parallel DC$. This implies that $N$ is the midpoint of $BC$, so $MN$ is a line joining the midpoints of $BD$ and $BC$, hence $DC = 2$ cm. That $MN \parallel DC$ also implies that $\angle BDC = 60°$, so triangle $BDC$ is equilateral with side 2. Therefore

$$\text{area of triangle } BDC = \frac{1}{2} \times 2^2 \times \sin 60° = \sqrt{3} \text{ cm}^2,$$
$$\text{and area of sector } BMN = \frac{60}{360} \times \pi \times 1^2 = \frac{\pi}{6} \text{ cm}^2.$$

Therefore the required area is $(\sqrt{3} - \frac{\pi}{6})$ cm$^2$. Hence (A).


**Solution 42:** Let the radius of the medium circle be $R$ cm, so that $AB = BO = R$, and let the radius of the smaller circle be $r$ cm, while the radius of the large circle is $AO = 2R$. Now, the small circle touches the large circle internally, hence, $ON = 2R - r$. The small circle also touches the medium circle externally hence $BN = R + r$. But, since $BN \perp AB$, we can apply Pythagoras' theorem in triangle $OBN$:

$$BN^2 + BO^2 = ON^2,$$
$$\text{therefore} \quad (R + r)^2 + R^2 = (2R - r)^2,$$
$$\text{hence} \quad 2R^2 - 6Rr = 0,$$
$$\text{so} \quad R(R - 3r) = 0.$$

This gives $R = 3r$. Considering triangle $OBN$, we have perimeter

$$OB + BN + NO = R + (R + r) + (2R - r)$$
$$= 4R = 8 \quad \text{(given)},$$

therefore $\quad R = 2, \quad$ so $r = \frac{1}{3}R = \frac{2}{3}$. Hence (A).

**Solution 43:**  In the diagram, let the radius of the circumcircle be $R$ and let $a, b$ and $c$ be triangle's sides. Let the heights from $A$ to $BC$, $B$ to $AC$, and $C$ to $AB$ be $h_1$, $h_2$ and $h_3$, respectively.

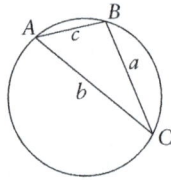

Now, the area of triangle is $A = \frac{1}{2}ab \sin C$. Therefore

$$\frac{ah_1}{2} = \frac{1}{2}ab \sin C, \quad \text{giving} \quad h_1 = b \sin C$$

$$\frac{bh_2}{2} = \frac{1}{2}ab \sin C, \quad \text{giving} \quad h_2 = a \sin C$$

$$\frac{ch_3}{2} = \frac{1}{2}ab \sin C, \quad \text{giving} \quad h_3 = \frac{ab}{c} \sin C.$$

The required sum is thus given by

$$h_1 + h_2 + h_3 = b \sin C + a \sin C + \frac{ab}{c} \sin C$$
$$= \frac{\sin C}{c} \times (bc + ac + ab).$$

The expression $\frac{\sin C}{c}$ should immediately remind us of the sine rule:

$$\frac{\sin C}{c} = \frac{\sin A}{a} = \frac{\sin B}{b} = \frac{1}{2R}.$$

where $R$ is the radius of the circumcircle (this last bit is usually forgotten by many people). But it is given that $R = 1$, hence $\frac{\sin C}{c} = \frac{1}{2}$. Now things are becoming clearer:

$$(a + b + c)^2 = a^2 + b^2 + c^2 + 2ab + 2bc + 2ac,$$
$$\text{therefore} \quad 9 = 5 + 2ab + 2bc + 2ac,$$
$$\text{hence} \quad 2 = ab + bc + ac.$$

This gives $h_1 + h_2 + h_3 = \frac{1}{2} \times 2 = 1$. Hence (A).

**Solution 44:**   The triangles $ABF$ and $GCF$ are congruent, because we have $\angle AFB = \angle GFC$ (vertically opposite angles), $\angle ABF = \angle GCF = 90°$, and $AB = CG$ (sides of the square). Therefore $F$ is the midpoint of $BC$. Now, convince yourself that triangles $AED$ and $FEB$ are similar (three angles equal). But $AD = 2FB$ (since $ABCD$ is a square and we have shown that $F$ is the midpoint of $BC$). Therefore $AE = 2EF$, so that $AE = \frac{2}{3}AF$ and $EF = \frac{1}{3}AF$.

Now to find $FK$, consider triangle $BCG$. Since $BF = FC$, $GF$ is its median, and similarly $BM$ is also a median. But these medians $BM$ and $GF$ intersect at $K$, so we have $FK = \frac{1}{3}FG = \frac{1}{3}AF$, by the congruence of triangles $ABF$ and $GCF$. Therefore, finally, $AE : EF : FK = \frac{2}{3}AF : \frac{1}{3}AF : \frac{1}{3}AF = 2 : 1 : 1$. Hence (B).

**Solution 45:**   Let $K$, a point on $DC$, be the circle's centre. Produce $KC$ to meet the circle's circumference at $M$, and drop perpendiculars from $K$ to meet $AB$ at $L$ and from $A$ to meet $DM$ at $F$.

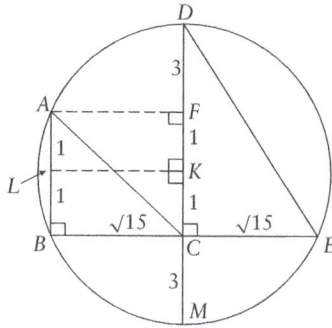

Now, since the perpendicular from the centre must bisect any chord of a circle, we have $AL = LB = 1$ (because we are given $AB = 2$). Therefore $FK = KC = 1$ too (opposite sides of rectangles). Then, since $KM = 4$ (radius), we have $CM = KM - KC = 4 - 1 = 3$. Next, by the principle of intersecting chords, we have $BC^2 = DC \cdot CM = 5 \cdot 3$, so $BC = \sqrt{5 \cdot 3} = \sqrt{15}$.

Now $AF = \sqrt{15}$ and $DF = 3$. Hence, using Pythagoras' theorem in triangle $AFD$,

$$AD^2 = AF^2 + DF^2$$
$$= 15 + 9 = 24,$$

therefore   $AD = 2\sqrt{6}$ cm.  Hence (D).

**Solution 46:**   Since $AG$ is a median of triangle $ABC$, area of triangle $ABG = $ area of triangle $BGC$. Now, consider the triangles $ABG$ and $AEB$. They have a common base $AB$ and the line $BG$ passes through their respective vertices. The common base theorem tells us that the ratio of their areas equals the ratio of the distances of their vertices from point $B$. That is,

$$\frac{\text{area } ABE}{\text{area } ABG} = \frac{BE}{BG}.$$

Our goal now is to find $\frac{BE}{BG}$. Since $BK = KC$, $AK$ is also a median of triangle $ABC$, hence $AK$ and $BC$ are medians intersecting at $E$. Hence $BE = \frac{2}{3}BG$. But area $ABG = \frac{1}{2}$ area $ABC$, hence

$$\frac{\text{area } ABE}{\text{area } ABC} = \frac{\text{area } ABE}{2 \times \text{area } ABG} = \frac{1}{2} \cdot \frac{2}{3} = \frac{1}{3}.$$

That is, $A_2 = 3A_1$. Hence (A).

**Solution 47:**   Let $Y$, $M$ and $Z$ be the points of tangency of the line to the three circles, so that $AY$, $CM$, $BZ$ are all perpendicular to $YZ$. Let $X$ and $L$ be points on $AY$ and $K$ be a point on $BZ$ such that $BX$, $CL$, $KC$ are all perpendicular to $AY$ (hence of course to $CM$ and $BZ$ also).

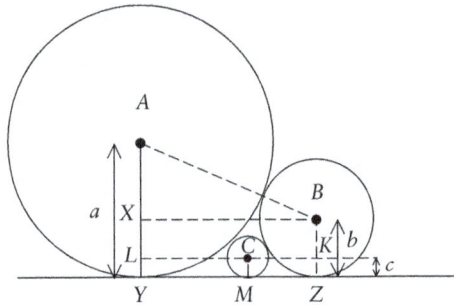

We see immediately that:

$$BC = b+c, \ BK = b-c, \ AX = a-b, \ AB = a+b,$$
$$AC = a+c \ \text{ and } \ AL = a-c.$$

We naturally think of using Pythagoras' theorem in some of the many right-angled triangles in that diagram. First, in triangle $ACL$:

$$\begin{aligned}
YM^2 = LC^2 = AC^2 - AL^2 \\
= (a+c)^2 - (a-c)^2 \\
= (a+c+a-c)(a+c-(a-c)) \\
= 4ac,
\end{aligned}$$

therefore   $YM = 2\sqrt{ac}.$

Next, in triangle $BCK$:

$$MZ^2 = KC^2 = BC^2 - BK^2$$
$$= (b+c)^2 - (b-c)^2$$
$$= (b+c+b-c)(b+c-(b-c))$$
$$= (2b)(2c),$$

therefore $\quad MZ = 2\sqrt{bc}.$

We can now write down the length of the common tangent $YZ$:

$$YZ = MY + MZ = 2(\sqrt{ac} + \sqrt{bc}).$$

Again, applying Pythagoras' theorem in triangle $AXB$:

$$XB^2 + AX^2 = AB^2,$$

therefore $\quad 4(\sqrt{ac} + \sqrt{bc})^2 + (a-b)^2 = (a+b)^2,$

so $\quad 4(ac + 2c\sqrt{ab} + bc) + a^2 + b^2 - 2ab = a^2 + b^2 + 2ab,$

hence $\quad 4(ac + 2c\sqrt{ab} + bc) = 4ab.$

This gives $\quad c(a + 2\sqrt{ab} + b) = ab,$

and so $\quad c(\sqrt{a} + \sqrt{b})^2 = ab,$

therefore $\quad c = \dfrac{ab}{(\sqrt{a} + \sqrt{b})^2}.$ Hence (A).

**Solution 48:** Label the angles $\angle QRP = \alpha$ and $\angle SRT = \beta$. Because $QRS$ is a straight line, $\alpha + \beta = 90$, hence $S\hat{T}R = \alpha$ and $Q\hat{P}R = \beta$.

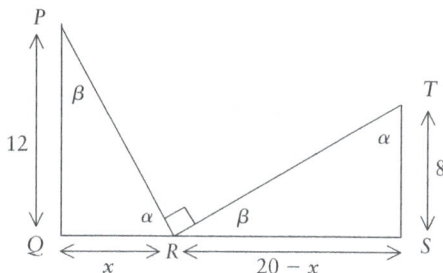

We now have similar triangles $QPR$ and $SRT$, so:

$$\frac{QP}{SR} = \frac{QR}{ST},$$

therefore $\quad \dfrac{12}{20-x} = \dfrac{x}{8},$

hence $\quad 8 \times 12 = x(20-x),$

giving the quadratic $\quad x^2 - 20x + 96 = 0.$

Solving gives $\quad x = \dfrac{20 \pm \sqrt{20^2 - 4 \times 96}}{2}$

$$= \frac{24}{2} \quad \text{or} \quad \frac{16}{2}.$$

Hence $x = 12$ or $8$, both of which are possible values for length $QR$. Since $12 - 8 = 4$, we have (A) as the correct answer.

**Solution 49:** Using the theorem that the three medians of a triangle all meet at a point two-thirds of the way along each, and using the given information, we are led to the diagram below:

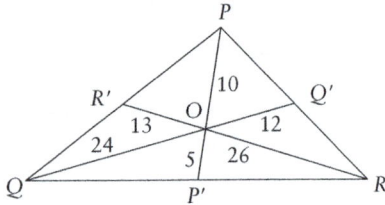

Looking at, for example, the triangle $OQR$, we see that two sides and one median are given, and it is intuitively clear that the third side $QR$ of the triangle is determined by these. But how can we calculate it? One good idea would be to see the triangle as half of a parallelogram, with the required $QR$ as diagonal. We achieve this by producing the line $OP'$ to a point $T$ so that $P'$ is the midpoint of $OT$. Then the diagonals of the quadrilateral $ORTQ$ bisect each other, so creating opposite pairs of similar triangles, and hence $ORTQ$ is a parallelogram (why?). Therefore $QT = OR = 26$ and $TR = QO = 24$.

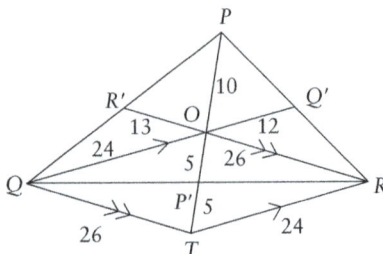

Now that we have some triangles in which all three sides are known, it is wise to check for standard forms. Observe that, in triangle $OQT$, $OT : OQ : QT = 5 : 12 : 13$, which is a Pythagorean triple. Hence, by the converse of Pythagoras' theorem, triangle $OQT$ is right-angled at $O$, and so of course is $OQP'$, to which we can then apply Pythagoras' theorem to obtain $QP'$. At this stage you should perhaps redraw that diagram to make it look right! You will be able to follow the rest of this solution much more easily. We look for right-angled triangles that contain segments of the perimeter, and apply Pythagoras' theorem to them. There are three such triangles, each with right angle at $O$: $OQP'$, $PQO$, $POQ'$. Hence

$$QR = 2QP' = 2\sqrt{QO^2 + OP'^2} = 2\sqrt{24^2 + 5^2} = 2\sqrt{601},$$

$$PQ = \sqrt{QO^2 + OP^2} = \sqrt{24^2 + 10^2} = 26,$$

$$PR = 2PQ' = 2\sqrt{PO^2 + OQ'^2} = 2\sqrt{10^2 + 12^2} = 4\sqrt{61}.$$

Hence the required perimeter $= 26 + 4\sqrt{61} + 2\sqrt{601}$.  Hence (C).

**Solution 50:**    For contrast with the algebraic proof and the dissection proof of Pythagoras' theorem given in Section 1.1.3, this is Euclid's *deductive* proof. This is the proof that so impressed the philosopher Thomas Hobbes that he is said to have fallen in love with geometry!! The diagram is the one on the left.

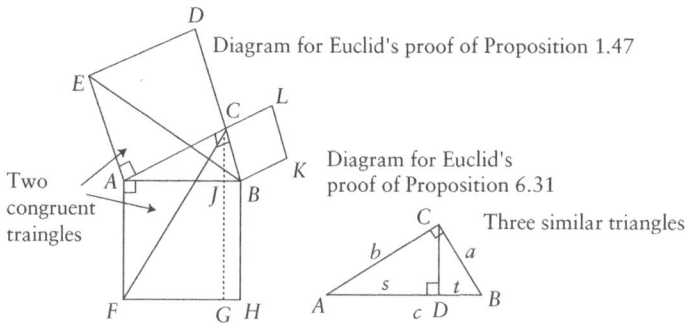

Diagram for Euclid's proof of Proposition 1.47

Two congruent traingles

Diagram for Euclid's proof of Proposition 6.31

Three similar triangles

First Euclid makes absolutely sure of things the diagram smuggles in, by carefully constructing the three squares on the sides, and proving that $BCD$ and $ACL$ are straight lines! (He uses a previous proposition to the effect that if two lines make adjacent angles with a third line that sum to two right angles then they are in the same straight line.) He goes on roughly as follows:

The triangles $ABE$ and $AFC$ are congruent, by Congruence Criterion (2): two sides and included angle. For $AB = AF$ (sides of the same square $AH$), $AE = AC$ (sides of the same square $AD$), and $B\hat{A}E = F\hat{A}C$, since each is the result of adding $\theta = B\hat{A}C$ to a right angle. Therefore these two triangles have the same area. But $\triangle ABE$ is half the area of the square $AD$ (stands on same base $AE$ and has same altitude $AC$), while $\triangle AFC$ is half the area of the rectangle $AG$ (stands on same base $AF$ and has same altitude $AJ$). Therefore the square $AD$ is equal in area to the rectangle $AG$.

In the same manner, it can be shown that the square $BL$ is equal in area to the rectangle $BG$. Therefore the sum of the two squares is equal in area to the sum of the two rectangles, which is the square on the hypotenuse $AB$. QED (*Quod Erat Demonstrandum*, as Euclid would say, except he would not say it in Latin, nor in English, but in Greek. 'That which was to be proved', is the usual translation, but we prefer: '$W^5$ – which was what was wanted'.)

**Solution 51:**  This (apparently easier) proof of Pythagoras' theorem appears later in Euclid in Book 6 as Proposition 6.31. But for Euclid and his contemporaries, this proof uses ideas that are far more sophisticated. Indeed, we have placed these ideas in our section on 'advanced geometry of the triangle' where the proof in the previous problem needs only ideas in our 'basic geometry' section. For this new proof is based on the theory of similarity, which had been thrown into confusion and crisis by the discovery of incommensurables (see Section 1 of Toolchest 2). For how could you talk of *the proportion of two lines* if there might not be any rational (*thinkable*) number to express such proportion? However, the great Eudoxus, a pupil of the philosopher Plato, came to the rescue by creating a new theory of proportion for geometrical magnitudes, quite independent of numbers. Euclid expounded Eudoxus' theory in Book 5, and then at last the theory of similarity could be applied with a clear mathematical conscience in Book 6 to great effect. Here is the proof of Pythagoras' theorem (although Euclid gives the result for 'any similar rectilinear figure' similarly described on the sides of the triangle). Referring to the diagram on the right above:

Draw the perpendicular $CD$ to the line $AB$. Since their angles are all equal (we skip that bit here as it is easy), the three triangles $\triangle ABC$, $\triangle ACD$, $\triangle CBD$, are all similar. Therefore

$$\frac{c}{a} = \frac{a}{t} \quad \text{and} \quad \frac{c}{b} = \frac{b}{s}, \quad \text{hence } a^2 + b^2 = tc + sc = (t+s)c = c^2.$$

# 2 Algebraic inequalities and mathematical induction

By the end of this topic you should be able to

(i) Use the method of mathematical induction to prove theorems.
(ii) Prove and use the inequality $x + \frac{1}{x} \geq 2$ for $x > 0$.
(iii) Prove and use Weierstrass' inequality.
(iv) Prove and use the Cauchy–Schwarz inequality.
(v) Prove and use the Arithmetic–Geometric (AM–GM) inequality.
(vi) Understand the logic of the discriminant of a quadratic equation, and use it to solve problems involving roots of such an equation.
(vii) Understand the properties of the modulus function, and solve inequalities involving it.

Here are three appetizer problems that you should be able to solve when you have worked through this Toolchest, although you may of course give them a go now. The answers to the first two can easily be found by trial and error, or by guesswork, but the tools in this Toolchest supply the means of proving them beyond doubt. The third problem looks more fearsome, but is a beauty! For any positive integer $n$ we define $n! := 1 \times 2 \times 3 \times \cdots \times (n - 1) \times n$, and we call this product '$n$-factorial'. If you start thinking about 'What, to the power of $n$, will always be greater than such a huge thing as

$n$ factorial?' you will begin to feel the fascination of it. This result can be proved by method of induction, but can be solved more economically using the important inequality of the means (AM–GM inequality). The solutions to these three problems are given at the end of the Toolchest.

**Appetizer Problem 1:**   *Prove that $n! > 2^n$ for all integers $n \geq 4$.*

**Appetizer Problem 2:**   *I have a set of four positive numbers whose sum is 12. What is the maximum product?*

(A) 96      (B) 81      (C) 256      (D) 144      (E) 60

**Appetizer Problem 3:**   *Prove that, for all positive integers $n$,*

$$\left(\frac{n+1}{2}\right)^n \geq n!$$

## 2.1  The method of induction

We shall use this very important method to prove many of the results here and elsewhere in this book. The method of induction normally applies to the set of numbers called the **natural** (or counting) numbers, $\mathbb{N}$.

$$\mathbb{N} = \{1, 2, 3, 4, \ldots\}.$$

Some people include the number zero, denoted 0, in the natural numbers; that is simply a matter of convention and we shall consistently adopt the definition given above. The set of natural numbers is also called the set of *positive integers*. It is crucial that you grasp clearly the distinctions between this set and some other sets: the *integers*, the *real numbers*, the *non-negative real numbers* and the *positive real numbers*. The definitions and illustrations below should make the difference between these sets clear.

### Number systems

**Integers**   An integer is a whole number, and can be positive or negative. Zero is also (by convention) an integer. The set $\mathbb{Z}$ of integers can be denoted in a number of ways:

$$\mathbb{Z} = \{\ldots, -4, -3, -2, -1, 0, 1, 2, 3, 4, \ldots\}$$
$$= \{0, 1, -1, 2, -2, 3, -3, 4, -4, \ldots\}.$$

**Real numbers**   A real number is, geometrically speaking, any point on the number line (infinite in both directions). Amongst the real numbers are certain familiar and easily expressible ones, called **rational** numbers. These are the numbers expressible as fractions $\frac{a}{b}$, where $a$ and $b$ are integers. These include the integers themselves, of course, by taking $b = 1$. We often denote the set of rational numbers by $\mathbb{Q}$ and the set of real numbers by $\mathbb{R}$, and we say the rationals form a subset of the reals: $\mathbb{Q} \subset \mathbb{R}$. This 'inclusion' relation is actually a 'proper inclusion' – there are also real numbers which are not rational numbers, and we call these **irrational** numbers. Why such a rude name? It carries an echo of the original (and natural) human response to the discovery of such numbers. The name has come down to us from the ancient Greeks who first discovered that the diagonal-to-side ratios for such simple figures as the square, the cube and the regular pentagon, cannot be expressed as fractions. (That is, there does not exist a 'common measure' – a small unit which will measure each of the diagonal and the side. The Greeks referred to such a pair of line segments as 'incommensurable'.) They treated all proportions henceforth from a purely geometrical standpoint and rejected entirely the 'irrational' as not deserving the name of number! Other cultures were more accepting while still not understanding, and the earliest known irrational numbers (still among the most common you are likely to meet) are the square roots of $2, 3, 5$, or indeed the square roots of any non-square positive integer.

Using the brilliant invention of place value notation extended to fractions (we have the ancient Babylonians to thank for this) combined with the Hindu–Arabic decimal system (including the vital symbol for zero), every real number can be represented as a decimal. This decimal representation is now universally taught to quite young children, but became widely accepted and understood only after 1585 when Simon Stevin published his book *De Thiende*, in Flemish, rapidly translated into French as *La Disme* ('The Tenths'). The decimal representation of a real number must be one of three kinds. If it is either *terminating* or *endless and periodic*, then the number is rational, for it is expressible as the sum of a finite geometric series, if it terminates, and as the sum to infinity of a geometric series, if it is of the endless periodic kind. (See Toolchest 6.) If the decimal representation is endless and *non-periodic*, the number is irrational. To see this, we need only consider what happens when we apply the algorithm for expressing the fraction $\frac{a}{b}$ as a decimal. If it is already in its lowest terms, there are only $b$ possible remainders on dividing by $b$: $0, 1, 2, \ldots, b - 1$. If the remainder is ever 0, the decimal terminates; otherwise, after a finite number (less than $b$) of steps we will have one of the remainders repeated. From that point the cycle of digits in the decimal expansion will repeat, endlessly.

**Non-negative real numbers**    On the geometrical number line, every number is a point dividing the line into two sets – those numbers to the left and those to the right. Using this as an intuitive basis, we can define the usual ordering $x \leq y$ (equivalently, $y \geq x$) on the real numbers, read as '$x$ is less than or equal to $y$' or '$y$ is greater than or equal to $x$'. The strict ordering $x < y$ (equivalently, $y > x$) holds precisely when $x \leq y$ *and* $x \neq y$, and is read as '$x$ is strictly less than $y$'. These order relations satisfy certain important laws that are the basis for all arithmetic inequalities. A modern approach is to define the order relation by means of a set of *axioms of order*. We shall list some of these in Section 2 of this Toolchest. If $x$ is a real number such that $x \geq 0$, then $x$ is called a non-negative number. We can use interval notation to write the set of non-negative numbers as $[0, +\infty)$. For example, 0 is non-negative, $\frac{1}{6}$ is non-negative, 10 is also non-negative, but $-\frac{1}{2}$ and $-10$ are **negative numbers**. Using interval notation, and the usual mathematical symbol $\in$ for 'belongs to' or 'is an element of', we can write $-10 \in (-\infty, 0)$, meaning: 'the number minus ten belongs to the set of negative numbers'.

**Positive numbers**   If $x$ is a real number such that $x > 0$, then $x$ is called a positive number. This means that 0 is not a positive number, but $\frac{1}{2}$ is a positive number, so is 7, and so on. Using interval notation, the set of positive numbers is $(0, +\infty)$. The natural numbers are just the positive integers; we write: $\mathbb{N} = \mathbb{Z} \cap (0, +\infty)$.

### Induction analogue: getting a physical picture of the principle

Most of us have played with bricks as children – either real ones or smaller wooden toy ones. One favourite playground trick was to stand the bricks on their ends in a row as shown in the diagrams and the pictures.

Before 'the push'

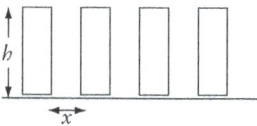

After 'the push'

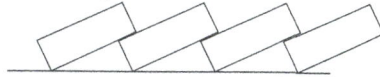

Pushing the end brick triggers off a 'chain reaction' and all the bricks will fall. However, this chain reaction will only occur if both the following conditions hold:

(i) at least one brick is given a push;
(ii) the distance $x$ between bricks satisfies $x < h$, where $h$ is the length of each brick (the height they stand above the level ground, assuming them to be rectangular and identical).

We think, of course, of ideal bricks, ignoring physical considerations like friction. We could have a sequence of bricks indefinitely long and, provided the two above conditions are met, **all** bricks will fall over.

With mathematical induction, our 'bricks' are the natural numbers and the 'chain reaction' is the result/theorem/contention we are supposed to prove, so that we are basically attempting to prove that some given statement, depending on $n$, is true for **all** natural numbers $n$ greater than some 'first number' $n_0$, which corresponds to the brick given the push.

If we succeed in finding any (preferably the smallest) natural number $n_0$ for which some statement $\mathcal{P}(n)$ with natural number variable $n$ is true (we write '$\mathcal{P}(n_0)$ is true') we have then achieved the equivalent of condition (i) in our 'brick' analogy.

To achieve the equivalent of condition (ii), we suppose that there is a certain number $k$ for which the statement $\mathcal{P}(k)$ is true and then we try to show that it must follow that $\mathcal{P}(k + 1)$ is also true. That is, in the 'brick' analogy, we are assured that if one brick falls then the next one will fall. Postulating that natural numbers are analogous to an infinite row of bricks, we would have then succeeded in showing that the chain reaction will occur, that is, $\mathcal{P}(n)$ is true for all natural numbers $n$ such that $n \geq n_0$. This is the rationale for the method of induction. We can appeal to induction because it is a fundamental postulate (or axiom) about what the natural numbers are. To be precise:

The **Principle of Mathematical Induction (PMI)** states:

If

(1) $\mathcal{P}(n_0)$ is true for the particular natural number $n_0$; and
(2) whenever $\mathcal{P}(k)$ is true for some number $k \geq n_0$ then $\mathcal{P}(k + 1)$ is true too;

then $\mathcal{P}(n)$ holds for all natural numbers $n \geq n_0$.

To prove a theorem: '$\mathcal{P}(n)$ holds for all natural numbers sufficiently large', using mathematical induction, we must first show that $\mathcal{P}(n_0)$ is true for some particular natural number $n_0$; this is often called 'step 1' and provides a 'basis for induction'. Unless we can find such a basis there is no point in attempting the second step, and we may be wasting our time – the statement may be false for all $n$.

Secondly, we must make the assumption (often called 'the inductive hypothesis') that $\mathcal{P}(k)$ is true for some number $k \geq n_0$, and deduce from this assumption that $\mathcal{P}(k + 1)$ (the formula when $k + 1$ replaces $k$) is true too. This deductive step is often called 'step 2'.

Clearly, the results of steps 1 and 2 together imply that $\mathcal{P}(n_0 + 1)$ is true, which again implies that $\mathcal{P}(n_0 + 2)$ is true, and so on. We state our conclusion like this: 'By the Principle of Mathematical Induction (or just by the PMI) $\mathcal{P}(n)$ is true for all $n \in \mathbb{N}$ (that is, for all natural numbers $n$) such that $n \geq n_0$.'

**Example:**   Prove that, for all natural numbers $n$,

$$1 + 2 + 3 + \cdots + n = \frac{n(n + 1)}{2}.$$

**Solution:**  It is good practice to 'visualize' mathematical results if possible: in this case, 'a triangular number is half a rectangular number'. Make some sketches of $n$-by-$(n+1)$ rectangles of dots, for $n = 2, 3, 4$, until you grasp the meaning of this statement. Now we set out to prove it by induction. Let $P(n)$ be the statement: $1 + 2 + 3 + \cdots + n = \frac{n(n+1)}{2}$. Note that we are using this shorthand notation to denote the *whole equation*, not just the left-hand side or the right-hand side. First, let us check the validity of $P(1)$. This means looking at left-hand and the right-hand sides of the equation and checking whether they are equal.

$$P(1) : \text{LHS} = 1, \quad \text{while RHS} = \frac{1 \times (1+1)}{2} = 1.$$

Therefore $P(1)$ is true.

For the second step, assume that $P(k)$ is true, that is: $1 + 2 + 3 + \cdots + k = \frac{k(k+1)}{2}$ is true for some $k \in \mathbb{N}$. Then

$$\begin{aligned}
\text{LHS of } P(k+1) &= 1 + 2 + 3 + \cdots + k + (k+1) \\
&= \frac{k(k+1)}{2} + (k+1) \quad \text{(by inductive hypothesis)} \\
&= \frac{k(k+1)}{2} + \frac{2(k+1)}{2} \\
&= \frac{(k+1)(k+2)}{2} = \text{RHS of } P(k+1).
\end{aligned}$$

We have shown that 'if $P(k)$ is true then $P(k+1)$ is also true', that is:

$$P(k) \text{ is true} \Rightarrow P(k+1) \text{ is true}.$$

Hence, by the Principle of Mathematical Induction, $P(n)$ is true for all $n \geq 1$, that is, simply for all natural numbers.

**Example:**  Prove that, for all natural numbers $n$,

$$1(1!) + 2(2!) + 3(3!) + \cdots + n(n!) = (n+1)! - 1.$$

**Solution:**  Recall the meaning of $n!$ given at the start of this Toolchest. Let $P(n)$ denote the above equation. Then

$$\text{LHS of } P(1) = 1(1!) = (1+1)! - 1 = \text{RHS of } P(1).$$

We have shown that $P(1)$ is true. For the second step, assume that $P(k)$ is true, that is:

$$1(1!) + 2(2!) + 3(3!) + \cdots + k(k!) = (k+1)! - 1$$

is true, for some $k \in \mathbb{N}$. Then

$$\text{LHS of } P(k+1) = 1(1!) + 2(2!) + \cdots + k(k!) + (k+1)(k+1)!$$
$$= \big((k+1)! - 1\big) + (k+1)(k+1)!$$
$$\text{(by inductive hypothesis)}$$
$$= \{(k+1)! + (k+1)!(k+1)\} - 1$$
$$= \{(k+1)!(1+k+1)\} - 1$$
$$= (k+2)! - 1 = \text{RHS of } P(k+1).$$

Hence $P(k+1)$ is true whenever $P(k)$ is true. Since $P(1)$ is true then $P(2)$, $P(3)$, and so on, are also true. That is, by the Principle of Mathematical Induction, $P(n)$ is true for all $n \in \mathbb{N}$.

**Example:**   Prove that $7^n - 1$ is always divisible by 6, for all $n \in \mathbb{N}$.

**Proof:**   Define the function $F(n) = 7^n - 1$. (Notice carefully that $F$ denotes a certain function of $n$, not an equation or a statement.) The statement that we want to investigate is

$$P(n): \text{'6 divides } F(n)\text{'},   \text{ or } \text{ '}F(n) \text{ is a multiple of 6'.}$$

Investigating $P(1)$ :   $F(1) = 7^1 - 1 = 6 \times 1$,   so that   6 divides $F(1)$, and $P(1)$ is true.

For the second step, suppose $P(k)$ is true for some $k \in \mathbb{N}$, i.e. $F(k) = 7^k - 1$ is a multiple of 6. This is the inductive hypothesis. Then

$$F(k+1) = 7^{k+1} - 1$$
$$= 7 \cdot 7^k - 1$$
$$F(k+1) - F(k) = 7 \cdot 7^k - 1 - (7^k - 1)$$
$$= 6 \cdot 7^k$$
$$\text{therefore}   F(k+1) = F(k) + 6 \cdot 7^k.$$

Now the RHS is a multiple of 6, because $F(k)$ is a multiple of 6 by our inductive hypothesis, so $F(k+1)$ must also be a multiple of 6. We have succeeded, then, in showing that if $P(k)$ is true then $P(k+1)$ is also true. But $P(1)$ is true, so that, by the Principle of Mathematical Induction, $P(n)$ is true for all $n \in \mathbb{N}$.

Henceforth, we shall often use the symbol $\forall$ to mean 'for all', as in: $P(n) \; \forall n \in \mathbb{N}$.

An alternative method of proof of the preceding result is to use the known factorization:

$$a^n - b^n = (a - b)(a^{n-1} + a^{n-2}b + \cdots + b^{n-1}).$$

Putting $a = 7$, $b = 1$, it becomes obvious that $7^n - 1^n$ is a multiple of 6 for all $n \in \mathbb{N}$. This is an example of 'direct proof'.                    ◇

**Example:**   Prove that $6^n - 5n + 4$ is divisible by 5, $\forall n \in \mathbb{N}$.

**Proof:**   Let $P(n)$ be the statement: '$6^n - 5n + 4$ is divisible by 5'. Let us investigate the statement $P(1)$ which is the sentence: '$6 - 5 + 4 = 5 \times 1$ is divisible by 5'. We see that $P(1)$ is true.

Now let $f(n) = 6^n - 5n + 4$, and let us assume that there exists some $k \in \mathbb{N}$ for which $f(k) = 6^k - 5k + 4$ is a multiple of 5, i.e. for which the statement $P(k)$ is true. This is the inductive hypothesis. Then

$$f(k+1) = 6^{k+1} - 5(k+1) + 4$$
$$= 6 \cdot 6^k - 5k - 5 + 4.$$
$$\text{So } f(k+1) - f(k) = (6 \cdot 6^k - 5k - 1) - (6^k - 5k + 4)$$
$$= 5 \cdot 6^k - 5 = 5(6^k - 1).$$

Now the RHS is a multiple of 5, so that, since $f(k)$ is a multiple of 5 by our inductive hypothesis, it follows that $f(k+1)$ is also a multiple of 5. We have proved that if $P(k)$ is true then $P(k+1)$ is true. Since $P(1)$ is true, it follows by the Principle of Mathematical Induction that $P(n)$ is true for all $n \in \mathbb{N}$.

Observe that we can also give a direct proof of the preceding result using congruence notation (see Theorems 1 and 2 in Toolchest 4) as follows:

$$6^n - 5n + 4 \equiv (1^n - 0 + 4) \ (\text{mod } 5) \ \ (\text{since } 6 \equiv 1 \ (\text{mod } 5),$$
$$5n \equiv 0 \ (\text{mod } 5))$$
$$\equiv 5 \ (\text{mod } 5)$$
$$\equiv 0 \ (\text{mod } 5).$$

That is, $6^n - 5n + 4$ leaves zero remainder when divided by 5, hence $6^n - 5n + 4$ is a multiple of 5, for all $n \in \mathbb{N}$.                    ◇

**Remark:**   It is perfectly possible to give a valid proof of results such as our two previous examples without introducing the notation $P(n)$ for the statement about $n$, or the functional notation $F(n)$, $f(n)$. This is partly a matter of taste, and as you gain confidence you may omit these notational devices or use whichever presentation you prefer. Here is a brief inductive proof of the last example:

Consider the assertion '$6^n - 5n + 4$ is divisible by 5'. When $n = 1$ this is true, since $6^1 - 5 \cdot 1 + 4 = 5$. Assuming that the assertion holds for $n = k$, we have

$$6^{k+1} - 5(k+1) + 4 = 6 \cdot 6^k - 5k - 5 + 4 = 5(6^k - 1) + (6^k - 5k + 4),$$

which will then also be divisible by 5, by our assumption. Therefore the assertion holds for all natural numbers.

**Example:** Prove the following result about the sum to $n$ terms of a geometric series with first term $a$ and common ratio $r$: For any $n \in \mathbb{N}$, and any $a, r \in \mathbb{R}$, with $r \neq 1$, we have

$$a + ar + ar^2 + \cdots + ar^{n-1} = \frac{a(r^n - 1)}{r - 1}.$$

**Proof:** Let $a, r$ be any real numbers, with $r \neq 1$. Let $\mathcal{P}(n)$ denote the equation above. Now

$$\text{LHS of } \mathcal{P}(1) = a = \frac{a(r - 1)}{r - 1} = \text{RHS of } \mathcal{P}(1).$$

Thus $\mathcal{P}(1)$ is true. Next, for the inductive hypothesis, assume that $\mathcal{P}(k)$ is true for some $k \in \mathbb{N}$. That is,

$$a + ar + ar^2 + \cdots + ar^{k-1} = \frac{a(r^k - 1)}{r - 1}.$$

Then

$$\text{LHS of } \mathcal{P}(k + 1) = a + ar + ar^2 + \cdots + ar^{k-1} + ar^k$$

$$= \frac{a(r^k - 1)}{r - 1} + ar^k \quad \text{(by the inductive hypothesis)}$$

$$= \frac{a(r^k - 1)}{r - 1} + \frac{(r - 1)ar^k}{r - 1}$$

$$= \frac{ar^k - a + rar^k - ar^k}{r - 1}$$

$$= \frac{rar^k - a}{r - 1}$$

$$= \frac{a(r^{k+1} - 1)}{r - 1} = \text{RHS of } \mathcal{P}(k + 1).$$

We have shown that if $\mathcal{P}(k)$ is true then $\mathcal{P}(k+1)$ is also true. We have also shown that $\mathcal{P}(1)$ is true, so that $\mathcal{P}(2), \mathcal{P}(3), \ldots$ are true, and by the PMI $\mathcal{P}(n)$ is true $\forall n \in \mathbb{N}$. We have proved this for all real $a$ and for all $r \neq 1$. Notice that when $a = 0$ the result is trivial.                                               ◇

We are now ready to use the method of induction to prove the validity of some of the useful and famous inequalities that follow in this Toolchest.

## 2.2 Elementary inequalities

Perhaps the most basic and useful inequality of all is this:

$$\text{For any real number } a, \ a^2 \geq 0.$$

Can we prove such a basic statement? Yes, it can be proved logically from the fundamental algebraic structure of the real numbers and the axioms of order for the real numbers. The algebraic structure is codified in axioms asserting that the two operations $+, \cdot$ are commutative and associative, have identities 0 and 1, respectively, and have inverses; also that they interact according to a distributive law. We shall not state these algebraic 'field axioms' explicitly here, for they are part of the commonly accepted core of school arithmetic. It is worth pointing out, however, that such basic arithmetic laws as the 'law of signs': 'minus-times-minus-equals-plus', and the fact that $a \cdot 0 = 0$ for all real numbers $a$, can be regarded as *theorems* as they can be proved from the field axioms. However, we will simply assume them along with the rest.

It will be helpful to give a list of some order axioms upon which all proofs of inequalities shall be based. Our choice is based on convenience; it is not claimed that these are logically independent or complete.

**Axiom 1:** *Trichotomy law*
  For any $a, b$ exactly one of the following holds: $a < b$, $a = b$, $a > b$.
**Axiom 2:** *Transitive law*
  If $a < b$ and $b < c$ then $a < c$.
**Axiom 3:** *Preservation of inequality by addition*
  If $a < b$ then $a + c < b + c$ for all $c \in \mathbb{R}$.
**Axiom 4:** *Preservation of inequality by positive multiplication/division*
  If $a < b$ and $c > 0$, then $ac < bc$ and $\frac{a}{c} < \frac{b}{c}$.
**Axiom 5:** *Reversal of inequality by negative multiplication/division*
  If $a < b$ and $c < 0$, then $ac > bc$ and $\frac{a}{c} > \frac{b}{c}$.

The following important elementary inequalities can be proved from the above.

1. $a^2 \geq 0$ for all $a \in \mathbb{R}$.

    This is proved by breaking it up into three cases, using the trichotomy law, and showing that the inequality holds in each case:

    If $a < 0$ then $a^2 > a \cdot 0 = 0$, by Axiom 5.
    If $a = 0$ then $a^2 = 0^2 = 0$.
    If $a > 0$ then $a^2 > a \cdot 0 = 0$, by Axiom 4.

2. If $0 < a < 1$ and $m$ is any positive integer, then

$$0 < a^m < 1.$$

This inequality can be proved easily 'by induction on $m$'. Given $0 < a < 1$ (so that the inequality holds for $m = 1$), assume that we have a positive integer $k$ such that $0 < a^k < 1$. Multiplying this by $a$, and using Axiom 4, gives

$$0 < a^{k+1} < a \quad \text{(since } a > 0\text{)},$$
$$\text{hence } 0 < a^{k+1} < 1 \quad \text{(since } a < 1\text{)}.$$

We have proved that if it's true for $k$ then it's true for $k + 1$; but it's true for $m = 1$, so that by the Principle of Mathematical Induction, it's true for all $m \in \mathbb{N}$.

3. If $0 < a < b$ and $r$ is any positive rational number, then

$$a^r < b^r \quad \text{and} \quad a^{-r} > b^{-r}.$$

We shall prove this just for positive integers $r$, by induction. It is trivially true for $r = 1$. Suppose that it holds for value $r = k \geq 1$, that is, $a^k < b^k$. Then we have

$$a^{k+1} = a \cdot a^k < a \cdot b^k, \quad \text{using Axiom 4 with the inductive hypothesis,}$$
$$< b \cdot b^k, \quad \text{using Axiom 4 with the previous result,}$$
$$= b^{k+1}.$$

Thus we have deduced from the inductive hypothesis that the inequality holds also for $r = k + 1$. By the PMI, it must hold for all $r \in \mathbb{N}$.

4. **Bernoulli's inequality.** If $a > 0$ then $(1 + a)^n \geq 1 + na$ for any natural number $n$.

One proof method is by induction. Equality holds for $n = 1$, and, if we assume it holds for $k \geq 1$ then

$$(1 + a)^{k+1} = (1 + a)(1 + a)^k$$
$$> (1 + a)(1 + ka), \quad \text{by the inductive hypothesis,}$$
$$= 1 + (k + 1)a + ka^2,$$
$$\geq 1 + (k + 1)a, \quad \text{because } ka^2 \geq 0,$$
$$\text{using inequality (1) and order axioms;}$$

so we have deduced that it holds also for $k + 1$. The PMI then finishes the job. This inequality also follows from the binomial theorem (see Toolchest 7).

## 2.3 Harder inequalities

**Inequality 1:** $x + \frac{1}{x} \geq 2$, $\forall x > 0$.

We start with the observation, following from elementary inequality no. 1, that, for all $x \in \mathbb{R}$,

$$(x-1)^2 \geq 0,$$

$$\text{therefore} \quad x^2 - 2x + 1 \geq 0,$$

$$\text{hence} \quad x^2 + 1 \geq 2x.$$

Notice that the last step uses Axiom 3. Then, since $x > 0$ implies $\frac{1}{x} > 0$ too, we have, by Axiom 4:

$$\frac{x^2+1}{x} \geq \frac{2x}{x},$$

$$\text{therefore} \quad x + \frac{1}{x} \geq 2.$$

This inequality comes in many guises and is quite popular with Olympiad examiners. It is clear that equality is attained for $x = 1$, so 2 is actually the minimum value of the function $f(x) = x + \frac{1}{x}$, for $x > 0$. It will help you to appreciate the situation if you sketch a rough graph of this function, drawing first the graphs of $y = x$, and $y = \frac{1}{x}$ before sketching the graph of their sum.

**Example:** Prove that if $a, b, c, d > 0$ then

$$\frac{cd(a^2+b^2) + bd(a^2+c^2)}{abcd} \geq 4.$$

**Solution:**

$$\frac{cd(a^2+b^2) + bd(a^2+c^2)}{abcd} = \frac{cd(a^2+b^2)}{abcd} + \frac{bd(a^2+c^2)}{abcd}$$

$$= \frac{a^2}{ab} + \frac{b^2}{ab} + \frac{a^2}{ac} + \frac{c^2}{ac}$$

$$= \frac{a}{b} + \frac{b}{a} + \frac{a}{c} + \frac{c}{a}.$$

Put $x = \frac{a}{b}$ so that $\frac{1}{x} = \frac{b}{a}$, and substitute in Inequality 1, observing that $x > 0$ because $a, b > 0$. We obtain $\frac{a}{b} + \frac{b}{a} \geq 2$. Similarly $\frac{a}{c} + \frac{c}{a} \geq 2$; adding these two equations gives the required result.

**Example:** If $0 < \theta < 90°$ and we want $\tan\theta + \cot\theta \geq k$, what is the greatest whole number value of $k$ for which this is true for all $\theta$ in the specified range?

(A) 1      (B) 2      (C) 3      (D) 4      (E) 5

**Solution:**   $\cot\theta = \dfrac{1}{\tan\theta}$, so that what we want is

$$\tan\theta + \frac{1}{\tan\theta} \ge k.$$

Since $\tan\theta > 0$ for $0 < \theta < 90°$, we can use Inequality 1 to deduce immediately that $k = 2$ is the greatest lower bound of the LHS, since 2 is the minimum value, attained by setting $\theta = 45°$. Hence (B).

**2. Weierstrass inequality:**   If $a_1, a_2, \ldots, a_n$ are positive real numbers, then, for $n \ge 2$:

$$(1 + a_1)(1 + a_2) \cdots (1 + a_n) > 1 + (a_1 + a_2 + \cdots + a_n).$$

**Proof:**   Let $\mathcal{P}(n)$ be the inequality:   $(1 + a_1)(1 + a_2) \cdots (1 + a_n) > 1 + (a_1 + a_2 + \cdots + a_n)$.

Consider  $\mathcal{P}(2): (1 + a_1)(1 + a_2) > 1 + a_1 + a_2$

i.e.  $1 + a_1 + a_2 + a_1 a_2 > 1 + a_1 + a_2.$

This inequality is true, for all positive $a_i$, since $a_1 a_2 > 0$.
    Assume that $\mathcal{P}(k)$ is true for some $k \in \mathbb{N}$, that is:

$$(1 + a_1)(1 + a_2) \cdots (1 + a_k) > 1 + a_1 + a_2 + \cdots + a_k.$$

Then

$$
\begin{aligned}
\text{LHS of } \mathcal{P}(k+1) &= (1 + a_1)(1 + a_2) \cdots (1 + a_k)(1 + a_{k+1}) \\
&= ((1 + a_1)(1 + a_2) \cdots (1 + a_k))(1 + a_{k+1}) \\
&> (1 + (a_1 + a_2 + a_3 + \cdots + a_k))(1 + a_{k+1}) \\
&\quad \text{(by inductive hypothesis)} \\
&> 1 + (a_1 + a_2 + \cdots + a_k) + a_{k+1} \\
&\quad \text{(since } (a_1 + \cdots + a_k)a_{k+1} > 0) \\
&= 1 + a_1 + a_2 + \cdots + a_k + a_{k+1} = \text{RHS of } \mathcal{P}(k+1).
\end{aligned}
$$

Hence, if $\mathcal{P}(k)$ is true then it follows that $\mathcal{P}(k + 1)$ is true. Since $\mathcal{P}(2)$ is true, $\mathcal{P}(n)$ is true $\forall\, n \ge 2$, by the PMI.    ◇

**3. Cauchy–Schwarz inequality:**   If $(a_1, a_2, \ldots, a_n)$ and $(b_1, b_2, \ldots, b_n)$ are two $n$-tuples of real numbers, then

$$(a_1 b_1 + a_2 b_2 + \cdots + a_n b_n)^2 \le (a_1^2 + a_2^2 + \cdots + a_n^2)(b_1^2 + b_2^2 + \cdots + b_n^2).$$

This is more economically expressed using summation notation (explained in Section 2 of Toolchest 6):

$$\left(\sum_{i=1}^{n} a_i b_i\right)^2 \leq \left(\sum_{i=1}^{n} a_i^2\right)\left(\sum_{i=1}^{n} b_i^2\right).$$

Equality occurs if, and only if, there is some real number $\lambda$ such that $b_i = \lambda a_i$, $\forall i$. That is, in vector notation, $\underline{b} = \lambda \underline{a}$.

**Proof:**

**Method 1 (Buniakowski's proof):** We start with the observation that, for any $a_i$, $b_i$ and $x$:

$$\sum_{i=1}^{n}(a_i x - b_i)^2 \geq 0$$

$$\text{i.e. } \sum_{i=1}^{n}(a_i^2 x - 2a_i b_i x + b_i^2) \geq 0$$

$$\text{i.e. } x^2 \sum_{i=1}^{n} a_i^2 - 2x \sum_{i=1}^{n} a_i b_i + \sum_{i=1}^{n} b_i^2 \geq 0$$

$$\text{i.e. } Ax^2 - 2Bx + C \geq 0,$$

where $A = \sum_{i=1}^{n} a_i^2$, $B = \sum_{i=1}^{n} a_i b_i$, $C = \sum_{i=1}^{n} b_i^2$.

Clearly, equality will hold if and only if all the non-negative squares are zero, that is, there is some value $x = \lambda$ such that $a_i \lambda = b_i$, $\forall i$.

Now, since this quadratic is non-negative for every value of $x$ (there is at most one root), its discriminant must be less than or equal to zero. (See Section 4 of this Toolchest for detailed discussion of discriminant.) That is

if $Ax^2 - 2Bx + C \geq 0 \ \forall \, x$ then $(-2B)^2 - 4AC \leq 0$, or $B^2 \leq AC$.

This is the inequality we set out to prove. The displayed statement clearly holds for strict inequalities also (no roots). Equality, $B^2 = AC$, will hold precisely when there is a root $x = \lambda$, and then $b_i = \lambda a_i$, $\forall i$, as shown above.

**Method 2:**

$$(a_1^2 + a_2^2 + \cdots + a_n^2)(b_1^2 + b_2^2 + \cdots + b_n^2) - (a_1 b_1 + a_2 b_2 + \cdots + a_n b_n)^2$$

$$= \sum_{i \neq j}^{n}(a_i^2 b_j^2 + a_j^2 b_i^2 - 2a_i b_j a_j b_i)$$

$$= \sum_{i \neq j}^{n}(a_i b_j - a_j b_i)^2 \geq 0.$$

Hence the required inequality. Equality will hold if and only if all the non-negative squares are zero, that is, $\forall\, i \neq j$, $a_i b_j = a_j b_i$, that is, $\frac{b_i}{a_i} = \frac{b_j}{a_j} = \lambda$ whenever denominators are non-zero, and $b_i = 0$ is zero whenever $a_i = 0$. This gives the required condition for equality.    ◇

**Example:**  If $a, b$ and $c$ are real numbers such that $a^2 + b^2 + c^2 = 1$, then what is the maximum value of $ab + bc + ac$?

(A) $\frac{1}{2}$    (B) $\frac{3}{4}$    (C) $\frac{1}{2}$    (D) 1    (E) Cannot be determined

**Solution:**  By the Cauchy–Schwarz inequality with $n = 3$ applied to the ordered triples $(a, b, a)$ and $(b, c, c)$

$$
\begin{aligned}
(ab + bc + ac)^2 &\leq (a^2 + b^2 + a^2)(b^2 + c^2 + c^2) \\
&= (1 - c^2 + a^2)(1 - a^2 + c^2) \\
&= (1 + x)(1 - x) \quad \text{(where } x = a^2 - c^2) \\
&= 1 - x^2 \leq 1.
\end{aligned}
$$

Now we know that 1 is an upper bound, but is it actually attained? Clearly it is attained by setting $a = b = c = \sqrt{\frac{1}{3}}$. Hence (D) is the correct answer.

**4. The AM–GM inequality:**  *Arithmetic Mean $\geq$ Geometric Mean.*

If $a_1, a_2, \ldots, a_n$ are positive numbers, then their **arithmetic mean** $A$, and their **geometric mean** $G$, are defined by:

$$
A = \frac{1}{n}(a_1 + a_2 + \cdots + a_n), \quad G = \sqrt[n]{(a_1 a_2 \ldots a_n)}
$$

and the inequality we have stated says that $A \geq G$.

**Proof:**

**Method 1 (by induction):**  This can be proved by induction on the number $n$ of positive numbers, starting with the observation that it is trivially true for $n = 1$ (and true for $n = 2$ by $0 \leq (\sqrt{a_1} - \sqrt{a_2})^2 = a_1 + a_2 - 2\sqrt{a_1 a_2}$). However, we shall prove it by induction on the number $m$ of members of the given set of positive numbers which differ from their geometric mean.

First we show that $A$ and $G$ both lie between $\min\{a_1, a_2, \ldots, a_n\}$ (i.e. the smallest value in the given set) and $\max\{a_1, a_2, \ldots, a_n\}$ (i.e. the greatest value in the given set), and that inequalities are strict if the numbers in the set are not all equal. For real numbers the operations of addition and multiplication are order independent (or *commutative*), so $A$ and $G$ are unchanged if the given numbers are named in any different order. Thus we may assume that

$$
a_1 \leq a_2 \leq \cdots \leq a_n
$$

so that

$$a_1 = \frac{na_1}{n} \le \frac{a_1 + a_2 + \cdots + a_n}{n} \le \frac{na_n}{n} = a_n$$

and $a_1 = (a_1^n)^{\frac{1}{n}} \le (a_1 a_2 \dots a_n)^{\frac{1}{n}} \le (a_n^n)^{\frac{1}{n}} = a_n$.

Therefore we have $a_1 \le A \le a_n$ and $a_1 \le G \le a_n$, and clearly equality can hold in any of the four places only if all $a_i$ are equal, in which case $A = G$.

Let $P(m)$ be the statement: 'Given any set of numbers such that at most $m \ge 0$ of them differ from their geometric mean, we have $A \ge G$'.

Consider $P(0)$: we must have $a_1 = a_2 = \cdots = a_n = G$ and $A = (nG)/n = G$, so that $A = G$. Thus the statement is true when $m = 0$. (Note that we are here extending the usual inductive method to allow $m$ to belong to the set of non-negative integers.)

Assume that $P(k)$ is true, that is, $A \ge G$ for every set of real numbers for which at most $k$ differ from the geometric mean, where $k \ge 0$. Our goal is to show that the statement will also hold for $k + 1$.

Consider a set $S$ of $n$ numbers $a_1, a_2, \ldots, a_n$ such that at most $k + 1$ of them differ from the geometric mean $G$. We may assume (as shown above) that $S$ is ordered by size: $a_1 \le a_2 \le \cdots \le a_n$. We may also assume that $a_1, \ldots, a_n$ are not all equal (for if so none would differ from $G$ and we would have $A = G$), it follows that

$$a_1 < G < a_n \quad \text{hence} \quad (G - a_1)(G - a_n) < 0.$$

Define a new set $S' = \{a_1', a_2, \ldots, a_{n-1}, a_n'\}$, as follows: $a_1' = G$, and $a_n' = a_1 a_n / G$. The new geometric mean $G'$ is still $G$, and clearly at most $k$ members differ from $G$, so that, by the inductive hypothesis,

$$(a_1' + a_2 + \cdots + a_{n-1} + a_n')/n \ge G.$$

Hence the arithmetic mean $A$ of the set $S$ is given by

$$A = \frac{1}{n}(a_1 + a_2 + \cdots + a_{n-1} + a_n)$$

$$= \frac{1}{n}\left((a_1' + a_2 + \cdots + a_{n-1} + a_n') + (a_1 - a_1') + (a_n - a_n')\right)$$

$$\ge G + \frac{1}{n}(a_1 - G) + \frac{1}{n}(a_n - a_1 a_n / G)$$

$$= G - \frac{1}{nG}(G - a_1)(G - a_n)$$

$$> G.$$

We have shown that $P(k + 1)$ is true whenever $P(k)$ is true, and we earlier showed that $P(0)$ is true, hence by the PMI it is true for all $n \in \mathbb{N}$.

**Method 2 (using the elementary functions):**   It can be proved by methods of mathematical analysis that

$$e^x \geq 1 + x \quad \forall x \geq 0 \tag{2.1}$$

or equivalently

$$x \geq \ln(1 + x) \quad \forall x \geq 0. \tag{2.2}$$

Now, if we take $x_i = a_i/A - 1$ and apply (2.2) for each $i = 1, 2, \ldots, n$, addition of the resulting inequalities gives:

$$\sum_{i=1}^{n} x_i \geq \sum_{i=1}^{n} \ln(x_i + 1). \tag{2.3}$$

Now the LHS $= \displaystyle\sum_{i=1}^{n} x_i = \sum_{i=1}^{n} \left(\frac{a_i}{A} - 1\right) = \left(\sum_{i=1}^{n} \frac{a_i}{A}\right) - n = n - n = 0,$

and the RHS $= \displaystyle\sum_{i=1}^{n} \ln(x_i + 1) = \sum_{i=1}^{n} \ln\left(\frac{a_i}{A} - 1 + 1\right)$

$$= \sum_{i=1}^{n} \ln\left(\frac{a_i}{A}\right) = \ln\left(\frac{a_1 a_2 \ldots a_n}{A^n}\right). \tag{2.4}$$

Thus (2.3) becomes $\quad 0 \geq \ln\left(\dfrac{a_1 a_2 \ldots a_n}{A^n}\right),$

therefore $\quad 1 = e^0 \geq \dfrac{a_1 a_2 \ldots a_n}{A^n}$

hence $\quad A \geq \sqrt[n]{(a_1 a_2 \ldots a_n)} = G.$

◇

**Example:**   Prove that $a^3 + b^3 + c^3 \geq 3abc$.

**Solution:**   The AM–GM inequality gives

$$\frac{a^3 + b^3 + c^3}{3} \geq \sqrt[3]{a^3 b^3 v^3},$$

therefore $\quad a^3 + b^3 + c^3 \geq 3abc$.

**Example:**   Three real numbers $x, y, z$ such that $x + y + z = 2$ and $x^2 + y^2 + z^2 = 4$ were chosen from the set of all real numbers. Show that each of $x, y$ and $z$ lie in the interval $[-\frac{2}{3}, 2]$. Are the extreme values attainable?

**Solution:**   The given system of equations is **symmetric** in $x, y$ and $z$, that is, any permutation of the three variables does not change the equations.

This means we need only prove restrictions on $x$, and they will hold also for $y$ and $z$. Therefore we aim at an inequality involving just the one variable $x$:

$$x + y + z = 2 \Rightarrow y + z = 2 - x$$

$$\text{and} \quad x^2 + y^2 + z^2 = 4 \Rightarrow y^2 + z^2 = 4 - x^2.$$

Now, using elementary inequality no. 1, for any $y$ and $z \in \mathbb{R}$, we have

$$(y - z)^2 \geq 0,$$

$$\text{therefore} \quad y^2 + z^2 \geq 2zy,$$

$$\text{hence} \quad 2y^2 + 2z^2 \geq y^2 + z^2 + 2zy,$$

$$\text{so} \quad 2(y^2 + z^2) \geq (y + z)^2,$$

$$\text{and finally} \quad y^2 + z^2 \geq \frac{1}{2}(y + z)^2.$$

Note that we could have quickly obtained the same inequality from the Cauchy–Schwarz inequality applied to the ordered pairs $(y, z), (1, 1)$. This inequality involving $y, z$ can now be translated into one involving just $x$, which can be solved easily, with the help of a sketch of a quadratic graph:

$$4 - x^2 \geq \frac{1}{2}(2 - x)^2$$

$$\text{that is} \quad 3x^2 - 4x - 4 \leq 0$$

$$\text{that is} \quad (3x + 2)(x - 2) \leq 0, \text{ or } x \in \left[-\frac{2}{3}, 2\right].$$

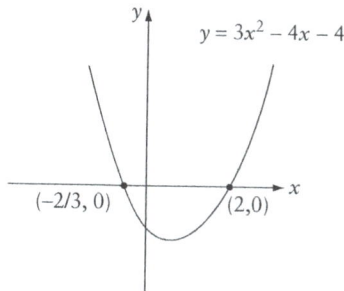

$y = 3x^2 - 4x - 4$

$(-2/3, 0)$   $(2,0)$

To see whether the extreme values are attainable or not, try $x = 2$, giving $y + z = 0$ and $y^2 + z^2 = 4$, which is obviously possible, as $z = 0$, $y = 2$ and $z = 2$, $y = 0$ are solutions. Next, trying $x = -\frac{2}{3}$ gives $y + z = \frac{8}{3}$, and

$y^2 + z^2 = \frac{32}{9}$,

therefore    $\left(\frac{8}{3} - z\right)^2 + z^2 = \frac{32}{9}$,

so that    $2z^2 - \frac{16z}{3} + \frac{32}{9} = 0$,

and hence    $18z^2 - 48z + 32 = 0$.

Now the discriminant of this equation is: $\Delta = b^2 - 4ac = (-48)^2 - 4 \cdot 18 \cdot 32 = 0$, so there is a real root for $z$ (and the same for $y$ by symmetry). Therefore $x = -\frac{2}{3}$ is attainable. Hence $x$ (and by symmetry $y$ and $z$ too) can take either of the extreme values 2 and $-\frac{2}{3}$.

**Example:**   If $a$ and $b$ are positive real numbers such that $a + b = 1$, prove that

$$\left(a + \frac{1}{a}\right)^2 + \left(b + \frac{1}{b}\right)^2 \geq \frac{25}{2}.$$

**Solution:**   If there was no restriction on $a$ and $b$, we could use Inequality 1 to get each bracket greater than or equal to 2, hence the whole expression bounded below by 8. The required bound, with the restriction taken into account, is stronger than this.

$$\left(a + \frac{1}{a}\right)^2 + \left(b + \frac{1}{b}\right)^2 = a^2 + \frac{1}{a^2} + b^2 + \frac{1}{b^2} + 4$$

$$= (a + b)^2 - 2ab + \left(\frac{1}{a} + \frac{1}{b}\right)^2 - \frac{2}{ab} + 4$$

$$= 1 - 2ab + \frac{1 - 2ab}{a^2b^2} + 4. \qquad\qquad (2.5)$$

However, using Inequality 1, we have

$$(a - b)^2 \geq 0,$$

therefore    $a^2 - 2ab + b^2 \geq 0$,

and so    $a^2 + 2ab + b^2 \geq 4ab$,

that is    $(a + b)^2 \geq 4ab$,

$$\text{giving}\quad \left(\frac{a+b}{2}\right)^2 \geq ab, \tag{2.6}$$

$$\text{hence}\quad ab \leq \frac{1}{4}. \tag{2.7}$$

Observe that we could have reached (2.6) directly by applying the AM–GM inequality to $a, b$.

We can use the order axioms and the given fact that $a, b$ are positive to make the following deductions from (2.7):

$$ab \leq \frac{1}{4} \Rightarrow 2ab \leq \frac{1}{2} \Rightarrow -2ab \geq -\frac{1}{2};$$

$$ab \leq \frac{1}{4} \Rightarrow \frac{1}{ab} \geq 4 \Rightarrow \frac{1}{a^2b^2} \geq 16.$$

Finally, using these inequalities in (2.5) gives

$$\left(a+\frac{1}{a}\right)^2 + \left(b+\frac{1}{b}\right)^2 \geq \left(1-\frac{1}{2}\right) + 16\left(1-\frac{1}{2}\right) + 4 = \frac{25}{2}.$$

*Isn't that every scientist and mathematician's wish? –*
*to have an infinitely knowledgeable and effortlessly brilliant device*
*you hurl questions to and it churns out nicely packaged answers!*

## 2.4   The discriminant of a quadratic expression

Given the quadratic equation $ax^2 + bx + c = 0$, the well-known formula for the two possible values for $x$ satisfying this equation is

$$x = \frac{-b \pm \sqrt{b^2 - 4ac}}{2a}.$$

The basis for deriving this formula, as you might already know, is the creation of a perfect square:

$$a\left(x + \frac{b}{2a}\right)^2 = ax^2 + bx + a\left(\frac{b^2}{4a^2}\right) = ax^2 + bx + c + \frac{b^2 - 4ac}{4a}.$$

Now, putting $ax^2 + bx + c = 0$ in this equation, derive the formula yourself. The value $b^2 - 4ac$ is so important for predicting the nature of solutions to the equation that mathematicians have given it a name: it is called the *discriminant*.

One of three things can happen with the discriminant. It can be greater than, equal to, or less than zero.

**Case 1:**    $b^2 - 4ac > 0$.

In this case, there are two values $x$ can take for which $ax^2 + bx + c = 0$, and these are both real numbers, namely:

$$x = \frac{-b + \sqrt{b^2 - 4ac}}{2a} \quad \text{and} \quad x = \frac{-b - \sqrt{b^2 - 4ac}}{2a}.$$

This means that if you were to plot the (parabolic) graph of $y = ax^2 + bx + c$, it would cross the $x$-axis at two distinct points, as shown below:

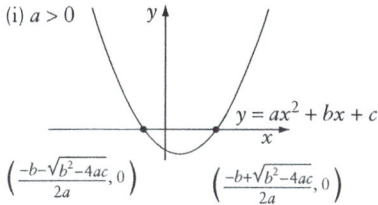

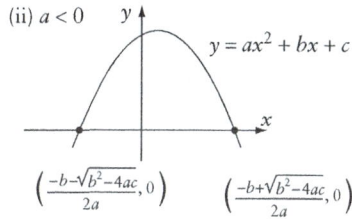

(i) $a > 0$    $y = ax^2 + bx + c$    $\left(\frac{-b-\sqrt{b^2-4ac}}{2a}, 0\right)$    $\left(\frac{-b+\sqrt{b^2-4ac}}{2a}, 0\right)$

(ii) $a < 0$    $y = ax^2 + bx + c$    $\left(\frac{-b-\sqrt{b^2-4ac}}{2a}, 0\right)$    $\left(\frac{-b+\sqrt{b^2-4ac}}{2a}, 0\right)$

**Case 2:**    $b^2 - 4ac = 0$.

In this case, there is only one value of $x$ for which $ax^2 + bx + c = 0$, and this single value is called a repeated root. The graph of such a function is tangent to the $x$-axis at this point as shown below:

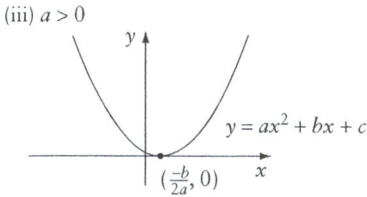

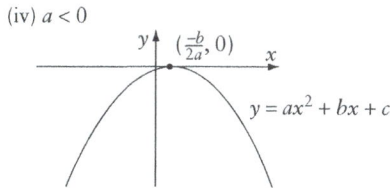

(iii) $a > 0$    $y = ax^2 + bx + c$    $\left(\frac{-b}{2a}, 0\right)$

(iv) $a < 0$    $\left(\frac{-b}{2a}, 0\right)$    $y = ax^2 + bx + c$

**Case 3:**    $b^2 - 4ac < 0$.

Have you ever thought what $\sqrt{-9}, \sqrt{-16}$, or $\sqrt{-2}$ might mean? This case challenges us to think about this. A mathematician called Rafael Bombelli overcame this problem (over 300 years ago) – because he found that these so-called *imaginary* numbers could be useful in finding real solutions to cubic equations! This resulted in mathematicians gradually accepting and introducing the complex numbers as a very useful enlargement of our repertoire of numbers.

In our present context, we say that the roots are not real – they are *complex* (or we still use the word imaginary) and the graph never meets the $x$-axis. In other words, the function takes positive values for all values of $x$, if $a > 0$, and takes negative values for all values of $x$, if $a < 0$.

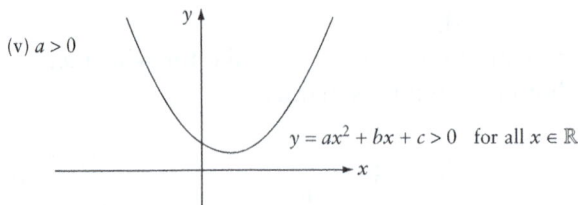

(v) $a > 0$

$y = ax^2 + bx + c > 0$ for all $x \in \mathbb{R}$

We can combine Case I and Case II to get the result:

$ax^2 + bx + c$ has real roots iff (if and only if) $b^2 - 4ac \geq 0$.

**Example:**   For the equation $(1 - 3k)x^2 + 3x - 4 = 0$ to have real roots, the largest integral value that $k$ may have is

(A) $-2$     (B) $-1$     (C) $0$     (D) $1$     (E) $2$

**Solution:**

$$\text{We require } b^2 - 4ac \geq 0$$
$$\text{that is } 9 - 4(1 - 3k) \cdot (-4) \geq 0$$
$$9 + 16(1 - 3k) \geq 0$$
$$25 \geq 45k$$
$$\frac{25}{48} \geq k.$$

Since $\left[\frac{25}{48}\right] = 0$, the required largest value is $k = 0$. Hence (C).

NB: $y = [x]$ is the 'integer part' function, and $[x]$ refers to the integer part of $x$, that is, the greatest integer less than or equal to $x$. For example, $[2.1] = 2$, $[2.8] = 2$, $[-3.1] = -4$, $[-3.8] = -4$, and so on.

## 2.5   The modulus function

We define the modulus function $f(x) = |x|$ as,

$$|x| = \begin{cases} x & \text{if } x \geq 0 \\ -x & \text{if } x < 0. \end{cases}$$

For example, $|-3| = -(-3) = 3$, $|3| = 3$, and so on.

### Properties of the modulus

1. $|x| \geq 0$, $|x| = 0$ if and only if $x = 0$.
2. $x \leq |x|$, $|x|^2 = x^2$.
3. $|xy| = |x| \cdot |y|$.
4. $|x + y| \leq |x| + |y|$, $|x - y| \geq |x| - |y|$.

Most of these properties are proved directly from the definition, by treating the two cases $x < 0$, $x \geq 0$ separately. The proof of the last: the so-called **triangle law**, depends upon properties 1–3:

$$|x + y|^2 = (x + y)^2 = x^2 + 2xy + y^2 = |x|^2 + 2xy + |y|^2$$
$$\leq |x|^2 + 2|x||y| + |y|^2 = (|x| + |y|)^2.$$

The second part of property 4 follows from the first by applying it to the two numbers $x - y$, $y$. The connection with the geometrical triangle law in Toolchest 1 is easily seen by putting $x = a - b$, $y = b - c$ to obtain the algebraic triangle law in a different form:

$$|a - c| \leq |a - b| + |b - c|.$$

**Example:** Solve the inequality $|2x - 7| \leq x$.

**Solution:** There is a graphical method: draw the graphs of the two functions $y = x$ and $y = |2x - 7|$, and observe for which $x$ the one graph is on top of the other. The second graph is obtained by drawing the graph of $y = 2x - 7$ and reflecting in the $x$-axis the portion that is beneath the axis. The diagrams show that the solution is an interval of real numbers:

$$\frac{7}{3} \leq x \leq 7.$$

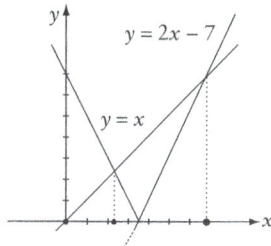

The alternative way is to use the definition of the modulus function, breaking the domain of $x$ into two sets divided by the critical point $\frac{7}{2}$ at which $y = |2x - 7| = 0$.

For $2x \geq 7$ the inequality is $2x - 7 \leq x$, giving $\frac{7}{2} \leq x < 7$;

and for $2x < 7$ the inequality is $7 - 2x \leq x$, giving $\frac{7}{3} \leq x < \frac{7}{2}$.

Putting the two sets together, we obtain the full interval as before.

**Example:** Which real numbers satisfy the inequality $|x| > |x^2 - 4|$?

**Solution:**

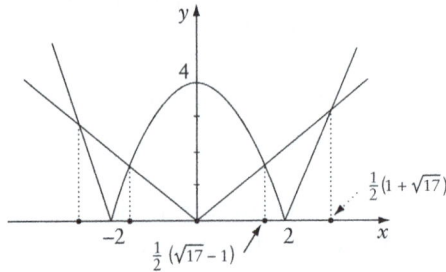

The sketch illustrates the two graphs $y = |x|$ and $y = |x^2 - 4|$, and there are two intervals $-\frac{1}{2}(\sqrt{17} + 1) < x < -\frac{1}{2}(\sqrt{17} - 1)$, and $\frac{1}{2}(\sqrt{17} - 1) < x < \frac{1}{2}(\sqrt{17} + 1)$, over which the $y$ value of the first graph is greater than that of the second. These values of $x$ are found by solving the quadratics that result from finding the points of intersection of the graphs of $y = x$ and $y = 4 - x^2$, and the points of intersection of $y = x$ and $y = x^2 - 4$. Hence (E).

**Example:**   Show that, for all real numbers $x$, $\left|x + \frac{1}{x}\right| \geq 2$, with equality at $x = \pm 1$.

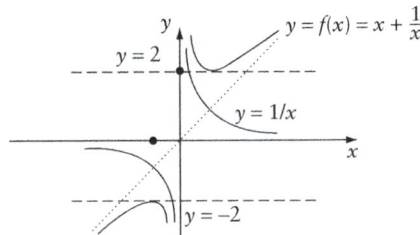

**Solution:**    This extends Inequality 1 to negative numbers. One approach is to use Inequality 1. Observing that $y = f(x) = x + \frac{1}{x}$ is an odd *function* $(f(-x) = -f(x)$, see Toolchest 5), you can sketch the graph for negative values of $x$ by reflecting twice, first in y-axis, then in x-axis. This will show that, for $x < 0$, we have $x + \frac{1}{x} \leq -2$. You can prove this inequality in similar fashion to the way we proved Inequality 1, but starting with $(x + 1)^2 \geq 0$ in place of $(x - 1)^2 \geq 0$, and recalling Axiom 5 when you divide through by negative $x$. Equality takes place if and only if $x = -1$. Both results can be elegantly combined by using the modulus function, and there is a direct proof using the second property above:

$$\left|x + \frac{1}{x}\right|^2 = \left(x + \frac{1}{x}\right)^2 = x^2 + \frac{1}{x^2} + 2 = \left(x - \frac{1}{x}\right)^2 + 4 \geq 4.$$

Now take positive roots. Equality will hold if and only if $x = \frac{1}{x}$, that is, $x^2 = 1$, or $x = \pm 1$.

**Example:** At how many points is the function $y = |\,|\,|x| - 3| - 2| + 1$ not differentiable? (That is, at how many points does it have a corner, so no gradient, no tangent?)

(A) None   (B) 1   (C) 3   (D) 5   (E) 7

**Solution:** The graph of the function $y = f(x)$ will help us answer the question, and this is best obtained by analysing the function step by step, 'from the inside outwards'. We start by sketching the graph of $y = g(x) = |x| - 3$, by translating the graph of $y = |x|$ three units downwards.

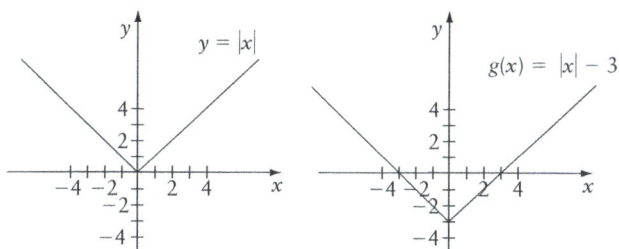

$y = |x|$

$g(x) = |x| - 3$

Next we sketch the graph of $y = h(x) = |\,|x| - 3| = |g(x)|$ by reflecting, in the $x$-axis as the mirror, the negative part (which lies below the $x$-axis) of the graph of $g(x)$. Next we find the graph of the function $y = m(x) = |\,|x| - 3| - 2 = |h(x)| - 2$, by translating the graph of $h(x)$ two units downwards.

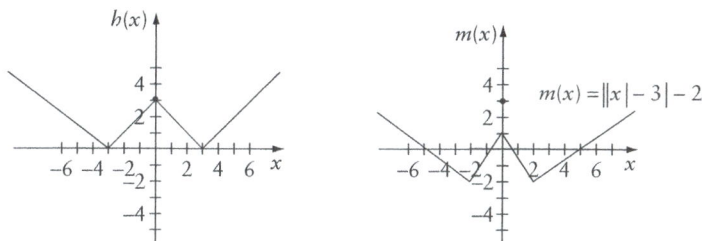

$h(x)$

$m(x)$

$m(x) = |\,|x| - 3| - 2$

To obtain the graph of $y = p(x) = |\,|\,|x| - 3| - 2| = |m(x)|$, we reflect the negative bits of the graph of $y = m(x)$ about the $x$-axis to obtain the left-hand diagram below. Then, finally, to obtain the graph of

$$f(x) = p(x) + 1 = |\,|\,|x| - 3| - 2| + 1,$$

we shift the graph of $y = p(x)$ one unit upwards to get

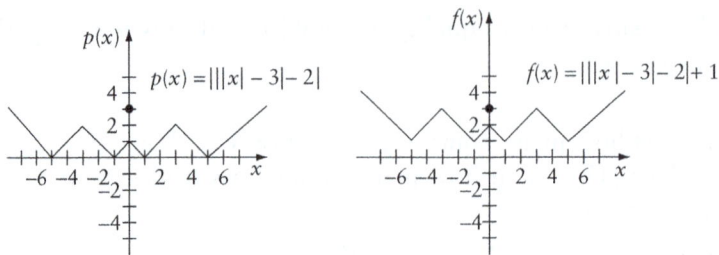

Intuitively, a function is differentiable at a point $(x, y)$ if a tangent can be drawn to the graph of the function at that point. This is not a rigorous definition of differentiability at a point. We say that a function $f(x)$ is differentiable at a point $(x_0, f(x_0))$ iff the following limit exists:

$$\lim_{x \to x_0} \left( \frac{f(x) - f(x_0)}{x - x_0} \right).$$

But the intuitive geometric idea of tangent is enough for us to solve our problem: functions are not differentiable at 'sharp' points like $A, B, C, \ldots$ etc., and from the graph above we have seven of these. There is no tangent to the graph of $f(x)$ at each of the points $A, B, \ldots, G$. Hence (E).

Here are the solutions to the three appetizer problems you met at the very beginning of the Toolchest.

**Solution of Appetizer Problem 1:**   **Proof:**   Let $\mathcal{P}(n)$ be the assertion: $n! > 2^n$. Since $1! = 1 \neq 2^1 = 2$,   $2! = 2 \neq 2^2 = 4$,   $3! = 6 \neq 2^3 = 8$, we see that the assertion is false for $n = 1, 2, 3$. But $\mathcal{P}(4)$ is the assertion: $4! = 24 > 2^4 = 16$, which is true.

Next, assume that $\mathcal{P}(k)$ is true for some $k \in \mathbb{N}$, with $k \geq 4$. That is $k! > 2^k$. This is the inductive hypothesis. Then

$$\begin{aligned}
(k+1)! &= (k+1)k! \\
&> (k+1)2^k, \quad \text{by inductive hypothesis,} \\
&> 2 \cdot 2^k, \quad \text{since } k \geq 4, \\
&= 2^{k+1}.
\end{aligned}$$

Thus, if $\mathcal{P}(k)$ is true then $\mathcal{P}(k+1)$ will also be true, provided that $k \geq 4$. But $\mathcal{P}(4)$ is true, hence, by the PMI, the result holds for all $n \geq 4$.

**Solution of Appetizer Problem 2:**   Let the numbers be $a_1, a_2, a_3$ and $a_4$, then

$$3 = \frac{12}{4} = \frac{a_1 + a_2 + a_3 + a_4}{4} \geq \sqrt[4]{a_1 a_2 a_3 a_4}, \quad \text{by the AM–GM inequality.}$$

This shows us that the greatest value of $a_1 a_2 a_3 a_4$ is obtained when equality occurs, i.e. when $a_1 = a_2 = a_3 = a_4 = 3$. Therefore the maximum product is $3^4 = 81$. Hence (B).

This result can be generalized, using the general AM–GM inequality: If the sum of $n$ positive numbers is given, their product is maximum when the numbers are all equal.

**Solution of Appetizer Problem 3:**   We show first how to prove that

$$\left(\frac{n+1}{2}\right)^n \geq n!, \quad \text{for all } n \in \mathbb{N}$$

by induction. Then, we show how it can be proved more rapidly, using the heavy machinery of the AM–GM inequality.

**Proof by induction:**   Since $\frac{1+1}{2} = 1 \geq 1!$, the statement holds for $n = 1$. Assume it holds for $n = k$; that is $\left(\frac{k+1}{2}\right)^k \geq k!$. First, we experiment with how one can get from this to the corresponding expression for $k+1$, and we easily reduce the problem to showing that $(n+1)^n \geq 2n^n$ for all positive integers $n$, and this is equivalent to $\left(\frac{n+1}{n}\right)^n \geq 2$. But this follows from the Binomial Theorem (see Toolchest 7). For

$$\left(1+\frac{1}{n}\right)^n = 1 + n \cdot \frac{1}{n} + \cdots \geq 2.$$

Now we can give the full demonstration of the inductive step:

$$\left(\frac{k+2}{2}\right)^{k+1} \geq 2\left(\frac{k+1}{2}\right)^{k+1} \quad \text{(by the result just shown)}$$

$$= 2\left(\frac{k+1}{2}\right)^k \left(\frac{k+1}{2}\right)$$

$$\geq k!(k+1) \quad \text{(by the inductive hypothesis)}$$

$$= (k+1)!.$$

**Proof by AM–GM inequality:**   We use the result of our first induction example, that

$$1 + 2 + 3 + \cdots + n = \frac{n(n+1)}{2}.$$

The AM–GM inequality gives

$$\frac{1+2+3+\cdots+n}{n} \geq \sqrt[n]{1 \times 2 \times 3 \times \cdots \times n} = \sqrt[n]{n!},$$

therefore   $\dfrac{n(n+1)}{2n} \geq \sqrt[n]{n!}$

so that   $\left(\dfrac{n+1}{2}\right)^n \geq n!$

## 2.6  Problems

**Problem 1:**  *The lengths of the sides of a triangle are $b + 1$, $7 - b$, and $4b - 2$. The number of values of $b$ for which the triangle is isosceles is:*

(A) 0   (B) 1   (C) 2   (D) 3   (E) None of these

**Problem 2:**  *How many acute angled triangles exist in which each angle is an integral number of degrees and the smallest angle is a quarter of the size of the largest angle? (Note: Consider all similar triangles to be identical.)*

(A) 1   (B) 2   (C) 3   (D) 4   (E) 5

**Problem 3:**  *If $a^2 + b^2 = 1$ and $c^2 + d^2 = 1$, what is the maximum value of $ac + bd + 1$?*

(A) 1.5   (B) $\sqrt{2}$   (C) 2   (D) $\sqrt{1.5}$   (E) $\sqrt{3}$

**Problem 4:**  *What is the smallest value of the positive integer $n$ such that $n^{300} > 3^{500}$?*

(A) 6   (B) 7   (C) 8   (D) 244   (E) 343

**Problem 5:**  *How many solutions of the form $(x, y)$, where $x$ and $y$ are integers, does the following system of inequalities have?*

$$(i)\quad x^2 - y < -1 \quad \text{and} \quad (ii)\quad x^2 + y < 5$$

(A) 2   (B) 3   (C) 4   (D) 5   (E) More than 5

**Problem 6:**  *Suppose $a$, $b$ and $c$ are integers satisfying the following conditions:*

$$(1)\quad a - b + c = 3$$
$$(2)\quad a > 10 \text{ and } b \geq 8$$
$$(3)\quad b^2 + c^2 < 80.$$

*What is the largest possible value of $a$?*

(A) 0   (B) 10   (C) 11   (D) 14   (E) 15

**Problem 7:**  *Given that $0 < \theta < 180°$ and that $2\sec\theta - \cos\theta\cot\theta \geq k$, find the maximum value of $k$.*

(A) $\sin\theta$   (B) 2   (C) 1   (D) 4   (E) 3

**Problem 8:**  *In a right-angled triangle ABC, the area is $4\ cm^2$ and the hypotenuse is $c$. What is the smallest value $c$ can take?*

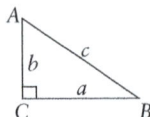

(A) 4  (B) 8  (C) 16  (D) $\sqrt{2}$
(E) Cannot be determined from information

**Remark:**  This problem, expressed in purely analytical terms, required us to minimize a function $f(a,b) = a^2 + b^2$ of two variables subject to a constraint $ab = 8$. It is a simple case of the general problem of minimizing a function subject to constraints on its variables, solved in advanced calculus by the 'method of Lagrange multipliers'. In our problem this method would seek a 'stationary point' of the function $F(x,y) = x^2 + y^2 - \lambda xy$, yielding (using calculus) two equations $2x - \lambda y = 0 = 2y - \lambda x$, which together with the constraint condition $xy = 8$ are sufficient to give us the solutions $\lambda = \pm 2, \ x = y = \pm 2\sqrt{2}$.

**Problem 9:**  *What is the smallest value of $3(a^2+b^2+c^2)-(a+b+c)^2$, where $a,b,c$ are real numbers?*

(A) 5  (B) 4  (C) 3  (D) 0  (E) 1

**Problem 10:**  *The products mn and rs are both 24. Both m and r have values between 1 and 24. If m is less than r, then s is*

(A) less than 24 and less than $n$
(B) greater than 24 and less than $n$
(C) equal to $n$
(D) greater than 24 and greater than $n$
(E) less than 24 and greater than $n$

**Problem 11:**  *For what values of k does the pair of simultaneous equations below have real solutions?*

$$x - ky = 0 \quad and \quad x^2 + y = -1.$$

(A) $-\frac{1}{2} \le k \le \frac{1}{2}$  (B) $-\frac{1}{4} \le k \le \frac{1}{4}$  (C) $k \le \frac{1}{2}$  (D) $0 \le k \le \frac{1}{2}$  (E) $k \le \frac{1}{4}$

**Problem 12:**  *Let $\log_a b + \log_b a = c$. The greatest whole number less than or equal to c for all $a, b > 1$ is:*

(A) 1  (B) 2  (C) 3  (D) 4  (E) 5

**Problem 13:**  *Let $a, b, c$ and $d$ be positive real numbers. Prove that*

$$\frac{1}{a} + \frac{1}{b} + \frac{4}{c} + \frac{16}{d} \ge \frac{64}{a+b+c+d}.$$

**Problem 14:**  *Let $a, b$ and $c$ be real numbers such that $a^2 + b^2 + c^2 = 1$. It is known that real numbers $P_1$ and $P_2$ can be found such that*

$$P_1 \le ab + bc + ac \le P_2.$$

*What is the smallest value of $P_2 - P_1$?*

(A) $\frac{3}{2}$  (B) 1  (C) $\frac{1}{2}$  (D) 2  (E) $2\frac{1}{2}$

## 2.7  Solutions

**Solution 1:**    The necessary and sufficient condition that three points drawn in a plane should satisfy in order to define a triangle is the *strict triangular inequality*, that the longest side should be strictly less than the sum of the other two sides, hence each side should be strictly less than the sum of the remaining two sides. We shall use this to find the range of values of $b$ for which it is possible to draw any triangle with the given lengths. The strict triangle inequality requires that

$$AB + BC > AC, \text{ that is } b + 1 + 4b - 2 > 7 - b, \text{ so that } b > \frac{4}{3},$$

$$AC + BC > AB, \text{ that is } 7 - b + 4b - 2 > b + 1, \text{ so that } b > -2,$$

$$AC + AB > BC, \text{ that is } 7 - b + b + 1 > 4b - 2, \text{ so that } b < \frac{5}{2}.$$

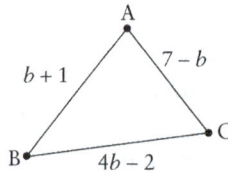

Taking these inequalities simultaneously leads us to the conclusion that the three sides can form a triangle if and only if

$$\frac{4}{3} < b < \frac{5}{2} \quad (\text{then also } b > -2).$$

Denote this interval by $I = (\frac{4}{3}, \frac{5}{2}) = (1\frac{1}{3}, 2\frac{1}{2})$.

We next consider the possible values of $b \in I$ for which the triangle $ABC$ is isosceles.

**Case I:** $AC = BC$: $7 - b = 4b - 2$ hence $-b - 4b = -2 - 7$, i.e. $b = \frac{9}{5} \in I$.
Hence $ABC$ can be isosceles with $AC = BC$.
**Case II:** $AB = AC$: $b + 1 = 7 - b$ hence $b = 3 \notin I$.
Hence $ABC$ cannot be isosceles with $AB = AC$.
**Case III:** $AB = BC$: $b + 1 = 4b - 2$ hence $b = 1 \notin I$.
Hence $ABC$ cannot be isosceles with $AB = BC$. There is only **one** value of $b$, namely $b = 1\frac{4}{5}$, for which $ABC$ is isosceles. Hence (B).

**Solution 2:**    Let the smallest angle be $x$, so that the largest angle is $4x$. Now, since the angle sum of a triangle is $180°$, the third angle will be $180 - x - 4x = 180 - 5x$, and since all the angles are acute we have

$$x \leq 180 - 5x \leq 4x \leq 90. \tag{2.8}$$

Now (2.8) can be separated into three inequalities:

| I | II | III |
|---|---|---|
| $x \leq 180 - 5x$ | $180 - 5x \leq 4x$ | $4x \leq 90$ |
| thus $6x \leq 180°$ | hence $9x \geq 180°$ | so $x < 22.5°$ |
| therefore $x \leq 30°$ | hence $x \geq 20°$ | |

Since all these are satisfied precisely when

$$20° \leq x \leq 22.5°,$$

and since $x$ assumes only whole number values, there are three solutions: $x = 20°$, $21°$, $22°$. Each of these gives rise to exactly one valid triangle, up to similarity (which means counting all similar triangles as one). Therefore (C) is the correct choice.

**Solution 3:**  After the experience of the examples in this Toolchest you should quickly think of using the following two inequalities:

$$a^2 + c^2 \geq 2ac, \text{ since } (a - c)^2 \geq 0,$$
$$b^2 + d^2 \geq 2bd, \text{ since } (b - d)^2 \geq 0.$$
$$\text{Adding, } a^2 + b^2 + c^2 + d^2 \geq 2(ac + bd),$$
$$\text{therefore} \quad 2 \geq 2(ac + bd)$$
$$\text{hence} \quad 1 \geq ac + bd.$$

Thus, an upper bound for $ac + bd$ is 1. But this value can be realized by setting $a = c = 1$ and $b = d = 0$, for instance, so it is actually the maximum value of $ac + bd$. Therefore the maximum value of $ac + bd + 1$ is $1 + 1 = 2$. Hence (C).

**Solution 4:**

$$\left(n^3\right)^{100} > \left(3^5\right)^{100} \Leftrightarrow n^3 > 3^5 = 243.$$

The answer now comes by simply evaluating a few cubes: $3^3 = 27$, $4^3 = 64$, $5^3 = 125$, $6^3 = 216 < 243$, $7^3 = 343 > 243$. Hence the required smallest value is $n = 7$, so (B) is the correct answer.

**Solution 5:**  Adding (i) and (ii) gives $2x^2 < 4$, so $x^2 < 2$, giving $x = 0$, $-1$ or $1$.

In the first case, $x = 1$, (i) gives $1 - y < -1$, hence $2 < y$, and (ii) gives $1 + y < 5$, hence $y < 4$. Together, these give $y = 3$, so we have a single solution $(1, 3)$. A similar analysis yields, for the case $x = 0$, the solutions $(0, 2)$, $(0, 3)$, $(0, 4)$, and, for the case $x = -1$, the solution $(-1, 3)$, so that we have a total of $1 + 3 + 1 = 5$ solutions. Hence (D).

**Solution 6:**  Firstly, from $b \geq 8$ we obtain $b^2 \geq 64$, and $-b^2 \leq -8$. Hence $c^2 < 80 - b^2 < 80 - 64 = 16$, and so $c \in \{-3, -2, -1, 0, 1, 2, 3\}$.

Further, since $64 \le b^2 < 80$ and $b$ is an integer, we must have $b = 8$. Then $10 < a = 3 + b - c = 3 + 8 - c \le 11 - (-3)$, by taking the lowest possible value of $c$. Thus $a \le 14$, hence (B).

**Solution 7:**   Consider the LHS:

$$2 \sec \theta - \cos \theta \cot \theta = \frac{2}{\sin \theta} - \cos \theta \cdot \frac{\cos \theta}{\sin \theta} = \frac{2 - \cos^2 \theta}{\sin \theta}$$

$$= \frac{1 + \sin^2 \theta}{\sin \theta} = \sin \theta + \frac{1}{\sin \theta}.$$

Now putting $\sin \theta = x$, we can apply Inequality 1, for $0 < \theta < 180°$ ensures that $\sin \theta > 0$.

Alternatively, go from basics: $(x - 1)^2 \ge 0$ for all real $x$, so $x^2 + 1 \ge 2x$, and division by $x > 0$ gives $x + \frac{1}{x} \ge 2$. And for $x = 1$ the expression attains its minimum value 2. Thus the maximum value of $k$ is 2.   Hence (B).

**Solution 8:**   A natural way to start is to express both the fixed area of the triangle and the variable hypotenuse $c$ in terms of the legs $a, b$. For generality, let's suppose we do not know the area. It is $ABC = \frac{ab}{2} = A$, say, hence $2ab = 4A$. Also, $c^2 = a^2 + b^2$, by Pythagoras' theorem. Now, there is an inequality connecting $2ab$ and $a^2 + b^2$: it is $a^2 + b^2 \ge 2ab$, for all real numbers $a, b$. (Apply the AM–GM inequality to $a^2$, $b^2$, or simply deduce it from $(a - b)^2 \ge 0$.)

Hence we have $c^2 \ge 4A$ and so $c \ge 2\sqrt{A}$, and this must be true for any right-angled triangle. Putting $A = 4$ gives $c \ge 4$. We must check that $c$ really can take this value 4, that is, we must show that this lower bound is actually a minimum. Putting $c = 4$, the relevant equations for $a$ and $b$ are $a^2 + b^2 = 16$ and $ab = 8$. Squaring the latter, and substituting in the former will give a quadratic equation which yields the single solution (obviously valid by inspection) $(a, b, c) = (\sqrt{8}, \sqrt{8}, 4)$. Hence (A).

**Solution 9:**   For convenience, denote the expression by $F(a, b, c)$. Consider

$$
\begin{aligned}
F(a, b, c) &= 3a^2 + 3b^2 + 3c^2 - (a + b + c)^2 \\
&= 3a^2 + 3b^2 + 3c^2 - (a^2 + b^2 + c^2 + 2ab + 2ac + 2bc) \\
&= 2a^2 + 2b^2 + 2c^2 - 2ab - 2ac - 2bc \\
&= a^2 - 2ab + b^2 + a^2 - 2ac + c^2 + c^2 - 2bc + b^2 \\
&= (a - b)^2 + (a - c)^2 + (c - b)^2.
\end{aligned}
$$

If $a = b = c$ then $F = 0$, and, for all $a, b, c$, $F \ge 0$, being a sum of squares. Hence (D).

**Solution 10:**   Summarizing what we are given:

$$(1) \ 1 < m < r < 24, \quad (2) \ m < r, \quad (3) \ mn = rs = 24.$$

It follows from (3) that that $\dfrac{s}{n} = \dfrac{m}{r}$, but $\dfrac{m}{r} < 1$ (by (2)), so that $\dfrac{s}{n} < 1$, hence $s < n$.

Also from (3), $n = \dfrac{24}{m} < 24$, since $m > 1$ from (1). Hence $s < n < 24$, so that (A) is the correct answer.

**Solution 11:**   We will need the ideas associated with the discriminant of a quadratic equation (Section 4). Substituting $x = ky$ in the second equation gives $k^2y^2 + y + 1 = 0$, which has real roots if (and only if)   $1 - 4k^2 \geq 0$. By sketching the graph of the quadratic equation $y = 1 - 4k^2$, or by re-expressing the inequality as $(1 + 2k)(1 - 2k) \geq 0$, we see that $-\frac{1}{2} \leq k \leq \frac{1}{2}$. Hence (A).

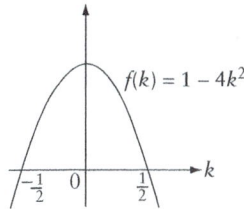

**Solution 12:**   The required insight here is that $\log_b a = \dfrac{1}{\log_a b}$. To see why, recall the definition: the logarithm of $a$ to the base $b$ is the power to which the base must be raised to give $a$. That is, if $x = \log_a b$, then $b = a^x$, hence $a = b^{\frac{1}{x}}$, hence $\frac{1}{x} = \log_b a$.

With this insight, our problem reduces to finding how small $c$ can be, where

$$c = x + \frac{1}{x}.$$

Now, since $a, b > 1$, we know (from that definition again) that $\log_a b > 0$, i.e. $x > 0$. Therefore we are dealing with Inequality 1 once more: if $x > 0$, $c = x + \frac{1}{x} \geq 2$ always, with equality for $x = 1$. The greatest whole number less than or equal to $c$ is 2.  Hence (B).

**Solution 13:**   Put $x = \dfrac{1}{a} + \dfrac{1}{b} + \dfrac{4}{c} + \dfrac{16}{d}$. Then

$$(a+b+c+d)x = 1 + \frac{a}{b} + \frac{4a}{c} + \frac{16a}{d} + \frac{b}{a} + 1 + \frac{4b}{c} + \frac{16b}{d}$$

$$+ \frac{c}{a} + \frac{c}{b} + 4 + \frac{16c}{d} + \frac{d}{a} + \frac{d}{b} + \frac{4d}{c} + 16$$

$$= 22 + \left(\frac{a}{b} + \frac{b}{a}\right) + 2\left(\frac{2a}{c} + \frac{c}{2a}\right) + 4\left(\frac{4a}{d} + \frac{d}{4a}\right)$$

$$+ 2\left(\frac{2b}{c} + \frac{c}{2b}\right) + 4\left(\frac{4b}{d} + \frac{d}{4b}\right) + 8\left(\frac{2c}{d} + \frac{d}{2c}\right).$$

Notice that each expression in brackets is not less than 2, for $x + \frac{1}{x} \geq 2$ for all $x > 0$ (this is Inequality 1). Therefore

$$(a + b + c + d)x \geq 22 + 2 + 2 \cdot 2 + 4 \cdot 2 + 2 \cdot 2 + 4 \cdot 2 + 8 \cdot 2 = 64.$$

**Solution 14:**   A useful connection between the given sum of squares and the expression to be bounded is $(a + b + c)^2$, and we know this is non-negative:

$$0 \leq (a + b + c)^2 = a^2 + b^2 + c^2 + 2ab + 2bc + 2ac$$
$$= 1 + 2(ab + bc + ac),$$

therefore    $-\dfrac{1}{2} \leq ab + bc + ac.$

If $a = 0$, $b = \frac{1}{\sqrt{2}}$, $c = \frac{1}{\sqrt{2}}$ then $ab + bc + ac = -\frac{1}{2}$, so the greatest possible value of $P_1$ is $-\frac{1}{2}$.

Now, to find the least possible value for the upper bound $P_2$, we recall that the Cauchy–Schwarz inequality gives us something on the right bigger than the square of *a sum of pairs* (which is what we are dealing with) on the left:

$$(a_1 b_1 + a_2 b_2 + a_3 b_3)^2 \leq (a_1^2 + a_2^2 + a_3^2)(b_1^2 + b_2^2 + b_3^2).$$

Setting $a_1 = a$, $b_1 = c$, $a_2 = a$, $b_2 = b$, $a_3 = b$, $b_3 = c$, (or, easier, just apply the Cauchy–Schwarz inequality to the two triples $(a, a, b)$, $(c, b, c)$) we get

$$(ac + ab + bc)^2 \leq (a^2 + a^2 + b^2)(c^2 + b^2 + c^2)$$
$$= (a^2 - c^2 + 1)(c^2 - a^2 + 1)$$
$$= (1 + x)(1 - x) \quad \text{(where } x = a^2 - c^2\text{)}$$
$$= 1 - x^2 \leq 1.$$

Equality occurs when $x = 0$, that is, $a^2 = c^2$. This means that the maximum value of $ac + ab + bc$ is 1, and so the least upper bound $P_2$ for the expression is 1. We have

$$-\frac{1}{2} \leq ac + ab + bc \leq 1,$$

with both bounds attained for certain values of the variables. Thus the smallest value of $P_2 - P_1$ is $1 - (-\frac{1}{2}) = \frac{3}{2}$. Hence (A).

# 3 Diophantine equations

*"At this very moment I am glad I am not Justitia – it doesn't appear to me that this Diophantine equations business is trivial. I mean, honestly, how do you keep simultaneous balance between a die and fountain!?"*

By the end of this topic you should be able to:

(i) Define the meaning of 'Diophantine equation'.
(ii) Solve Diophantine equations of the form $ax + by = \gcd(a, b)$.
(iii) Solve Diophantine equations of the form $ax + by = c$.
(iv) Set up and solve Diophantine equations of the form in (iii) from word problems.
(v) Solve some simple nonlinear Diophantine equations.

## 3.1 Introduction

As an appetizer, here is a typical problem of the kind Sections 1–5 should equip you to solve. You are invited to try it as soon as you wish. You may find it hard for now, but by the end of Section 5, where its solution is given, it should not look difficult to you.

**Appetizer Problem:**  *Yeukai's grandmother has given her $10, and asked her to buy the maximum possible total number of mangoes and oranges, using all the money, but getting more oranges than mangoes. Mangoes cost 7 cents each and oranges cost 13 cents each. What should she buy?*

The term 'Diophantine equation' refers to any equation in one or more variables whose solution(s) must come from a restricted set of numbers. The set might be the rationals, integers, non-negatives, etc. For example, suppose we have to find all $x, y \in \mathbb{Z}$ such that

$$21x + 7y = 11,$$

then we have to solve a *linear Diophantine equation* in two variables. In fact, the equation above has no integral solutions, for reasons you shall soon find out. In our discussion we shall restrict ourselves to solutions coming from the set of integers, unless otherwise stated. Diophantine equations are named after Diophantus of Alexandria, who lived around 250 CE. Not much is known about him, but he was an Egyptian who received a Greek education. Diophantus did much work on the exact solution of equations and gave considerable impetus to the slow development of algebraic symbolism that culminated in our modern economical symbolic notation. Before we can study the methods for solution of Diophantine equations, we must introduce some necessary concepts and procedures.

## 3.2    Division algorithm and greatest common divisor

This algorithm formalizes the procedure of 'division with remainders' in the integers. Given integers $a, b$ with $b > 0$, there exist unique $q, r \in \mathbb{Z}: a = qb + r$, with $0 \le r < b$; that is, $\dfrac{a}{b} = q + \dfrac{r}{b}$, where $q$ is the **quotient** and $r$ is the **remainder**.

If $r = 0$ we say that $b$ **divides** $a$, and we write $b|a$.

The **greatest common divisor** of $a$ and $b$, denoted by $\gcd(a, b)$, is the largest positive integer which divides both $a$ and $b$. For this to exist, at least one of the integers $a$ and $b$ must be non-zero, for 0 is divisible by any number. Let's define the gcd formally:

**Definition:**    Let $a, b \in \mathbb{Z}$ not both zero; then $\gcd(a, b)$ is the unique natural number $d$ such that:

(i)  both $d|a$ and $d|b$;
(ii) if $c|a$ and $c|b$, then $c|d$.

The natural question is: How do we find $\gcd(a, b)$, for any given $a, b \in \mathbb{Z}$?

One way is to factorize both, and then select the factors that appear in both. This can be very time-consuming for large numbers; even modern computers lack the speed to factorize very large numbers efficiently. (This is exploited in methods of safe encryption of information.) Another way is to use the following schema, given by Euclid. The basic idea is to divide larger by smaller, then divide smaller by remainder (which is smaller than it), and so on, until the division is exact.

## 3.3  Euclidean algorithm

Let $0 < b \leq a$. By the division algorithm, we have:

$$a = q_1 b + r_1, \quad 0 \leq r_1 < b.$$

If $r_1 = 0$, then $b|a$ so that $\gcd(a, b) = b$; if $r_1 \neq 0$ take $b$ and $r_1$ in the division algorithm to obtain

$$b = q_2 r_1 + r_2, \quad 0 \leq r_2 < r_1.$$

If $r_2 = 0$, stop: we have $\gcd(a, b) = r_1$; otherwise, continue this process until the remainder becomes zero. Suppose the zero remainder is obtained after $n + 1$ steps, thus:

$$a = q_1 b + r_1, \quad 0 < r_1 < b,$$
$$b = q_2 r_1 + r_2, \quad 0 < r_2 < r_1,$$
$$r_1 = q_3 r_2 + r_3, \quad 0 < r_3 < r_2,$$
$$\cdots \quad \cdots \quad \cdots$$
$$r_{n-2} = q_n r_{n-1} + r_n, \quad 0 < r_n < r_{n-1},$$
$$r_{n-1} = q_{n+1} r_n + 0.$$
$$\text{Now} \quad \gcd(a, b) = r_n.$$

You can satisfy yourself that $r_n$ really is $\gcd(a, b)$ by checking:

(i) $r_n|a$, $r_n|b$. Start with the last equation and move upwards, observing that $r_n|r_{n-1}$, hence $r_n|r_{n-2}$ (because it divides both terms on RHS), hence $r_n|r_{n-2}$, etc.
(ii) Any number that divides both $a$ and $b$ must divide $r_1$ (from first equation $r_1 = a - q_1 b$), hence must divide $r_2$ (from second equation) hence (eventually) must divide $r_n$.

Now let us use the Euclidean algorithm to find $\gcd(178, 312)$:

Step (i):  $312 = 1 \cdot 178 + 134$  (of course $0 < 134 < 178$)
Step (ii):  $178 = 1 \cdot 134 + 44$
Step (iii):  $134 = 3 \cdot 44 + 2$
Step (iv):  $44 = 22 \cdot 2 + 0.$

Therefore $\gcd(178, 312) = 2$. Note that if $x|y$ then $-x|y$, that is, divisors occur in pairs. So $\gcd(178, 312) = \gcd(-178, 312) = \gcd(178, -312) = \gcd(-178, -312)$. With these tools in our toolbag let us consider the solution of linear Diophantine equations in two variables.

## 3.4   Linear Diophantine equations

In this section we will show how to apply the Euclidean algorithm to find a solution for Diophantine equations of the form $ax + by = \gcd(a, b)$, and then we will show how to get a solution (if one exists) for the more general equation $ax + by = c$. Finally, we will obtain *all* solutions of such an equation, included in what we call the *general solution*.

**Example:**   Consider the question of finding a solution for the equation:

$$178x + 312y = \gcd(178, 312).$$

The first step is to find $\gcd(178, 312)$, using the Euclidean algorithm as above. We see that $\gcd(178, 312) = 2$. Working backwards now:

$$
\begin{aligned}
2 &= 134 - 3 \cdot 44 \quad \text{(from Step (iii))}\\
&= 134 - 3 \cdot (178 - 1 \cdot 134) \quad \text{(from Step (ii))}\\
&= 4 \cdot 134 - 3 \cdot 178\\
&= 4 \cdot (312 - 178) - 3 \cdot 178 \quad \text{(from Step (i))}\\
&= 4 \cdot 312 - 7 \cdot 178.
\end{aligned}
$$

We see that $x = -7$ and $y = 4$ is a solution. But is this pair the only solution?

Certainly not! It is easy to see that the extra terms we have inserted in the equation below will cancel out, whatever $t$ is, so that $x = -7 + 312t$ and $y = 4 - 178t$ is a solution, for any integral value of $t$:

$$178(-7 + 312t) + 312(4 - 178t) = 2.$$

This gives us an infinity of solutions. But does it include *all* solutions? No! For consider $x = -7 + 156t$ and $y = 4 - 89t$; this gives solution for any $t \in \mathbb{Z}$, testing in the same way, and for $t$ odd we get new solutions. Now we *do* have all the solutions. Indeed, we call this the **general solution**. Why we can be sure it includes all possible solutions will become clear below. We will divide our investigation into two parts: finding a particular solution if one exists, and finding the general solution.

### 3.4.1   Finding a particular solution of $ax + by = c$

By following the method exemplified in the above example, we can always find a solution for $ax + by = \gcd(a, b)$, but this does not mean we can always find a solution for equation $ax + by = c$. Consider the question of solving this linear Diophantine equation:

Find integral $x, y$ such that $3x + 4y = 9$.

Of the two following natural questions, we deal with (1) in this subsection and (2) in the next:

(1) Are there any solutions to begin with?
(2) For a given set of conditions (for example, we might want all numbers positive) is there more than one solution? Are there finitely many solutions? How can we be sure of having them all? If there are an infinite number is there a general way of representing the solution?

In our example $3x + 4y = 9$, applying the Euclidean algorithm, we have

$$4 = 1 \cdot 3 + 1$$
$$3 = 3 \cdot 1$$

so that $\gcd(3,4) = 1$. Yes, of course we knew that already, but those equations are good for more than that! We also have $1 = (-1)3 + (1)4$, so the pair $(-1, 1)$ is a solution for $3x + 4y = 1$. Now 9 is divisible by 1, so we can multiply the whole equation by 9 to get:

$$9\,((-1)3 + (1)4) = 9,$$
$$(9 \times (-1))3 + (9 \times 1)4 = 9,$$
$$3(-9) + 4(9) = 9,$$

giving a solution $(-9, 9)$ for $3x + 4y = 9$. It is clear that $(x, y) = (-1, 3)$ is another solution, and we shall show below how to find the general solution.

What if we had wanted solutions of $2x + 4y = 9$? We would have found $\gcd(2, 4) = 2$, and solution $(-1, 1)$ for $2x + 4y = 2$. But 9 is *not* divisible by 2, so we cannot multiply the whole equation as we did before. In fact. It is easy to see that *there is no solution* (in integers) for $2x + 4y = 9$, for the LHS is even, while the RHS is odd.

In general, if the equation $ax + by = c$ has a solution then $d = \gcd(a, b)$ must divide $c$. For it divides $a$ and $b$, hence it divides the LHS of the equation, so must divide the RHS. Let $c = rd$, some $r \in \mathbb{Z}$. Then if $(x_0, y_0)$ is any particular solution of the equation $ax + by = d$, we have $ax_0 + by_0 = d$ hence $a(rx_0) + b(ry_0) = rd = c$, so that $(rx_0, ry_0)$ is a solution of the equation $ax + by = c$.

The following theorem summarizes our answer to the question (1) of whether a solution exists.

**Theorem 1 (Solvability condition)**  *Consider the general linear Diophantine equation in two unknowns $x$, $y \in \mathbb{Z}$:*

$$ax + by = c, \quad a, b, c \in \mathbb{Z}.$$

*Let $d = \gcd(a, b)$. Then $ax + by = c$ has a solution if and only if $c$ is divisible by $d$.*

### 3.4.2   Finding the general solution of $ax + by = c$

First we look for a particular solution. We check if one is possible by using the Euclidean algorithm to find $d = \gcd(a, b)$; if $d$ does not divide $c$ no solution is possible, by the preceding theorem. Otherwise, if $c = rd$, we work backwards through the equations of the algorithm to find a solution $ax_0 + by_0 = d$, hence (multiplying by $r$) arriving at a solution of the given equation: $a(rx_0) + b(ry_0) = rd = c$. Next, we can get an infinite number of solutions (in fact *all* solutions) from:

**Theorem 2**   *(a) If $x_0$, $y_0$ is a solution of $ax + by = c$, then $x_0 + bt$, $y_0 - at$ is a solution, for any integer $t$, or indeed for any rational number $t$ such that $bt$, $at$ are integers. (b) Furthermore, if $x$, $y$ is **any** solution of $ax + by = c$, then it is of the form $x_0 + \frac{b}{d}t$, $y_0 - \frac{a}{d}t$, where $d = \gcd(a, b)$, $t \in \mathbb{Z}$.*

**Proof:**   (a) It is given that $ax_0 + by_0 = c$. Now

$$a(x_0 + bt) + b(y_0 - at) = ax_0 + by_0 + abt - bat$$
$$= ax_0 + by_0$$
$$= c.$$

(b) We have $ax + by = c$ and $ax_0 + by_0 = c$, hence (subtracting) $a(x - x_0) + b(y - y_0) = 0$, and therefore (dividing by $d$)

$$p(x - x_0) + q(y - y_0) = 0$$

where $p = \dfrac{a}{d}$, $q = \dfrac{b}{d}$, and $p$, $q$ are relatively prime (see Toolchest 4).

Thus $q$ must divide $x - x_0$, so we have an integer $t = \dfrac{x - x_0}{q}$, and also $y - y_0 = \dfrac{-p(x - x_0)}{q} = -pt$. Therefore we have $x = x_0 + qt$, $y = y_0 - pt$.   ◇

Consider now the problem of finding the general solution of the Diophantine equation we looked at above.

**Problem:**   Find *all* pairs $x, y \in \mathbb{Z}$ such that $3x + 4y = 9$.

**Solution:**   We showed that $(-9, 9)$ was a particular solution. And so $(-9 + 4t, 9 - 3t)$ is a solution, for any integer $t$. Putting $t = 2$, for example, gives us our solution $(-1, 3)$. By the Theorem 2 above we know (since $\gcd(3, 4) = 1$) that *all* solutions are included in this representation, which is therefore the general solution.

**Example:**   Find all positive solutions (in integers) for $19x + 99y = 1999$.

**Solution:**   Using Euclid's algorithm,

$$99 = 5 \cdot 19 + 4$$
$$19 = 4 \cdot 4 + 3$$
$$4 = 1 \cdot 3 + 1$$
$$3 = 3 \cdot 1$$

Hence $\gcd(19, 99) = 1$, which is a factor of 1999, so a solution exists. We can find one using the algorithm backwards:

$$1 = 4 - 3 = 4 - (19 - 4.4) = -19 + 5.4 = -19 + 5(99 - 5.19)$$
$$= (-26)19 + (5)99,$$

so $(-26, 5)$ is a solution. A simpler and more obvious solution is $(100, 1)$. Either of these can now be used to get all other solutions, thus:

$$x = 100 + 99t, \quad y = 1 - 19t, \quad t \in \mathbb{Z}.$$

(Can you find which $t$ gives $(-26, 5)$?) Now, we want $x, y > 0$. The benefits of finding a *parametric representation* of the solutions become evident, as we express in terms of the *parameter $t$* these two conditions $x > 0$, $y > 0$:

$$100 + 99t > 0, \ \text{ so that } \ t > -\frac{100}{99},$$

$$\text{and } \quad 1 - 14t > 0, \ \text{ so that } \ t < \frac{1}{49},$$

hence $t = -1$ or $t = 0$. Now $t = -1$ gives $(x, y) = (1, 20)$, and $t = 0$ gives the solution we obtained by inspection, namely $(x, y) = (100, 1)$. Thus, $(1, 20)$ is the only other positive solution.

## 3.5   Euclidean reduction, or 'divide and conquer'

There is a simple alternative algorithm which allows us to find the solution of the linear Diophantine equation $ax + by = c$ (if solvable, of course), without first having to exhibit a particular solution. The basic idea of this method, known as the method of Euclidean reduction, is to express the solution $(x, y)$ in the form $x = i + jt$ and $y = k + mt$, where $i, j, k, m \in \mathbb{Z}$. Consider the following problem:

**Example:**   Fungai steps into the supermarket to buy a candle. A candle costs 29 cents. She has a $2 note, and the shopkeeper has 5 cent and 6 cent coins only in his till. If Fungai is to get her change in these denominations, what is the largest product she can obtain after multiplying the number of 5 cent coins by the number of 6 cent coins in her change?

**Solution:**   Let the number of 6 cent coins be $x$ and the number of 5 cent coins be $y$, so that

$$6x + 5y = 171 \quad \text{with} \quad x, y > 0.$$

Now $\gcd(6,5) = 1$ and 1 divides 171 so a solution exists. We have

$$x = \frac{171 - 5y}{6} = 28 + \frac{3 - 5y}{6}.$$

Letting $\quad p = \dfrac{3 - 5y}{6} \in \mathbb{Z}: \quad x = 28 + p \quad \text{and} \quad y = \dfrac{3 - 6p}{5}$

$$= -p + \frac{3 - p}{5}.$$

Letting $\quad q = \dfrac{3 - p}{5} \in \mathbb{Z}: \quad y = -p + q \quad \text{and} \quad p = 3 - 5q,$

$$\text{so} \quad x = 28 + p = 28 + (3 - 5q)$$
$$= 31 - 5q,$$
$$\text{and} \quad y = -(3 - 5q) + q = 6q - 3.$$
$$\text{Thus} \quad (x, y) = (31 - 5q, \ 6q - 3).$$

We want $x = 31 - 5q > 0$ and $y = 6q - 3 > 0$, which yields $q < \frac{31}{5}$ and $q > \frac{3}{6} = \frac{1}{2}$, hence

$$q \in \{1, 2, 3, 4, 5, 6\}.$$

$$\text{if } q = 1, \quad \text{then } (x, y) = (26, 3) \quad \text{and} \quad xy = 78$$
$$\text{if } q = 2, \quad \text{then } (x, y) = (21, 9) \quad \text{and} \quad xy = 189$$
$$\text{if } q = 3, \quad \text{then } (x, y) = (16, 15) \quad \text{and} \quad xy = 240$$
$$\text{if } q = 4, \quad \text{then } (x, y) = (11, 21) \quad \text{and} \quad xy = 232$$
$$\text{if } q = 5, \quad \text{then } (x, y) = (6, 27) \quad \text{and} \quad xy = 162$$
$$\text{if } q = 6, \quad \text{then } (x, y) = (1, 33) \quad \text{and} \quad xy = 33.$$

From this we see that the largest product is 240.

*Man, don't you just want to thank the heavens for Euclidean reduction? Consider this: Suppose the number of Appletiser bottles in your fridge is the same as the number of houses in your street. Now if I take 100 cans from your fridge…*

*No, man! If you take my Appletisers, especially my cold Appletisers, from my fridge, you will wind-up in a wheel-chair!*

**Example:**   Solve $43x + 5y = 250$ where $x$ and $y$ are positive integers.

**Solution:**   We have $\gcd(43, 5) = 1$ and $1$ divides $250$, hence there is at least one solution. Now

$$x = \frac{250 - 5y}{43} = 5 + \frac{35 - 5y}{43}.$$

Letting $p = \dfrac{35 - 5y}{43} \in \mathbb{Z}$:   $x = 5 + p$ and $y = \dfrac{35 - 43p}{5}$

$$= 7 - 8p - \frac{3p}{5}.$$

Letting $q = \dfrac{3p}{5} \in \mathbb{Z}$:   $y = 7 - 8p - q$ and $p = \dfrac{5q}{3} = q + \dfrac{2q}{3}.$

Letting $r = \dfrac{2q}{3} \in \mathbb{Z}$:   $p = q + r,$   $q = \dfrac{3r}{2} = r + \dfrac{r}{2}.$

Letting $t = \dfrac{r}{2} \in \mathbb{Z}$:   $r = 2t,$   $q = r + t = 3t,$

$$p = q + r = 3t + 2t = 5t,$$

$$\text{hence } x = 5 + p = 5 + 5t,$$

$$\text{and } y = 7 - 8p - q = 7 - 8 \cdot 5t - 3t$$

$$= 7 - 43t.$$

$$\text{Thus } (x, y) = (5 + 5t,\ 7 - 43t).$$

We want $x, y > 0$ so that $5 + 5t > 0$ and $7 - 43t > 0,$ hence $t > -1$ and $t < \frac{7}{43}$. Since $t \in \mathbb{Z}$,  $t = 0$ gives the only solution, which is $(x, y) = (5, 7)$.

**Example:**   Pencils cost 1 cent each, pens 5 cents each and books 7 cents each. You buy 15 items from 75 cents. You want the number of pens to be greater than that of pencils – which is not to be zero, and you want the number of books not to be less than that of pens. How many pencils must you buy?

**Solution:**   Let there be $x$ pencils, $y$ pens, and $z$ books, so (i) $x < y \le z$; (ii) $x + y + z = 15$; (iii) $x + 5y + 7z = 75$.
   From equations (ii) and (iii), we have: $3x + y = 15$, so that $y = 15 - 3x$, which leads us to the table:

| $x$ | 1 | 2 | 3 | 4 | 5 | $\cdots$ | impossible values |
|---|---|---|---|---|---|---|---|
| $y$ | 12 | 9 | 6 | 3 | 0 | $\cdots$ | impossible values |
| $z$ | 2 | 4 | 6 | 8 | 10 | $\cdots$ | impossible values |

We inspect the table; only the third column makes it! That is, it satisfies condition (i), as well as (ii) and (iii). Therefore you must buy three pencils.

Here is the solution to the problem posed at the beginning of Section 1.

**Solution of Appetizer Problem:**   Let the number of the mangoes be $x$ and the number of oranges $y$ so that:

$$7x + 13y = 1000, \text{ with } 0 \leq x < y.$$

$$\text{Now } x = \frac{1000 - 13y}{7} = 142 - y + \frac{6 - 6y}{7}.$$

Letting $p = \dfrac{6 - 6y}{7} \in \mathbb{Z}: \ x = 142 - y + p \text{ and } y = \dfrac{6 - 7p}{6}$

$$= 1 - p - \frac{p}{6}.$$

Letting $q = \dfrac{p}{6} \in \mathbb{Z}: \ y = 1 - p - q \text{ and } p = 6q.$

$$\text{Thus } x = 142 - (1 - 7q) + 6q = 141 + 13q,$$

$$\text{and } y = 1 - 6q - q = 1 - 7q.$$

Now, we have the constraints $0 < x < y$, so

$$0 < 141 + 13q < 1 - 7q, \text{ hence } -10 \leq q < -7,$$

which gives as possibilities:

if $q = -10$, then $(x, y) = (11, 71)$, $x + y = 82$;
if $q = -9$, then $(x, y) = (24, 64)$, $x + y = 88$;
if $q = -8$, then $(x, y) = (37, 57)$, $x + y = 94$.

The maximum number of fruits is 94, so she must buy 37 mangoes and 57 oranges.

## 3.6   Some simple nonlinear Diophantine equations

There are many different kinds of nonlinear equations and no standard procedure that will cover all, so in this section we give you a representative set of examples to illustrate a variety of useful techniques. First, here is an appetizer – a typical problem of the kind you should be able to solve when you have worked through the examples. You are invited to try it now, although you may find it hard before you have worked through this section. You will find the solution at the end of the section.

**Appetizer Problem:**   *How many ordered pairs of integers $(x, y)$ are there which satisfy the following?*

$$0 < x < y \text{ and } \sqrt{1998} = \sqrt{x} + \sqrt{y}.$$

(A)  None   (B)  1   (C)  2   (D)  3   (E)  4

**Example:**  How many pairs of natural numbers $(x, y)$ are there for which $x^2 - y^2 = 64$?

**Solution:**  We are given $64 = x^2 - y^2 = (x - y)(x + y)$. Since $x$ and $y$ are natural numbers, and since $x > y$ from the given equation, both $x - y$ and $x + y$ are natural numbers, with $x - y < x + y$. Listing the factorizations of 64:

$$64 = 1 \times 64 = 2 \times 32 = 4 \times 16 = 8 \times 8.$$

Consider for instance $(x - y)(x + y) = 2 \times 32$, which leads to $x + y = 32$, $x - y = 2$, giving solutions $x = 17$, $y = 15$, so that $(17, 15)$ is one pair with the required property. Consider the other factorizations of 64, and satisfy yourself that only two pairs lead to a solution of $x^2 - y^2 = 64$.

**Example:**  How many pairs of natural numbers $(x, y)$ are there for which $x + y = xy$?

**Solution:**  Since $x$ and $y$ are non-zero real numbers, we can find a real number $k \neq 0$ such that $x = ky$. Hence, from the hypothesis that $x + y = xy$, we have:

$$ky + y = ky^2,$$
$$\text{therefore} \quad k + 1 = ky \quad (\text{since } y \neq 0),$$
$$\text{hence} \quad y = 1 + \frac{1}{k}, \quad (\text{since } k \neq 0).$$

Now we have $x = ky$ and $y = 1 + \frac{1}{k}$. But there is only one value of $k$ for which both $1 + \frac{1}{k}$ and $ky$ are natural numbers (where $y$ is a natural number), and this is $k = 1$. Thus $(2, 2)$ is the only ordered pair with the required property.

**Example:**  If $x$ and $y$ are natural numbers such that $x + y + xy = 9$, what is the largest value of $xy$?

**Solution:**  We set out to adjust the equation so that the LHS is a product $(x + s)(y + t)$. We require $sy = y$, $tx = x$, hence $s = t = 1$, and we need to add 1 to both sides:

$$x + y + xy = 9,$$
$$\text{hence} \quad x + 1 + y + xy = 10,$$
$$\text{giving} \quad (x + 1)(y + 1) = 10.$$

Both $x + 1$ and $y + 1$ are natural numbers. We now proceed to list the factorizations of 10 as a product of two natural numbers. We can choose $x > y$, since the given equation is symmetrical. Thus $(x + 1)(y + 1) =$

$10 \times 1 = 5 \times 2$. Now $x + y = 10$, $y + 1 = 1$ gives $(x, y) = (9, 0)$, but $0 \notin \mathbb{N}$ so the first factorization doesn't give a solution. Take $x + 1 = 5$, $y + 1 = 2$, which gives $(x, y) = (4, 1)$. The only value of $xy$ is 4, and hence it is (trivially) the largest value.

**Example:** How many ordered pairs $(x, y)$ are there which satisfy the equation

$$\frac{1}{x} + \frac{1}{y} = \frac{1}{12}, \quad x, y \in \mathbb{N}?$$

**Solution:** We use a similar trick to that of the previous example, transforming the given equation step by step into an equivalent equation of the form $(x + s)(y + t)$:

$$\frac{1}{x} + \frac{1}{y} = \frac{1}{12}$$
$$12(x + y) = xy$$
$$xy - 12x - 12y = 0$$
$$xy - 12x - 12y + 144 = 144$$
$$(x - 12)(y - 12) = 144.$$

Factorize $144: 144 = 144 \times 1 = 72 \times 2 = 36 \times 4 = 24 \times 6 = 18 \times 8 = 16 \times 9 = 12 \times 12$, so there are 13 such ordered pairs. Do you see why?

**Example:** Determine the number of integral solutions $(x, y, z)$ with $x, y, z$ distinct, of

$$|x| \cdot |y| \cdot |z| = 12.$$

**Solution:** Factorizing gives $12 = 1 \cdot 2 \cdot 2 \cdot 3$, and thus, for $x, y$ and $z$ to be distinct, the possibilities for the three absolute values $|x|, |y|, |z|$ are (ignoring order) $(1, 2, 6)$ and $(1, 3, 4)$. Each of these can be ordered in six different ways, and then, for each of the six ways, we can have $0, 1, 2$, or 3 of the integers positive, making a total of eight ways. Thus altogether there are $2 \cdot 6 \cdot 8 = 96$ solutions to identify.

For example, from $(1, 2, 6)$ we have: $(1, 2, 6)$, $(1, 6, 2)$, $(2, 1, 6)$, $(2, 6, 1)$, $(6, 1, 2)$, $(6, 2, 1)$, and for each of these we have eight, as for the first one: $(1, 2, 6)$, $(-1, 2, 6)$, $(1, -2, 6)$, $(1, 2, -6)$, $(-1, -2, 6)$, $(1, -2, -6)$, $(-1, 2, -6)$, $(-1, -2, -6)$.

Here is the solution to the problem posed at the beginning of this section.

**Solution of Appetizer Problem:** Since $\sqrt{y} = \sqrt{1998} - \sqrt{x}$, squaring both sides gives

$$y = 1998 + x - 6\sqrt{2 \cdot 3 \cdot 37x} \quad \text{(using } (\sqrt{a} - \sqrt{b})^2 = a + b - 2\sqrt{ab}).$$

Now $y$ is an integer, so $x$ must be of the form $(2 \cdot 3 \cdot 37)k^2$, in order that $\sqrt{2 \cdot 3 \cdot 37x}$ takes an integral value. Clearly $k = 1$ is the only value that produces $x \, (= 222)$ positive and smaller than $y \, (= 888)$. Hence (B).

## 3.7  Problems

**Problem 1:**  *In how many different ways is it possible to stamp a letter to the value of 81 cents, using only 4 cent and 7 cent stamps?*

(A) 0  (B) 1  (C) 3  (D) 5  (E) 14

**Problem 2:**  *How many integral solutions are there of* $-20x + 16y = 12$, $|x| < 10, |y| < 10$.

(A) 3  (B) 0  (C) 10  (D) 4  (E) 1

**Problem 3:**  *Three tired sailors have collected a pile of coconuts, and decide to go to sleep and divide the coconuts evenly in the morning. One of them stays awake, throws a coconut to the monkeys, buries a third of those left, and goes to sleep. A second sailor wakes up, throws a coconut to the monkeys, buries a third of what's left and goes back to sleep. The third sailor awakens as it's getting light, throws one coconut to the monkeys, and divides the remainder into three equal piles. How many coconuts, at least, could they have had to start with?*

**Problem 4:**  *Show, quickly, that these three numbers are relatively prime:*

$$2\,500\,137\,802, \quad 1\,420\,515\,313, \quad 3\,920\,653\,117$$

**Problem 5:**  *Suppose a, b and c are distinct positive integers such that* $abc + ab + bc + ac + a + b + c = 1000$. *Find the value of* $a + b + c$.

(A) 46  (B) 31  (C) 49  (D) 28  (E) 27

**Problem 6:**  *Zorodzai does a piece of work in a certain integral number of hours. Tawanda does the same piece of work in a certain number of hours which is also an integer. Tawanda takes less time than Zorodzai, and one of those two integers is prime. What fraction of the work does Zorodzai do in an hour if when they work together on the same piece of work, they take 4 hours to complete it?*

(A) $\frac{1}{20}$  (B) $\frac{1}{5}$  (C) $\frac{1}{4}$  (D) $\frac{1}{9}$  (E)  Indeterminate

**Problem 7:**  *The number of ordered triplets* $(a, b, c)$ *such that* $a + 2b = c$ *and* $a^2 + b^2 = c^2$ *and* $a, b, c \in \mathbb{N}$ *is*

(A) 0  (B) 1  (C) 4  (D) 11  (E)  Infinitely many

**Problem 8:**   *How many ordered pairs $(a, n)$, $a, n \in \mathbb{N}$, are there satisfying the equation*

$$n! + 10 = a^2?$$

*(Recall that we define n! as $1 \times 2 \times 3 \times \cdots \times n$., and call it 'n factorial'.)*

(A) 1  (B)  2  (C) 0  (D) infinitely many  (E)  5

**Problem 9:**   *If $ax + 3y = 5$ and $2x + by = 3$ represent the same line then $a + b$ equals*

(A)  5  (B) $\frac{77}{15}$  (C) $\frac{19}{15}$  (D) $\frac{31}{5}$  (E) $\frac{77}{10}$

**Problem 10:**   *Tendai received a $100 gift voucher to be spent at a large stationery store. He aims to purchase 20 items, using all the money. The items he is interested in are pens at $5 each, notebooks at $7 each and rubbers at $2 each. In how many ways can he spend his money if he buys at least one of each of these three items?*

(A) 0    (B) 1  (C) 2  (D) 3  (E) 6

**Problem 11:**   *The diagram shows a rectangular border made up of an integral number of squares of side equal to the width of the border. We require the (shaded) border to be equal in area to the unshaded rectangular region within the border. What is the smallest number of squares needed to make up the border of the rectangle?*

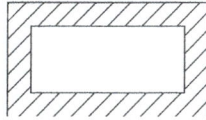

(A) 48  (B) 24  (C) 64  (D) 36  (E) 72

## 3.8  Solutions

**Solution 1:**   To find a particular solution $(x_0, y_0)$ of the linear Diophantine equation $4x + 7y = 81$, we could be lucky and spot one, or, more systematically, we apply Euclid's algorithm to the pair $4, 7$:

$$7 = 1 \cdot 4 + 3$$
$$4 = 1 \cdot 3 + 1$$
$$3 = 3 \cdot 1.$$

As we knew already, $\gcd(4, 3) = 1$, but the equations also give us a solution for $4x + 7y = 1$: working backwards, $1 = 4 - 3 = 4 - (7 - 4) = 4(2) + 7(-1)$, so $(x, y) = (2, -1)$ is a solution. Now multiply through by 81 to get a solution for the given equation: $81 = 4(2 \times 81) + 7(-1 \times 81)$, hence $(x_0, y_0) = (162, -81)$. Since $\gcd(4, 3) = 1$, the general solution is now given by $(162 - 7t, \ 4t - 81)$, where $t$ is an integer chosen to make each number of the solution pair non-negative: $162 \geq 7t, 4t \geq 20$, so that $20 < t \leq 23$. Thus we have three solutions, corresponding to $t = 21, 22, 23$: $(15, 3)$, $(8, 7)$, $(1, 11)$. Hence (C).

Of course you can also do this by 'Euclidean reduction' as in the examples of Section 5. Another approach, basically similar to Euclidean reduction, uses the *congruence arithmetic* introduced in Toolchest 4. Any solution $x$ of $4x + 7y = 81$ must satisfy $4x \equiv 81 \pmod 7 \equiv 4 \pmod 7$, hence $x \equiv 1 \pmod 7$. Thus, the solutions are all of the form $x = 1 + 7k$. Substituting this in the given equation, $7y = 81 - 4(1 + 7k) = 77 + 28k$, so $y = 11 - 4k$. Therefore the general solution is $(x, y) = (1 + 7k, \ 11 - 4k)$, where $k$ is any integer making each of the pair non-negative. Now $k = 0, 1, 2$ provides us with the same three solutions we found by the first method.

**Solution 2:** Since $\gcd(20, \ 16) = 4$, and $4$ divides $12$, we know there is a solution. First (ignoring minus sign for convenience) we seek a solution $(x, y)$ for the equation $5x + 4y = 3$. Applying Euclid's algorithm to the pair $5, 4$:

$$5 = 1 \cdot 4 + 1$$

$$4 = 4 \cdot 1.$$

As expected, $\gcd(5, 4) = 1$, and the equations give us the solution $(1, -1)$ for $5x + 4y = 1$. Now, multiplying by 3, we have the solution $(3, -3)$ for the equation $5x + 4y = 3$, and hence for the equation $20x + 16y = 12$. Using Theorem 2, the general solution of this equation is therefore $(3 - 4k, \ -3 + 5k)$, and hence the general solution of the given equation $-20x + 16y = 12$ is $(x, y) = (4k - 3, -3 + 5k)$, where $k$ can be any integer. For both $|x| < 10$ and $|y| < 10$, it is necessary and sufficient to have $-10 < 4k - 3 < 10$ and $-10 < 5k - 3 < 10$, that is $-\frac{7}{5} < k < \frac{13}{5}$, and this will be so only for the four values $k = -1, 0, 1, 2$. Hence (D).

Now do this by Euclidean reduction, and also explore the congruence notation of Toolchest 4, and try using the approach demonstrated in the previous problem. To solve the equation $5x + 4y = 3$, you look for numbers $x$ satisfying $5x \equiv 3 \bmod 4$). Of the values $x = 0, 1, 2, 3$, only $x = 3$ satisfies the congruence (i.e. leaves remainder 3 when divided by 4), hence the solutions are $x = 3 + 4k$, with corresponding $y = -5k - 3$.

**Solution 3:** Working backwards from three equal piles of $k$ nuts in the morning, there must have been, when the second sailor went back to sleep, $3k + 1$ nuts; and when the first went to sleep there were $\frac{3}{2}(3k + 1) + 1$ nuts. They must therefore have started with $x = \frac{3}{2}\left(\frac{3}{2}(3k + 1) + 1\right) + 1$ nuts. Thus we must solve, for positive integers $x, k$, the Diophantine equation

$4x - 27k = 19$. Applying Euclid's algorithm to the pair $27, 4$:

$$27 = 6 \cdot 4 + 3$$
$$4 = 1 \cdot 3 + 1$$
$$3 = 3 \cdot 1.$$

Naturally, $\gcd(27, 4) = 1$, and the equations give us a solution $(x, k) = (7, 1)$ for $4x - 27k = 1$, from: $1 = 4 - 3 = 4 - (27 - 6 \cdot 4) = 4(7) - 27(1)$. Thus we have a solution $(x, k) = (133, \ 19)$ for $4x - 27k = 19$. Now the general solution, by Theorem 2, is $(x, k) = (133 + 27t, \ 19 + 4t)$, and we want the smallest positive value of $k$. This will occur when $t = -4$, giving $k = 3$, and the original pile of coconuts as $x = 25$. At this point, you should check that this works, by starting with 25 nuts and replaying the night's events. Of course, it would have been fairly obvious in the morning that some were missing since the pile would have been reduced from 25 to 9, but they would no doubt each have kept quiet and publicly blamed the monkeys!

Try to do this one also by the congruence method displayed in the previous two problems.

**Solution 4:** It was part of the demonstration of the Euclidean algorithm that if $a, b, r$ are any positive integers and $a = qb + r$, then $\gcd(a, b) = \gcd(b, r)$. For any common factor of $a$ and $b$ must divide $r = a - qb$, so is a common factor of $b$ and $r$. And conversely, any common factor of $b$ and $r$ must divide $a = qb + r$, so will be a common factor of $a$ and $r$. Applying this idea to our problem, any common factor of our three numbers

$$a = 2\,500\,137\,802, \quad b = 1\,420\,515\,313, \quad c = 3\,920\,653\,117$$

will also divide

$$c - a = 3\,920\,653\,117 - 2\,500\,137\,802 = 1\,420\,515\,315 = d$$

and so will also divide

$$d - b = 1\,420\,515\,315 - 1\,420\,515\,313 = 2.$$

But $b$ and $c$ are odd numbers, so the only possible common factor is 1.

**Solution 5:** As in the examples, we look for some sort of factorization. We observe that

$$(a + 1)(b + 1)(c + 1) = abc + ac + ab + bc + a + b + c + 1$$
$$= 1001$$
$$= 7 \times 11 \times 13.$$

Since $a, b, c > 0$, none of $a + 1$, $b + 1$, $c + 1$ can be 1. Thus $a$, $b$, $c$ must take the values $7 - 1$, $11 - 1$, $13 - 1$ in some order. Therefore their sum is $6 + 10 + 12 = 28$. Hence (D).

**Solution 6:**   Let one of them take $m$ hours to do the work and the other take $n$ hours to do the same piece of work and let $n$ be prime. So, in an hour, the one does $(\frac{1}{m})^{th}$ of the work and the other $(\frac{1}{n})^{th}$ of the work. So, together they do $\frac{1}{m} + \frac{1}{n}$ of the work in an hour's time i.e. $\frac{m+n}{mn}$. Consider the following table:

| Fraction of work done | Time taken |
|---|---|
| $\dfrac{m+n}{mn}$ | 1 hour |
| $\dfrac{m+n}{mn} \times \dfrac{mn}{m+n}$ | $\dfrac{mn}{m+n}$ |
| 1 whole | $\dfrac{mn}{m+n}$ |

This means that they take $\frac{mn}{m+n}$ hours to complete the work when they have concerted their efforts. So put $\frac{mn}{m+n} = 4$. This gives

$$4m + 4n = mn,$$

$$\text{hence} \quad n^2 = mn - 4m - 4n + n^2$$

$$= m(n - 4) + n(n - 4)$$

$$= (m + n)(n - 4).$$

Since $n$ is a prime positive integer and $m$ is a positive integer, this is an integral factorization of the square prime $n^2$, and the only possibilities are (1) $1 \times n^2$, and (2) $n \times n$.

   Case (1): $1 \times n^2$
   Subcase (a): $m + n = 1$, $n^2 - n + 4 = 0$. We have no real solutions since the discriminant is negative: $(-1)^2 - 4(4) = -15 < 0$. (See Toolchest 2, Section 5.)
   Subcase (b): $m + n = n^2$, $n - 4 = 1$, so $n = 5$ and $m = 20$. Since 5 is prime we do indeed have a solution. Now Zorodzai takes more time than Tawanda, hence she does $\left(\frac{1}{20}\right)$th of the work in an hour. Hence (A) is a solution.
   We leave you to show that this is the only solution: show that Case (2): $n \times n$ gives no other solution.

**Solution 7:**   Let

$$a + 2b = c \tag{3.1}$$
$$a^2 + b^2 = c^2 \tag{3.2}$$

Squaring equation (3.1) gives $a^2 + 4ab + 4b^2 = c^2 = a^2 + b^2$ by (3.2), hence $4ab + 3b^2 = 0$, i.e. $b(4a + 3b) = 0$. So

$$b = 0 \tag{3.3}$$
$$\text{or } 4a = -3b. \tag{3.4}$$

Equation (3.3) contradicts the fact that the numbers are natural numbers, and (3.4) does the same. Hence such numbers do not exist, and (A) is the correct choice.

**Solution 8:**   The key to this lies in recognizing that the number $n!$ contains every number less than or equal to $n$ as a factor, so that, whatever factor we want, we can have it for big enough $n$. Thus, if $n \geq 4$, then $n!$ is a multiple of 4. Now, since $n!$ and 10 are even numbers, $a^2$ is also an even number. But a perfect square which is even is a multiple of 4, so $a^2$ is a multiple of 4. Now the equation is saying that the difference of two multiples of 4 ($n!$ and $a^2$) gives a non-multiple of 4 (i.e 10). But this can never happen, for: $4p - 4q = 4(p - q)$. We have a contradiction. Hence, for $n \geq 4$, the equation has no solution. Putting $n = 0$, $n = 1$ and $n = 2$, gives $a^2 = 11$, 11 and 12, respectively, and these are obviously not solutions. But $n = 3$ gives $a^2 = 16$, which is therefore the only solution. Hence (A).

**Solution 9:**

$$ax + 3y = 5 \text{ gives } 3y = -ax + 5, \text{ hence } y = -\frac{a}{3}x + \frac{5}{3},$$
$$\text{and } 2x + by = 3 \text{ gives } by = -2x + 3, \text{ hence } y = -\frac{2}{b}x + \frac{3}{b}.$$

For these two equations to represent the same line, we must have (equating gradients and intercepts):

$$-\frac{a}{3} = -\frac{2}{b} \text{ and } \frac{5}{3} = \frac{3}{b},$$

which yield $ab = 6$, and $5b = 9$, hence $b = \frac{9}{5}$.

Solving: $a \cdot \frac{9}{5} = 6$ gives $a = \frac{10}{3}$, so that $a+b = \frac{9}{5} + \frac{10}{3} = \frac{77}{15}$. Hence (B).

**Solution 10:**   Let the number of pens bought be $p$, the number of notebooks bought be $n$, and the number of rubbers $r$. All his \$100 is spent, hence

$$5p + 7n + 2r = 100. \tag{3.5}$$

He aims to purchase 20 items, hence

$$p + n + r = 20. \tag{3.6}$$

Now $(3.5)-2\times(3.6)$ gives us $3p + 5n = 60$ and hence $p = 20 - \frac{5n}{3} > 0$.

Possible values are $n = 3, 6, 9$, giving $p = 15$, $r = 2$, $p = 10$, $r = 4$, and $p = 5$, $r = 6$, respectively. Hence we have three ways, so that (D) is the correct response.

**Note:**    In being assured that the **only** solutions for $n$ are the multiples of 3, we have reasoned that (since $p$ is integral) 3 divides $5n$ and hence 3 divides $n$. We are here using a simple but important bit of number theory, that if prime number $p$ divides a product $ab$, then it must either divide $a$ or $b$. Observe that this is *not* true for nonprimes: try $p = 4$, $a = 2$, $b = 6$. (See Toolchest 4 for more details.)

**Solution 11:**    Let the inner rectangle be bordered by $a \times b$ squares, so that the area of the large rectangle is composed of $a + 2$ by $b + 2$ squares. Label the diagram appropriately, and it is clear that the shaded border area is $4 + 2b + 2a$, and the unshaded rectangular area is $ab$.

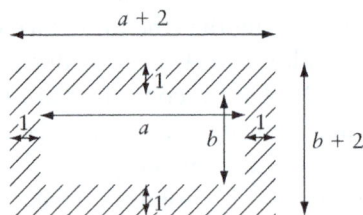

For these to be equal, we get the following Diophantine equation, which can be expressed equivalently in factorized form:

$$ab = 4 + 2b + 2a$$
$$ab - 2b - 2a = 4$$
$$ab - 2b - 2a + 4 = 4 + 4$$
$$b(a - 2) - 2(a - 2) = 8$$
$$(b - 2)(a - 2) = 8 = 4 \times 2.$$

Hence one of $a - 2$ and $b - 2$ should be 2 and the other should be 4, giving 48 squares altogether. The product $8 \times 1$ could also be considered but this leads to a total of 60 squares, and $48 < 60$. Hence (A).

Alternatively, since the shaded and unshaded areas are equal and their sum equal to the area of the rectangle, we could have set $ab = \frac{1}{2}(a+2)(b+2)$. This leads us to $(a - 2)(b - 2) = 8$, and thus to the same results as before.

# 4  Number theory

By the end of this topic you should be able to:

  (i) Use the tests of divisibility as a problem solving tool.
 (ii) Use the congruence notation and do congruence arithmetic.
(iii) Use Fermat's Little Theorem and Wilson's theorem.
(iv) Use the Unique Factorization Theorem (also known as the Fundamental Theorem of Arithmetic).
 (v) Use the Chinese Remainder Theorem.

As appetizers, here are two typical problems of the kind you should be able to solve when you have worked through Sections 1–4, and might find even easier after Section 5. You are invited to try them as soon as you wish. The solution of each is given at the end of Section 4, and a streamlined approach to Appetizer Problem 1 is given at the end of Section 5.

**Appetizer Problem 1:**  *Find the remainder when $2^{4901}$ is divided by* 11.

(A) 1   (B) 3   (C) 6   (D) 10   (E) 2

**Appetizer Problem 2:**  *What are the last two digits in the number* $11^{111}$?

(A) 01   (B) 11   (C) 21   (D) 31   (E) 41

## 4.1   Divisibility, primes and factorization

We collect here some simple facts about divisibility that will be used frequently. In talking about divisibility of numbers, they are always understood to be integers. The idea of divisibility and the notation is introduced at the beginning of Toolchest 3.

We say that $b$ **divides** $a$, and we write $b|a$, if there is an integer $c$ such that $a = bc$. Thus, $4|36$ because $36 = 4 \cdot 9$; and $(-3)|(18)$ because $18 = (-3) \cdot (-6)$. Any number $b$ divides $0$ because $0 = b \cdot 0$. Note carefully that '$b|a$' is a *statement* about $a$ and $b$, while $c = a/b = a \cdot b^{-1}$ is a *number*, called the **quotient** of $a$ by $b$. Note also that '$b$ divides $a$' is equivalently expressed as '$b$ is a **factor** of $a$', and as '$a$ is a **multiple** of $b$'.

A **prime** number is a natural number greater than 1 which is divisible only by 1 and itself. The first few primes are $\{2, 3, 5, 7, 11, 13, 17, \ldots\}$. Eratosthenes showed how to find many primes by using his *sieve*. Arrange the first 1000 (or so) numbers in an orderly table, then ring the number 2 and cross off every even number; ring the next uncrossed number 3, and cross off every third number after that; ring the next uncrossed number 5, and cross off all remaining multiples of 5; etc. (If your table is constructed systematically you will notice that there is a geometrical pattern in the multiples at each stage. Indeed, if you colour *all multiples* at any stage by using a distinctive colour, you will have created an attractive design – a visualization of the corresponding mutiplication table.) When you have completed the crossing off process for your table, you will be left with only the ringed prime numbers.

Finding patterns in the distribution of primes, or generating formulas for primes, has challenged people for many centuries. The 'frequency' of the primes falls off as we go further: to find 100 non-primes in a row, look at $101! + 2$, $101! + 3$, $\ldots$, $101! + 101$. (Recall that $101! = 1 \cdot 2 \cdot 3 \cdot 4 \cdots \cdot 100 \cdot 101$.) But there is always a greater prime to find, as proved by Euclid. His clever and elegant idea is to take any collection of $n$ distinct primes, multiply them together and add 1: the new number $N = (p_1 p_2 p_3 \cdots p_n) + 1$ is *not* divisible by any of those primes (why?), so is either a new prime or is divisible by some new prime. The 'why' in that proof really depends on unique factorization:

Every natural number $N > 1$ can be factored uniquely (we shall state the theorem precisely later) into a product of prime numbers, and is therefore divisible by each of these primes, and by any product of a subset of them.

The **greatest common divisor** of two positive integers $a$ and $b$, denoted by $\gcd(a, b)$, is the largest positive integer which divides both $a$ and $b$. (This is also sometimes called the highest common factor, or $\text{hcf}(a, b)$.) The **least common multiple** of $a$ and $b$, denoted by $\text{lcm}(a, b)$, is the smallest positive integer which is divisible by both $a$ and $b$. The $\gcd(a, b)$ and $\text{lcm}(a, b)$ are each easily discovered once one has the prime factorizations of $a$ and $b$.

Since $1176 = 2^3 \cdot 3 \cdot 7^2$ and $1260 = 2^2 \cdot 3^2 \cdot 5 \cdot 7$, we have:

$\gcd(1176, 1260) = 2^2 \cdot 3 \cdot 7 = 84$, and
$\text{lcm}(1176, 1260) = 2^3 \cdot 3^2 \cdot 5 \cdot 7^2 = 17640$.

If a product $ab$ is divisible by a prime $p$, then either $p|a$ or $p|b$. For example, if $3|(5 \cdot n)$ then (since 3 is a prime, and does not divide 5) we know $3|n$.

If, for any integer $d$, we have $d|a$ and $d|b$, then $d|(pa \pm qb)$, for any integers $p, q$. For example, if 3 is known to divide $17 + 7n$, then it cannot divide $n$. For if it did divide $n$ it would have to divide $17 = (17 + 7n) - 7n$, which is not true.

Two integers are called **relatively prime** if they have no common divisor except 1. For positive integers $a, b$, this is equivalent to each of $\gcd(a, b) = 1$, $\text{lcm}(a, b) = ab$.

## 4.2  Tests for divisibility

You may be conversant with some of the tests for divisibility. We list a good number of these rules below, without proof. (You are challenged to prove them.) The sign $\Leftrightarrow$ is to be read as 'implies and is implied by', or 'if and only if', or 'is necessary and sufficient for'.

Let $N = (a_n a_{n-1} \cdots a_1 a_0)_{10} = a_n 10^n + a_{n-1} 10^{n-1} + \cdots + a_1 10 + a_0$ be a positive integer written in base 10.

1. $N$ is divisible by 2 $\Leftrightarrow$ the last digit $a_0$ (i.e the units digit) is even.
2. $N$ is divisible by 3 $\Leftrightarrow$ the sum of all the digits is divisible by 3.
3. $N$ is divisible by 4 $\Leftrightarrow$ the number formed by the last two digits is divisible by 4.
4. $N$ is divisible by 5 $\Leftrightarrow$ the last digit is 0 or 5.
5. $N$ is divisible by 6 $\Leftrightarrow$ the sum of all digits is divisible by 3 and the units digit is even.
6. $N$ is divisible by 7 $\Leftrightarrow$ $M$ is divisible by 7, where $M$ is the smaller number derived from $N$ by subtracting twice the last digit from the number formed by the rest of the digits.
7. $N$ is divisible by 8 $\Leftrightarrow$ the number formed by the last three digits is divisible by 8.
8. $N$ is divisible by 9 $\Leftrightarrow$ the sum of all the digits is divisible by 9.
9. $N$ is divisible by 10 $\Leftrightarrow$ the units digit is 0.
10. $N$ is divisible by 11 $\Leftrightarrow$ the difference between the sum of digits in the odd places (where the places are numbered starting from the right) and the sum of the digits in the even places (where again the numbering

starts from the right) is a multiple of 11. For example, 979 is a multiple of 11: observe that $(9 + 9) - (7) = 11$; and 8679 is a multiple of 11: observe that $(8 + 7) - (9 + 6) = 0 = 11 \times 0$, etc.

11. $N$ is divisible by 12 $\Leftrightarrow$ the number formed by the last two digits is divisible by 4 and sum of the digits is divisible by 3.

12. $N$ is divisible by 25 $\Leftrightarrow$ the number formed by the last two digits is divisible by 25.

13. $N$ is divisible by 125 $\Leftrightarrow$ the number formed by the last three digits is divisible 125.

**Example:**    Find all positive divisors of $N = 47\,292$.

**Solution:**    As an even number, $N$ is divisible by 2. Since 92 is divisible by 4, so is $N$, and since 292 is not divisible by 8, neither is $N$. The sum of the digits is 24 which is divisible by 3 but not by 9, so this holds for $N$ also. Since the last digit is neither 0 nor 5, $N$ is not divisible by 5. Checking for divisibility by 7, we perform transformations $N \to M$ as in the test:

$$47\,292 \to 4\,729 - (2 \times 2) = 4725 \to 472 - 10 = 462 \to 46 - 4 = 42.$$

Since 7 divides 42, it divides $N$. Since $(9 + 7) - (2 + 2 + 4) = 8$ is not divisible by 11, neither is $N$. So far we have $N = 2^2 \cdot 3 \cdot 7 \cdot 563$. We still have to check divisibility of 563 by the primes 13, 17, 19 and 23, and (since 563 is relatively small) this can be done quickly by routine division. The results will be negative, so we know 563 is a prime factor. Why? Because $25^2 = 625 > 563$, so any factorization $563 = a \cdot b$ would have one of the two factors less than 25. We are done.

**Example:**    The number $N$ is formed by writing all the two digit numbers from 18 to 95 consecutively, i.e. $N = 181920\ldots92939495$. Find the greatest power of 3 that is a factor of $N$

(A) 3    (B) 9    (C) 27    (D) 243    (E) 81

**Solution:**    The first step is to find the sum of the digits of $N$. A method of evaluating the sum in which we reduce the chances of getting tangled up, is this:

Observing that the digit 1 occurs 20 times in the numbers from 18 to 95, 2 occurs 18 times, and so on, we arrive at the sum of 753, for which $3^1$ is the greatest power of 3 that divides it, hence (A).

## 4.3    The congruence notation: finding remainders

Consider now Appetizer Problem 1 given at the begining of this toolchest – finding the remainder when $2^{4901}$ is divided by 11. It would be a very nice thing to have a quick way of finding the integers $k$ and $r$ such that

$$2^{4901} = 11k + r, \quad \text{where } 0 \le r < 11.$$

The required remainder is $r$. It is not a good idea to try and evaluate $2^{4901}$; in fact, you will not live long enough, nor have enough paper to write it down! We cannot expect to find $k$. But there are methods that give $r$ without having to work out $2^{4901}$ or $k$.

Appetizer Problem 2 is (in slightly different words) asking for the remainder when $11^{111}$ (another huge number) is divided by 100.

One method of approaching each of the appetizer problems is illustrated in a number of the examples and problems of this toolchest; it involves observation of the cycles of the remainders. This is, however, a 'rhinoceros', or brute force, method. It can be laborious when considering remainders after division by large integers.

A more efficient method for solving these and even more intricate problems, is Gauss's **congruence notation**. The great advantage of this method is that we need deal only with small numbers (the remainders). In Appetizer Problem 1, we will not have to find the value of $k$ in order to get our target value $r$.

**Definition:**   Let $a, b$ and $m$ be integers, with $m > 0$. We say $a$ **is congruent to** $b$ **modulo** $m$ if $m$ divides $a - b$. We write this as $a \equiv b(\text{mod } m)$.

You may be familiar with the observation that, if two different integers leave the same remainder on division by an integer $m > 0$, then their difference is a multiple of $m$, i.e. $m$ divides their difference. For example, because 11 and 111 leave the same remainder when divided by 10,

$$111 - 11 = (11(10) + 1) - (1(10) + 1) = 10 \times 10,$$

a multiple of 10. In general, if $a$, $b$ leave the same remainder $r$ on division by $m$, then

$$a = q_1 m + r, \; b = q_2 m + r \quad \text{so that } a - b = (q_1 - q_2)m.$$

The remainder cancels out when the difference between the two numbers is taken. And conversely, it is clear that if the difference is a multiple of $m$, then the two numbers have the same remainder. Hence we have:

$a \equiv b(\text{mod } m)$ if and only if $a$ and $b$ leave the same remainder when divided by $m$.

For another example:

$$15 \equiv 1(\text{mod } 7), \quad \text{since } 15 - 1 = 2 \times 7,$$

that is, each of 15 and 1 leave remainder 1 when divided by 7. The following congruences are also true:

$$8 \equiv 2(\text{mod } 3), \quad \text{since } 8 - 2 = 3 \text{ is divisible by } 3$$

$$40 \equiv 7(\text{mod } 11), \quad \text{since } 40 - 7 = 33 \text{ is divisible by } 11$$

$$10 \equiv 0(\text{mod } 5), \quad \text{since } 10 - 0 = 10 \text{ is divisible by } 5$$

$$-7 \equiv 17(\text{mod } 8), \quad \text{since } -7 - 17 = -24 \text{ is divisible by } 8.$$

The relation of 'congruence modulo $m$' is just a particular example of the more general concept of an **equivalence relation** $R$, on a set.

If $A$ is a set of objects (think of numbers, or people), then a **relation $R$ on set $A$** is some rule or relationship associating or 'attaching' a particular $a$ to a particular $b$, where $a$, $b \in A$. That is, given the relation $R$ on $A$, we can tell, for any given pair $(a, b)$, whether $a$ is, or is not, in the relation $R$ to $b$. Think of the relation 'daughter' on the set of all people in the country Utopia; of two people: $a$ is, or is not, the daughter of $b$. If $a$ is related to $b$ under $R$, we write: $aRb$.

Some relations on people, like '$a$ is *daughter* of $b$' (or *aunt*, or *parent*) are not one-to-one: here, there may be more than one $b$ to which a given $a$ is related, and there may be many instances of $a$ related to a given $b$. Or there may be none, as for a man, or a virgin, under the relation *is the mother of*. A relation for which every $a$ has some unique $b$ is called a *function*; for example, in most societies, *wife* is a function on the set of married women, with $b$ (her husband) in the set of men. Some relations are such that they divide up the set $A$ into distinct classes of related objects. For example, the relations: *has the same name as*, or *has the same birthday as*, or *has the same blood group as* each divide the set of people into disjoint groups of associated (or 'congruent') people. These are the **equivalence classes** under that particular relation, and we call such relations **equivalence relations**.

In the light of all that, we may, given some integer $m$, define the equivalence relation $R$ on the set of integers by:

$$aRb \text{ if: } m|(b - a),$$

that is, $a, b$ belong to the same equivalence class modulo $m$ if and only if $m$ divides the difference of $a, b$, if and only if $a \equiv b(\text{mod } m)$. This way of looking at congruence modulo $m$ emphasizes the fact that it is a relation that associates each $a \in \mathbb{Z}$ with some $b \in \mathbb{Z}$ (not unique – there are an infinite number of them). A little thought will convince you that this is in fact an equivalence relation, dividing the integers up into $m$ classes, each distinguished by the remainder (one of the numbers $0, 1, 2, \ldots, m - 1$) that its members have when divided by $m$. We are going to call these the **residue classes** of the integer $m$.

**Some properties of congruence modulo $m$**

Recall that by $p|q$ ('$p$ divides $q$') we mean $q = kp$, for some $k \in \mathbb{Z}$.

**Theorem 1**  *If $a \equiv b(\mathrm{mod}\ m)$, then, for any $x \in \mathbb{Z}$,*

(1) $(a + x) \equiv (b + x)(\mathrm{mod}\ m)$,
(2) $(a - x) \equiv (b - x)(\mathrm{mod}\ m)$,
(3) $(ax) \equiv (bx)(\mathrm{mod}\ m)$,
(4) $(a^n) \equiv (b^n)(\mathrm{mod}\ m)$, $\forall n \in \mathbb{N}$.

**Proof:**  Recall that $a \equiv b(\mathrm{mod}\ m)$ means $m|(a - b)$.

(1) $(a + x) - (b + x) = a - b$, so $m|\big((a + x) - (b + x)\big)$, hence $(a + x) \equiv (b + x)(\mathrm{mod}\ m)$.
(2) $(a - x) - (b - x) = a - b$, so $m|\big((a - x) - (b - x)\big)$, hence $(a - x) \equiv (b - x)(\mathrm{mod}\ m)$.
(3) $(ax) - (bx) = x(a - b)$, so $m|\big((ax) - (bx)\big)$, hence $(ax) \equiv (bx)(\mathrm{mod}\ m)$.
(4) $a^n - b^n = (a - b)(a^{n-1} + a^{n-2}b + a^{n-3}b^2 + \cdots + b^{n-1})$, so $m|\big(a^n - b^n\big)$, hence $(a^n) \equiv (b^n)(\mathrm{mod}\ m)$.     ◇

**Example:**  The number of integers $n$ between 1 and 2000 (inclusive) for which $2^n + 1$ is divisible by 3 is:

(A) 300   (B) 600   (C) 1000   (D) 100   (E) 500

**Solution:**  Starting with the obvious congruence:

$$2 \equiv (-1)(\mathrm{mod}\ 3),$$
$$\text{therefore}\quad 2^n \equiv (-1)^n(\mathrm{mod}\ 3),$$
$$\text{hence}\quad 2^n + 1 \equiv \big((-1)^n + 1\big)(\mathrm{mod}\ 3).$$
$$\text{Thus}\quad 2^n + 1 \equiv 0(\mathrm{mod}\ 3), \text{ for all odd } n,$$
$$\text{while}\quad 2^n + 1 \equiv 2(\mathrm{mod}\ 3), \text{ for all even } n.$$

That is, $2^n + 1$ is divisible by 3 for all odd $n$, and is not divisible by 3 for all even $n$. Therefore the required number equals the number of odd numbers between 1 and 2000 inclusive, which is 1000. Hence (C).

Alternatively, starting with the familiar identity

$$a^n - b^n = (a - b)(a^{n-1} + a^{n-2}b + \cdots + ab^{n-2} + b^{n-1}),$$

and setting $a = 2$ and $b = -1$, we obtain

$$2^n - (-1)^n = [2 - (-1)]P = 3P, \text{ where } P \in \mathbb{Z}.$$

$$\text{Hence}\quad 2^n + 1 = \big(2^n - (-1)^n\big) + \big(1 + (-1)^n\big)$$
$$= 3P + \big(1 + (-1)^n\big),$$

which shows that $2^n + 1$ is divisible by 3 when $n$ is odd, but gives a remainder of $1 + 1 = 2$ when $n$ is even.

## 4.4  Residue classes

Under congruence modulo $m$, the set of integers is partitioned into exactly $m$ classes, as explained earlier. These classes correspond to the possible remainders $r$ on division by $m$, i.e. $r = 0, 1, \ldots, m - 1$, and each class has an infinite membership. For the class giving remainder $r$, the members are

$$qm + r, \quad \text{where } q = 0, \pm 1, \pm 2, \ldots$$

These classes are called the 'residue classes' with respect to $m$. Observe that, for a given $m$, each integer is a member of exactly one of these $m$ classes, and note the word 'residue' has the same meaning as 'remainder'.

### The basic arithmetic of residue classes: congruence arithmetic

**Theorem 2**   *Let* $a \equiv b (\mathrm{mod}\ m)$ *and* $c \equiv d (\mathrm{mod}\ m)$. *Then*

(1) $a + c \equiv (b + d)(\mathrm{mod}\ m)$,
(2) $a - c \equiv (b - d)(\mathrm{mod}\ m)$,
(3) $ac \equiv bd(\mathrm{mod}\ m)$.

**Proof:**   If $a \equiv b(\mathrm{mod}\ m)$ then $m | a - b$, which means we can find an integer $s$ for which

$$a - b = sm, \quad \text{or} \quad a = b + sm. \tag{4.1}$$

Similarly, if $c \equiv d(\mathrm{mod}\ m)$, then $m | c - d$, which means we can find an integer $t$ for which

$$c = d + tm. \tag{4.2}$$

Now, adding equations (4.1) and (4.2) gives:

$$a + c = (b + d) + m(s + t),$$

$$\text{therefore} \quad (a + c) - (b + d) = m(s + t),$$

$$\text{so that} \quad (a + c) \equiv (b + d)(\mathrm{mod}\ m).$$

This proves part (1). Next, taking (4.1)–(4.2), we prove part (2) similarly:

since $a - c = (b - d) + m(s - t)$ we have $(a - c) \equiv (b - d)(\mathrm{mod}\ m)$.

Finally, to prove part (3), we multiply (4.1) and (4.2):

$$ac = (b + sm)(d + tm)$$

$$= bd + btm + sdm + stm^2$$

$$= bd + m(bt + sd + stm)$$

$$= bd + mp, \text{ where } p \in \mathbb{Z},$$

$$\text{therefore} \quad ac \equiv bd(\mathrm{mod}\ m). \qquad \diamond$$

A complete reappraisal of what the rules are saying will be helpful. If we write the residue class of $a$ (modulo $m$) as $C_a$, then we can define the sum of two residue classes as:

$$C_a + C_b = C_{a+b}$$

and similarly the difference and the product:

$$C_a - C_b = C_{a-b}, \quad C_a \cdot C_b = C_{a \cdot b}.$$

Now, these sums are *well-defined*! This means that if you take any two respective representatives of two residue classes, and add/subtract/multiply them, the sum/difference/product will lie in a certain residue class (usually different for the sum, difference and product); and this class will be the same no matter what representatives you happened to choose!

Let us turn our attention back to the two appetizer problems we posed at the beginning of this toolchest and discussed on page 149.

**Solution of Appetizer Problem 1:** We make a table of the remainders on division by 11 of the first few powers of 2, and look for a pattern. Note that to find each successive remainder we need only double the previous remainder, so we don't need to evaluate high powers of 2.

| Number: | $2^1$ | $2^2$ | $2^3$ | $2^4$ | $2^5$ | $2^6$ | $2^7$ | $2^8$ | $2^9$ | $2^{10}$ | $2^{11}$ |
|---------|-------|-------|-------|-------|-------|-------|-------|-------|-------|----------|----------|
| (mod 11): | 2 | 4 | 8 | 5 | 10 | 9 | 7 | 3 | 6 | 1 | 2 |

We observe a cycle of order 10 – every tenth remainder is 1, every $(10k + 1)$th remainder is 2, where $k$ is any non-negative integer, etc. But $4901 \equiv 1 \pmod{10}$ so the remainder for that power is 2.

The fact that $2^{10} \equiv 1 \pmod{11}$ gives us another route to the answer, using congruence arithmetic, because we know that $1^n = 1$ for any real number $n$. Indeed, congruence arithmetic also gives us a slightly quicker way of arriving at this useful fact:

$$2^4 = 16 \equiv 5 \pmod{11}$$

$$2^8 = \left(2^4\right)^2 \equiv 5^2 \pmod{11} \equiv 3 \pmod{11}$$

$$2^{10} = 4(2^8) \equiv 4 \cdot 3 \pmod{11} \equiv 12 \pmod{11} \equiv 1 \pmod{11}.$$

Now we can use this as follows:

$$2^{4900} = (2^{10})^{490} \equiv 1^{490} \pmod{11} \equiv 1 \pmod{11},$$

so that $\quad 2^{4901} = 2 \cdot 2^{4900} \equiv 2 \cdot 1 \pmod{11} \equiv 2 \pmod{11}.$

Hence, the remainder on dividing $2^{4901}$ by 11 is 2, found without having to find how many times 11 goes into that huge number.

Here is the solution to Appetizer Problem 2 posed at the beginning of this Toolchest.

**Solution of Appetizer Problem 2:**   Considering the first powers of 11:

| Number | $11^1$ | $11^2$ | $11^3$ | $11^4$ | $11^5$ | $11^6$ | $11^7$ | $11^8$ | $11^9$ | $11^{10}$ | $11^{11}$ |
|---|---|---|---|---|---|---|---|---|---|---|---|
| Last two digits | 11 | 21 | 31 | 41 | 51 | 61 | 71 | 81 | 91 | 01 | 11 |

The pattern can be extrapolated. Thus, for example, when the exponent has form $1+10n$, the number will end in 11, and, since $111 = 1+(10 \times 11)$, the number $11^{111}$ also ends in 11. Hence (B).

This result was achieved by pattern recognition. We now give full justification, using congruence modulo 100. Starting with the third power of 11: $11^3 = 11 \cdot 11^2 \equiv 11 \cdot 21 (\text{mod } 100) \equiv 31 (\text{mod } 100))$. For the fourth power we need only multiply $11 \times 31$, etc. This method easily leads to $11^{10} \equiv 1 (\text{mod } 100)$, and then, $11^{1+10n} = 11 \cdot (11^{10})^n \equiv 11 \cdot 1^n (\text{mod } 100) \equiv 11 (\text{mod } 100)$.

## 4.5    Two useful theorems

Here is a problem you will be able to solve with the help of the two theorems of this section. You are invited to play around with it now, to whet your appetite for these two powerful tools. Its solution is given at the end of this section.

**Appetizer Problem:**   *Find the remainder when $p^6 - 1$ is divided by 504, where $p > 7$ is a prime number.*

We now state (without proof) these two important theorems in number theory, and illustrate their use.

**Theorem 3 (Fermat's Little Theorem)**   *If $p$ is a prime number and $n$ is relatively prime to $p$, then:*

$$n^{p-1} \equiv 1 (\text{mod } p), \text{ and also, } n^p \equiv n (\text{mod } p).$$

This means, of course, that $n^{p-1} - 1$ is a multiple of $p$, and so is $n^p - n$. The last assertion follows because $n^p - n = n(n^{p-1} - 1)$. For example, if we choose $n = 8$ and $p = 3$ then

$$n^{p-1} - 1 = 8^{3-1} - 1 = 63 = 3 \times 21, \text{ and } n^p - n = 8^3 - 8 = 8 \times 3 \times 21.$$

Recall that '$n$ is relatively prime to $p$' means that $n$ and $p$ do not have any common factors except 1. Since $p$ is a prime here, this simply means that $n$ is not a multiple of $p$, so that $n = kp + r$, where $k$ is an integer, and

$1 \leq r < p$. Because (using the Binomial Theorem of Toolchest 7) $n^p = (kp + r)^p \equiv r^p \pmod{p}$, it is clear that Fermat's Little Theorem is equivalent to the following (perhaps more memorable) statement:

If $p$ is a prime number and $1 \leq n < p$, then $n^p \equiv n \pmod{p}$, that is, $n^p$ has remainder $n$ when divided by $p$.

*Oh, isn't he adorable? ... Fermat's little one!*

**Theorem 4 (Wilson's Theorem)**    *If p is a prime number, then*

$$(p - 1)! + 1 \equiv 0 \pmod{p},$$

*that is, $(p - 1)! + 1$ is a multiple of p.*

Let's try it:

$$\frac{(2-1)!+1}{2} = \frac{1+1}{2} = 1 \in \mathbb{Z}$$

$$\frac{(3-1)!+1}{3} = \frac{2+1}{3} = 1 \in \mathbb{Z}$$

$$\frac{(5-1)!+1}{5} = \frac{25}{5} = 5 \in \mathbb{Z}$$

$$\frac{(7-1)!+1}{7} = \frac{720+1}{7} = 103 \in \mathbb{Z}.$$

**Example:** Find the remainder when $3^{1999}$ is divided by 47.

(A) 20   (B) 21   (C) 41   (D) 19   (E) 18

**Solution:** Since 47 is prime and 3 and 47 are relatively prime, Fermat's Little Theorem gives

$$3^{46} \equiv 1 (\text{mod } 47).$$

$$3^{1999} = (3^{46})^{43} \cdot 3^{21}, \quad (\text{because } 1999 = (46 \times 43) + 21)$$

$$\equiv (1)^{43} \cdot 3^{21} (\text{mod } 47)$$

$$\equiv 3^{21} (\text{mod } 47).$$

Now we are going to build up to that power of 21, starting with a smaller power and squaring successively, so that we never have to deal with really big numbers.

$$3^2 = 9 \equiv (-38)(\text{mod } 47),$$

$$3^4 \equiv (-38)^2 (\text{mod } 47) \equiv 1444(\text{mod } 47) \equiv 34(\text{mod } 47),$$

$$3^8 \equiv (3^4)^2 (\text{mod } 47) \equiv 34^2 (\text{mod } 47) \equiv 1156(\text{mod } 47)$$

$$\equiv 28(\text{mod } 47),$$

$$3^{16} \equiv 28^2 (\text{mod } 47) \equiv 784(\text{mod } 47) \equiv 32(\text{mod } 47).$$

Now,   $$3^{21} = 3^{16} \cdot 3^4 \cdot 3 \equiv 32 \cdot 34 \cdot 3(\text{mod } 47)$$

$$\equiv 3264(\text{mod } 47) \equiv 21(\text{mod } 47).$$

Thus, the remainder when $3^{1999}$ is divided by 47 is 21. Hence (B).

Note that there are alternative ways of doing this. For example:

$$3^5 = 243 \equiv 8(\text{mod } 47),$$

$$3^{10} \equiv 8^2 (\text{mod } 47) \equiv 64(\text{mod } 47) \equiv 17(\text{mod } 47),$$

$$3^{20} \equiv (17)^2 (\text{mod } 47) \equiv 289(\text{mod } 47) \equiv 7(\text{mod } 47),$$

$$3^{21} = 3^{20} \cdot 3 \equiv 7 \cdot 3(\text{mod } 47) \equiv 21(\text{mod } 47).$$

The situations in which Fermat's Little Theorem or Wilson's theorem are handy come in many different guises, and we ought therefore to be constantly on the lookout for possible applications. This does not mean, however, that we should take leave of our common sense and look upon these results as 'universal problem solvers' that will single-handedly solve all given problems. At times we have to resort to ideas from elsewhere in addition to the theorems, and it's often a creative combination of theorems and ideas that solves the problem.

**Example:** Find the remainder when $2^{1990}$ is divided by 1990. (You can tell when this problem was set!)

(A) 1024    (B) 1990    (C) 1991    (D) 1023    (E) 2199

**Solution:** Observe that $1990 = 199 \times 10$, and 199 is prime, hence Fermat's Little Theorem tells us that 199 divides $2^{199} - 2$, i.e. there exists some integer $k$ such that

$$2^{199} = 199k + 2,$$
$$\text{so} \quad 2^{1990} = (199k + 2)^{10}$$

$$= \sum_{i=0}^{10} \binom{10}{i} (199k)^i \cdot 2^{10-i}$$

$$= 2^{10} + 10 \cdot 2^9 \cdot (199k) + \frac{10 \cdot 9}{1 \cdot 2} 2^8 \cdot (199k)^2$$
$$+ \cdots + 10 \cdot 2 \cdot (199k)^9 + (199k)^{10}.$$

Here we used the Binomial Theorem (see Toolchest 7), and the summation notation (explained in Toolchest 6). Our equation tells us that $2^{1990} - 2^{10}$ is a multiple of 199. Also, $2^{1990} \equiv 2^{10} \pmod{10} \equiv 4 \pmod{10}$, because the last digit of $2^n$ has cycle $2, 4, 8, 6, 2, \ldots$, of period four, and $1990 \equiv 2 \pmod 4$. Hence $2^{1990} - 2^{10}$ is actually a multiple of $199 \cdot 10 = 1990$. In other words, the remainder when $2^{1990}$ is divided by 1990 is $2^{10} = 1024$. Hence (A).

**Example:** What are the last two digits of $3^{1999}$?

(A) 49    (B) 41    (C) 69    (D) 67    (E) 61

**Solution:** We get the last two digits by working with modulo 100. (To get the last $n$ digits we would work with modulo $10^n$.) Thus:

$$3^2 \equiv 9 \pmod{100},$$
$$3^4 \equiv 81 \pmod{100},$$

$$3^8 \equiv 81 \times 81 (\mathrm{mod}\ 100) \equiv 61(\mathrm{mod}\ 100),$$

$$3^{10} \equiv 9 \times 61(\mathrm{mod}\ 100) \equiv 49(\mathrm{mod}\ 100),$$

$$3^{20} \equiv 49^2(\mathrm{mod}\ 100) \equiv 1(\mathrm{mod}\ 100),$$

and this is the breakthrough. We have $1999 = 20 \cdot 99 + 19$, which shows that

$$3^{1999} = (3^{20})^{99} \cdot 3^{19}$$

$$\equiv (1)^{99} 3^{19} (\mathrm{mod}\ 100) \equiv (3^{10})(3^9)(\mathrm{mod}\ 100)$$

$$\equiv 49 \times 61 \times 3(\mathrm{mod}\ 100)$$

$$\equiv 67(\mathrm{mod}\ 100).\quad \text{Hence (D)}.$$

Here is the solution to the problem posed at the beginning of this section.

**Solution of Appetizer Problem:**  We will need a variety of tools to solve this one. A little experimentation will help us get started:

$$\text{For } p = 11: \quad p^6 - 1 = 1\,771\,560 = 504 \times 3515;$$

$$\text{for } p = 13: \quad p^6 - 1 = 4\,826\,808 = 504 \times 9577.$$

We may then suspect that $p^6 - 1$ is actually a multiple of 504 whenever $p$ is prime, i.e. the required remainder is zero. But we must prove this, for general $p$.

Now $504 = 2^3 \cdot 3^2 \cdot 7$, so that, for $p^6 - 1$ to be a multiple of 504 for every prime $p$, then it must be a multiple of $2^3$, $3^2$ and 7, and we will show just that.

Of course 7 is also prime, and so $p$ and 7 are relatively prime (they cannot be equal since $p > 7$). We then deduce by Fermat's Little Theorem that $p^{7-1} \equiv 1(\mathrm{mod}\ 7)$, i.e. $p^6 - 1$ is a multiple of 7.

Now, since $p$ is prime and $p > 2$, we know $p$ is odd, so $p - 1$ and $p + 1$ are both even. In fact they are consecutive even numbers, so that one is a multiple of 2 and another of 4, hence $(p + 1)(p - 1) = p^2 - 1$ is a multiple of $2 \times 4 = 8$. But we have

$$p^6 - 1 = (p^2)^3 - 1 = (p^2 - 1)(p^4 + p^2 + 1).$$

Here we use the factorization $a^n - b^n = (a - b)(a^{n-1} + a^{n-2}b + a^{n-3}b^2 + \cdots b^{n-1})$, which holds for all $a, b \in \mathbb{R}$ and $n \in \mathbb{N}$.

Therefore $p^6 - 1$ is a multiple of $p^2 - 1$, and is hence divisible by 8.

The product of the three consecutive integers $(p - 1)p(p + 1)$ is divisible by 3. To see this, consider what happens if the first $p - 1$ leaves remainder 1 or 2 on division by 3. Incidentally, since at least one of the three must be even, their product is divisible by $3! = 6$. (By similar reasoning, the product of any $n$ consecutive integers must be divisible by $n!$ – convince yourself that this is true.)

Returning to our product $(p - 1)p(p + 1)$, we know that $p$ is a prime number greater than 7, so is not divisible by 3, hence $(p^2-1) = (p-1)(p+1)$ must be divisible by 3.

We set out to use this fact, expressing $p^6 - 1$ in terms of $p^2 - 1$:

$$p^6 - 1 = (p^2 - 1)(p^4 + p^2 + 1) \quad \text{(using result above)}$$
$$= (p^2 - 1)((p^2)^2 - 2p^2 + 3p^2 + 1)$$
$$= (p^2 - 1)((p^2)^2 - 2p^2 + 1 + 3p^2)$$
$$= (p^2 - 1)((p^2 - 1)^2 + 3p^2).$$

Now, we showed earlier that $(p^2 - 1)$ is a multiple of 3, so that $(p^2 - 1)^2 + 3p^2$ is also a multiple of 3. Hence $(p^2 - 1) \times ((p^2 - 1)^2 + 3p^2)$ is a multiple of $3 \times 3 = 9$.

Finally, then, $p^6 - 1$ is a multiple of $7 \times 8 \times 9 = 504$, so that the required remainder is indeed zero!

At the end of Section 4 we gave answers to the appetizer problems posed at the beginning of the toolchest. Now we can shorten the solution to the first.

**Solution of Appetizer Problem 1:** Since 11 is prime and 2 is relatively prime to 11, Fermat's Little Theorem gives us immediately that $11^{10} \equiv 1 (\text{mod } 11)$. From this we arrive at the answer just as before.

## 4.6    The number of zeros at the end of $n!$

This and related problems are regulars on the international mathematics competition scene.

Consider finding the number of zeros at the end of 1999! Let us take out all the factors 2 and 5 for a start:

$$1999! = (2^p 5^q)r,$$

where $r$ and 10 are *relatively prime* – i.e. $r$ is not divisible by 2 or 5.

Consider the sequence:

$$1, 2, 3, 4, 5, 6, 7, 8, 9, 10, \ldots, 1998, 1999.$$

Powers of 2 obviously occur more frequently than powers of 5 amongst all those numbers from 1 to 1999, so we can quickly note that $p \geq q$. This means that the number of factors equal to 10 in the product, each of which needs a 5 and a 2, must be $q$, the number of fives, and this will be the number of zeros at the end.

We set out, therefore, to count the number of fives. Every fifth member of the above sequence is a multiple of 5, and we have

$$\left[\frac{1999}{5}\right] = 399 \text{ of these.}$$

Recall that $[x]$ denotes the greatest integer less than or equal to $x$, or 'the integer part of $x$'.

Also, every 25th member of the sequence is divisible by 25 and each of these will contribute an additional factor 5, so we have a total of

$$\left[\frac{1999}{25}\right] = 79 \text{ of these.}$$

Every 125th member of the sequence is divisible by 125 and so each will contribute yet another factor 5, so we have a total of

$$\left[\frac{1999}{125}\right] = 15 \text{ of these.}$$

Continuing in this way, we will see that

$$q = \left[\frac{1999}{5}\right] + \left[\frac{1999}{25}\right] + \left[\frac{1999}{125}\right] + \left[\frac{1999}{625}\right]$$
$$= 399 + 79 + 15 + 3 = 496.$$

Hence we have 496 zeros at the end of 1999!
It is now easy to generalize:

The number of zeros at the end of $n!$ is given by the finite sum:

$$q = \left[\frac{n}{5}\right] + \left[\frac{n}{5^2}\right] + \left[\frac{n}{5^3}\right] + \cdots .$$

We can make this a little more general by saying that the highest power of $p$ in $n!$ ($p \leq n$) is given by the finite sum

$$\left[\frac{n}{p}\right] + \left[\frac{n}{p^2}\right] + \left[\frac{n}{p^3}\right] + \cdots .$$

**Example:** How many zeros are there at the end of $(5^n)!$ where $n \in \mathbb{N}$?

(A) $n$    (B) $n - 1$    (C) $\frac{n(n-1)}{2}$    (D) $\frac{1}{4}(5^n - 1)$    (E) $\frac{1}{4}(5^{n-1} - 1)$

**Solution:**  Let the number of zeros be $q$. Then:

$$q = \left[\frac{5^n}{5}\right] + \left[\frac{5^n}{5^2}\right] + \cdots + \left[\frac{5^n}{5^{n-1}}\right] + \left[\frac{5^n}{5^n}\right]$$

$$= 5^{n-1} + 5^{n-2} + \cdots + 5 + 1 \quad \text{(a geometric series)}$$

$$= \frac{1}{4}(5^n - 1). \quad \text{Hence (D)}.$$

We now consider a very simple yet extremely powerful fact in number theory, recognized and proved by the ancient Greeks, and sometimes called the Fundamental Theorem of Arithmetic.

## 4.7  The Unique Factorization Theorem

Here is a problem to help you to fully appreciate the tools of this section. You are invited to try it now. Its solution is given at the end of the section.

**Appetizer Problem:**  *N is a natural number such that $\dfrac{N}{5}$ is a perfect square and $\dfrac{N}{2}$ is a perfect cube. The smallest value of N for which $\dfrac{N}{3^3}$ is another natural number is:*

(A) 2 916 000   (B) 729 000   (C) 250 000   (D) 1 458 000   (E) 364 500.

The fundamental theorem of arithmetic is that every natural number can be factored in essentially one way into a product of prime numbers. Here is the precise statement:

**Theorem 5   Unique Factorization Theorem**  *Any natural number N can be written in precisely one way in the form*

$$N = p_1{}^{z_1} p_2{}^{z_2} p_3{}^{z_3} \cdots p_k{}^{z_k}$$

*where $p_1, p_2, \ldots, p_k$ are prime, $p_1 < p_2 < \cdots < p_k$, and $z_1, z_2, \ldots, z_k$ are positive integers.*

Note that our strict ordering of the values $p_i$ means we do not count different orderings of their powers as distinct factorizations. Although we can write $15 = 3 \cdot 5 = 5 \cdot 3$ these are not different; in the statement of our theorem, only the first is registered. To give more insight into the usefulness of unique factorization, we demonstrate its use in proving that square roots of primes are irrational. (Note that there are a number of other ways of proving this fact.)

Recall (see Toolchest 2, Section 1) that a **rational number** $q$ is a real number which is a fraction, that is: $q = \frac{m}{n}$, where $m, n \in \mathbb{Z}$. Thus, $3.142 = \frac{3142}{1000}$,

$10 = \frac{10}{1}$, $3.3 = \frac{33}{10}$, $3.\dot{3} = \frac{10}{3}$ are rational. All numbers that are not rational are said to be **irrational**. For example, $\sqrt{2}$, $\sqrt{3}$, $\sqrt{5}$, $\sqrt{6}$, $\sqrt{7}$, $\sqrt{8}$, $\sqrt{10}$, $\pi$, $e$ are all irrational.

**Example:**   Prove that $\sqrt{2}$ is irrational.

**Solution:**   Suppose that $\sqrt{2}$ is in fact rational, and set

$$\sqrt{2} = \frac{a}{b}, \quad \text{with } a, b \in \mathbb{Z}.$$

Squaring both sides:   $b^2 = 2a^2$.

Now, this equation contradicts the unique factorization theorem, because (since there is an even number of twos in any square number) in the unique factorization of the RHS, 2 is raised to an odd power, and in the unique factorization of the LHS, 2 is raised to an even power. This contradiction tells us that our supposition was false – we cannot have $\sqrt{2} = \frac{a}{b}$, with $a, b \in \mathbb{Z}$. That is, $\sqrt{2}$ is irrational.

Here is the solution to the problem posed at the beginning of this section.

**Solution of Appetizer Problem:**   Let $N = 2^x 3^y 5^z$.

Firstly, $N/5 = 2^x 3^y 5^{z-1}$ is a perfect square implies all factors in the unique factorization appear an even number of times, hence $z$ is odd and $x$, $y$ are even.

Secondly, $N/2 = 2^{x-1} 3^y 5^z$ is a perfect cube implies that the power of each factor in the unique factorization is a multiple of 3, hence the smallest value $x$ can take is 4 and also $y$, $z$ are (positive) multiples of 3.

It is now clear that the smallest values $x$, $y$ and $z$ can take, satisfying all these conditions, are:

$$x = 4, \ y = 6, \quad \text{and} \quad z = 3.$$

Also, since $3 < 6$, $N/3^3 \in \mathbb{N}$. Therefore $N = 2^4 \cdot 3^6 \cdot 5^3 = 1\,458\,000$, hence (D) is correct.

## 4.8   The Chinese Remainder Theorem

We start by considering a typical problem that can be solved easily with the help of the Chinese Remainder Theorem. You are invited to try it before looking at the methods of solution, one given immediately afterwards, and another (using the theorem) at the end of the section.

**Appetizer Problem:**   *Brilliant Duck in Bertrand Carroll's fiction book 'Mathland' was indeed brilliant. She devised a quick way of counting her* *ducklings which was basically a residue evaluation process after division (separately) by 5, 3, and 11. She knew that:*

(i) *When counted by fives, 2 ducklings remained.*
(ii) *When counted by threes, 2 ducklings remained.*
(iii) *When counted by elevens, 3 ducklings remained.*

*What is the smallest possible number of ducklings she has?*

Please spend some time on this problem before reading on.

**Intuitive approach to Appetizer Problem:** Your thinking may have revealed that the number of ducklings is given by the number which occurs in each of the following arithmetic sequences:

$$2, \ 7, \ 12, \ 17, \ldots, \ 2 + 5p, \ldots$$
$$2, \ 5, \ 8, \ 11, \ldots, \ 2 + 3q, \ldots$$
$$3, \ 14, \ 25, \ 36, \ldots, \ 3 + 11r, \ldots$$

The smallest common number 47 can be found by developing those sequences a little, and occurs when $p = 9$, $q = 15$ and $r = 4$.

Therefore Brilliant Duck has 47 ducklings.

Such comparison of sequences can be a very handy tool for rapidly solving problems such as the Brilliant Duck problem. However, the method may be cumbersome, and is not general – it might fail altogether in some cases.

Examples of such situations include:

(i) The smallest common term may be given by very large values of $p$, $q$ and $r$ – so large that even the most gifted fortune-teller will not have a clue where to start!

(ii) Many more than three sequences (as in this example) may be necessary.
(iii) A combination of (i) and (ii) (i.e. a real nightmare!).

A more systematic approach to solving Brilliant Duck's and related problems is given by the following:

**Theorem 6 (The Chinese Remainder Theorem)** *Suppose we wish to find a number x which leaves:*

*a remainder of $r_1$ when divided by $d_1$,*
*a remainder of $r_2$ when divided by $d_2$,*

$$\vdots \qquad\qquad \vdots$$

*and a remainder of $r_n$ when divided by $d_n$,*

*where no two of the divisors $d_1, d_2, \ldots, d_n$ have any factors in common.*

*Let $D = d_1 d_2 \ldots d_n$, and $y_i = \dfrac{D}{d_i}$. Now, if we can find numbers $a_i$ such that:*

$$a_i y_i \equiv 1 \pmod{d_i} \quad \text{for each } i\colon 1 \le i \le n,$$

*then a solution to our problem is:*

$$x = a_1 y_1 r_1 + a_2 y_2 r_2 + \cdots + a_n y_n r_n = \sum_{i=1}^{n} a_i y_i r_i.$$

**Remarks:** We have used the summation notation explained in Toolchest 6. Note that there are infinitely many possible values of $x$ satisfying the conditions above; for if $x_0$ is a solution then so is $x_0 \pm kD$, $k \in \mathbb{Z}$. We shall, however, limit ourselves to the smallest positive value satisfying the required conditions.

**Proof:** We have to show that for each $i$ (running from 1 to $n$), $x$ leaves a remainder $r_i$ when divided by $d_i$.
Consider, for each $1 \le j \le n$ and $1 \le i \le n$,

$$m = \frac{a_i y_i r_i}{d_j} = \frac{a_i D r_i}{d_i d_j}.$$

If $i \ne j$, then $d_i d_j | D$ so that $m \in \mathbb{Z}$. However, if $i = j$, then $d_i d_j$ does not divide $D$, since none of $d_1, d_2, \ldots, d_n$ share any common factors, hence $m \notin \mathbb{Z}$.
This shows us that if $x = \sum_{i=i}^{n} a_i y_i r_i$ is divided by $d_j$, every term other than the $j$th term is a multiple of $d_j$ and hence leaves a remainder of zero when divided by $d_j$. Hence to obtain the remainder when $x$ is divided by $d_j$ we only need consider the $j$th term: $a_j y_j r_j$. Now, by choice of the numbers $a_i$,

$$a_j y_j \equiv 1 \pmod{d_j}$$

$$\text{hence} \quad a_j y_j r_j \equiv r_j \pmod{d_j}. \qquad\qquad \diamond$$

*I told you, didn't I? – we should always eat here each time we feel like eating out. Come in at the right time and you will find the proprietress giving away the remainders...*

We now use the Chinese Remainder Theorem to solve the Brilliant Duck problem above.

**Solution of Appetizer Problem:**   We have $D = d_1 \times d_2 \times d_3 = 5 \times 3 \times 11 = 165$, and from $y_i = \frac{D}{d_i}$, we have:

$$y_1 = \frac{165}{5} = 33, \quad y_2 = \frac{165}{3} = 55, \quad y_3 = \frac{165}{11} = 15.$$

It remains to find $a_1$ such that $33a_1 - 1$ is divisible by 5, $a_2$ such that $55a_2 - 1$ is divisible by 3, and $a_3$ such that $15a_3 - 1$ is divisible by 11. It should be easy to see that $a_1 = 2$, $a_2 = 1$ and $a_3 = 3$ give the desired results. It's now a simple matter to evaluate $x$, which is given by:

$$x = \sum_{i=1}^{n} a_i y_i r_i = a_1 y_1 r_1 + a_2 y_2 r_2 + a_3 y_3 r_3$$

$$= (2)(33)(2) + (1)(55)(2) + (3)(15)(3)$$

$$= 132 + 110 + 135 = 377.$$

Now, any number of the form $377 \pm 165k$ is a solution. However, to get the smallest possible solution, we set $k = 2$ to obtain: $377 - 330 = 47$, as before.

The working could be neatly laid out in tabular form as shown below:

| i | 1 | 2 | 3 | |
|---|---|---|---|---|
| $d_i$ | 5 | 3 | 11 | |
| $r_i$ | 2 | 2 | 3 | |
| $y_i$ | 33 | 55 | 15 | |
| $a_i$ | 2 | 1 | 3 | |
| $a_i y_i r_i$ | 132 | 110 | 135 | $\sum_{i=1}^{n} a_i y_i r_i = x = 377$ |

$$D = 5 \times 3 \times 11 = 165, \quad \text{so take } 377 - 165(2) = 47.$$

## 4.9  Problems

**Problem 1:**  *Which of the following will always be an odd number for all possible integral values of N?*

(A) $5N + 1$   (B) $6N^2 + 1$   (C) $5N - 1$   (D) $N^2 + 2N + 3$
(E) $7N^2 + 1$

**Problem 2:**  *If the natural numbers are arranged as in a triangular array below and the rows are as labelled:*

```
                          1                                row 1
                      2       3                            row 2
                  4    5     6   7                         row 3
              8   9   10   11    12  13  14  15            row 4
      16  17  18  19  20  21  22  ...        ...  31       row 5
```

*find the last number in the tenth row.*

(A) 1024   (B) 1022   (C) 1025   (D) 1023   (E) 1021

**Problem 3:**  *The positive integers are arranged in a triangular array as shown below. What is the first number in row 21 of this array?*

```
                    1
                2   3   4
            5   6   7   8   9
        10  11  12  13  14  15  16
```

(A) 401   (B) 400   (C) 402   (D) 441   (E) 440

**Problem 4:**   *The fraction $\frac{16}{64}$ is unusual. The digit 6 occurs on the top and the bottom and if it is 'cancelled' we are left with the correct answer $\frac{1}{4}$. Which of the fractions below has this same property?*

(A) $\frac{12}{24}$   (B) $\frac{23}{92}$   (C) $\frac{15}{45}$   (D) $\frac{19}{95}$   (E) $\frac{16}{96}$

**Problem 5:**   *Think of any whole number, double it and add five. Double this answer and then add two. Now take away the number you first thought of. Your answer will always be*

(A) even   (B) odd   (C) a multiple of 3   (D) a multiple of 5   (E) a multiple of 6

**Problem 6:**   *A positive whole number is said to be perfect if it equals the sum of its proper factors, e.g. 6 is perfect because $1 + 2 + 3 = 6$. The difference between the first two perfect numbers lies between:*

(A) 19 and 55   (B) 1 and 18   (C) 57 and 74   (D) 77 and 113
(E) 61 and 83

**Problem 7:**   *What is the eleventh number in the sequence below?*

$$1, 2, 3, 5, 8, 13, 21, 34, \ldots.$$

(A) 144   (B) 89   (C) 155   (D) 156   (E) 153

**Problem 8:**   *The whole numbers x and y are non-multiples of 3 greater than zero. Find the sum of the numbers that can be the remainders when $x^3 + y^3$ is divided by 9.*

(A) 3   (B) 7   (C) 2   (D) 9   (E) 5

**Problem 9:**   *How many digits has the number $8^{28} 5^{80}$?*

(A) 56   (B) 82   (C) 83   (D) 181   (E) 108

**Problem 10:**   *The integers $1, 2, 3, \ldots, 9$ are written on individual slips of paper and all are put into a hat. Mike chooses a slip at random, notes the integer on it, and replaces it in the hat. Honest then picks a slip at random and notes the integer written on it. Taurai then adds up Mike's and Honest's numbers. Which digit is most likely to be the units digit of this sum?*

(A) 0   (B) 1   (C) 2   (D) 8   (E) 9

**Problem 11:** *How many factors (including 1 and itself) does the number 576 have?*

(A) 25  (B) 26  (C) 21  (D) 20  (E) 30

**Problem 12:** *The number a is randomly chosen from the set $\{1, 2, 3, \ldots, 100\}$. The number b is similarly selected from the same set. What is the probability that the number $3^a + 7^b$ has its units digit equal to 8?*

(A) $\frac{1}{6}$  (B) $\frac{1}{8}$  (C) $\frac{3}{16}$  (D) $\frac{1}{5}$  (E) $\frac{1}{4}$

**Problem 13:** *Suppose $S_n$ is given by $(n^3 - 2^n)/10$. What is the fractional part of $S_{37}$?*

(A) 0.9  (B) 0.7  (C) 0.5  (D) 0.4  (E) 0.1

**Problem 14:** *Use Fermat's Little Theorem to find which of the numbers shown below divides the number X which is the product:*

$$X = (9\,999\,999\,999\,999\,999)(9\,999\,999\,999)$$

(A) 12  (B) 999  (C) 289  (D) 23  (E) 187.

**Problem 15:** *For how many positive integers n is the number $\dfrac{n^3 - 1}{5}$ prime?*

(A) None  (B) One  (C) Two  (D) Four  (E) Eight

**Problem 16:** *What is the last digit (units) of the number $42^{42}$?*

(A) 0  (B) 2  (C) 4  (D) 6  (E) 8.

**Problem 17:** *Find the sum of all the divisors of the number 1800.*

(A) 157  (B) 1 605  (C) 1 042  (D) 59  (E) 14 030

**Problem 18:** *There are less than 20 000 students at the University of Zimbabwe. Exactly 0.100100100 ... % of them study all of the time whilst exactly 27.272727 .... % of them study only at examination times. How many students are there at the University?*

(A) 10 989  (B) 10 681  (C) 11 297  (D) 7 981  (E) 13 997

**Problem 19:** *What is the smallest prime divisor of $5^{1999} + 6^{1999}$?*

(A) 2  (B) 5  (C) 7  (D) 11  (E) 13

**Problem 20:** *How many perfect squares are there between 1 and 100 which are of the form $c^2 + 5c + 6$, where c is a positive whole number?*

(A) 10  (B) 9  (C) 5  (D) 0  (E) 6

**Problem 21:** *The natural numbers from 1 to $X \times Y$ are written in increasing order, in the cells of a table which contains X rows and Y columns. It is* known that the number 20 is in row 3 while 41 is in row 5 and 105 is in the last row. Find the values of X and Y.

(A) $X = 12$, $Y = 8$   (B) $X = 10$, $Y = 9$   (C) $X = 11$, $Y = 9$
(D) $X = 10$, $Y = 8$   (E) $X = 12$, $Y = 9$

**Problem 22:**   How many ordered integral pairs $(a, b)$ make $\dfrac{a^3 + b^3}{a - b}$ a perfect square?

(A) 1   (B) 3   (C) 9   (D) None of these   (E) 0

**Problem 23:**   What is the greatest square number with 4 digits in base 6?

(A) 5555   (B) 5554   (C) 5501   (D) 5401   (E) 5553

**Problem 24:**   If $x, y$ are positive integers with $45x = y^2$, what is the smallest possible value of $x + y$?

(A) 20   (B) 45   (C) 9   (D) 15   (E) 5

**Problem 25:**   If $n$ is a positive integer, which of the following is always divisible by three?

(A) $(n + 1)(n + 4)$   (B) $n(n + 2)(n + 6)$   (C) $n(n + 2)(n + 4)$
(D) $n(n + 3)(n - 3)$   (E) $(n + 2)(n + 3)(n + 5)$

**Problem 26:**   What is the last digit in the sum: $2^{1996} + 3^{1996}$?

(A) 0   (B) 7   (C) 8   (D) 9   (E) 6

**Problem 27:**   What are the last two digits of $6^{2007} + 7^{2007}$?

(A) 59   (B) 21   (C) 79   (D) 23   (E) 69

## 4.10   Solutions

**Solution 1:**   Recall that any odd number can be written in the form $2a + 1$ where $a$ is an integer. So $6N^2 + 1 = 2(3N^2) + 1$ and since $N^2$ is an integer, $6N^2 + 1$ is always an odd number. Hence (C) is a solution.

It is easy to see that none of the others consistently yield odd numbers, by substituting a couple of trial values of $N$, e.g. $N = 0, 1, 2$, etc. Just one counterexample is sufficient.

**Solution 2:**   Observe that the first number in the row $n$ is $2^{n-1}$. This implies that the first number in row 11 is $2^{10} = 1024$. Hence, the last number in row 10 is $1024 - 1 = 1023$. Hence (D).

Another way of seeing this is that the total number of integers in the first 10 rows is: $1 + 2 + 2^2 + 2^3 + \cdots + 2^9 = 2^{10} - 1$ as the sum of a geometric series.

**Solution 3:**    Observe that the last number in row $n$ is $n^2$ (being the sum of the first $n$ odd numbers!). Now consider the last number in row 20. This is $20^2 = 400$, so that the first number in row 21 is $400 + 1 = 401$. Hence (A).

**Solution 4:**    $\frac{19}{95} = \frac{1}{5}$ as required. Hence (D).

Can you think of a way of finding *all* such peculiar fractions $n/m$ where $11 \le n, m \le 99$?

**Solution 5:**    Let $x$ be the number you start with. Move through the steps from $x$, to $2x + 5$, to $2(2x + 5)$, to $2(2x + 5) + 2$, to $(2(2x + 5) + 2) - x$ which is $3x + 12 = 3(x + 4)$, a multiple of 3 for all integer values of $x$. Hence (C).

**Solution 6:**    The next perfect number is found (by experimentation) to be 28, since $28 = 1 + 2 + 4 + 7 + 14$. Now $28 - 6 = 22$, so the answer is (A).

**Note:**    There is a formula, discovered by the Pythagoreans and proved in Euclid to generate perfect numbers: $2^{n-1}(2^n - 1)$ is perfect provided $2^n - 1$ is prime. Setting $n = 2$, $n = 3$ gives the perfect numbers 6 and 28. Setting $n = 4$ fails (why?), and setting $n = 5$ gives another perfect number. You can prove (do it!) that this will always give perfect numbers, by listing all the factors – greatly simplified by the expression in brackets being prime – and adding together the terms of the two geometric series thus obtained. However, although Euler proved much later that *all* even perfect numbers must be given by this formula, it is still not known whether there are any odd ones! Strictly speaking, then, even if you know the formula you still have to fall back on experimentation to check the numbers between 6 and 28.

**Solution 7:**    Observe that $3 = 1 + 2$, $5 = 2 + 3$, $8 = 5 + 3$, that is, letting $u_n$ be the $n$th term:

$$u_3 = u_1 + u_2$$
$$u_4 = u_2 + u_3$$
$$\cdots$$
$$u_9 = u_7 + u_8 = 21 + 34 = 55$$
$$u_{10} = u_8 + u_9 = 34 + 55 = 89$$
$$u_{11} = u_9 + u_{10} = 55 + 89 = 144, \quad \text{hence (A).}$$

Note: These are the famous Fibonacci numbers, first published by Leonardo Fibonacci in his 'Book of Counting': *Liber Abaci* in 1202, in connection with rabbit populations. For more on Fibonacci numbers, see Toolchest 6, Problem 5.

**Solution 8:**    A whole number non-multiple of 3 takes exactly one of the two forms $3m + 1$ or $3m + 2$, where $m$ is a whole number. (This means it leaves either 1 or 2 as remainder when divided by 3.) Hence, we have three cases to look at (we can ignore a fourth case, by symmetry):

| Case I | Case II | Case III |
|---|---|---|
| $x = 3m + 1$ | $x = 3m + 2$ | $x = 3m + 2$ |
| $y = 3n + 1$ | $y = 3n + 1$ | $y = 3n + 2$ |

where $m$ and $n$ are whole numbers greater than zero.

*Case I:*

$$x^3 + y^3 = 27m^3 + 27n^3 + 27m^2 + 27n^2 + 9m + 9n + 1 + 1$$
$$= 9(3m^3 + 3n^3 + 3m^2 + 3n^2 + m + n) + 2.$$

Therefore the remainder on division by 9 is 2.

*Case II:*

$$x^3 + y^3 = 27m^3 + 27n^3 + 54m^2 + 27n^2 + 36m + 9n + 8 + 1$$
$$= 9(3m^3 + 3n^3 + 6m^2 + 3n^2 + 4m + n + 1).$$

Therefore the remainder on division by 9 is 0.

*Case III:*

$$x^3 + y^3 = 27m^3 + 27n^3 + 54m^2 + 54n^2 + 36m + 36n + 8 + 8$$
$$= 9(3m^3 + 3n^3 + 6m^2 + 6n^2 + 4m + 4n + 1) + 7.$$

Therefore the remainder on division by 9 is 7.

Hence, the sum of all the positive remainders is $2 + 0 + 7 = 9$ making (D) the correct response. The solution could be shortened (in expression, not in method) by using congruence arithmetic. As before, we observe that $x$ or $y$ can be of the form $3a + k$ where $a \in \mathbb{Z}$ and $k = 1$ or 2. Then $x^3 = (3a)^3 + 3(3a)^2 k + 3(3a)k^2 + k^3 \equiv k^3 \pmod 9$. Thus, $x^3$, $y^3$ can be $1 \pmod 9$ or $8 \pmod 9$. We then have the following cases (ignoring a fourth case because of symmetry):

| $x$ | $y$ | $(x^3 + y^3) \pmod 9$ |
|---|---|---|
| $3a + 1$ | $3b + 1$ | 2 |
| $3a + 1$ | $3b + 2$ | 0 |
| $3a + 2$ | $3b + 2$ | 7 |
| | | 9 |

**Solution 9:**   Observing all the twos and fives, we think of extracting all the tens:

$$(8)^{28}(5)^{80} = (2^3)^{28}(5)^{80}$$
$$= (2^4)(2)^{80}(5)^{80}$$
$$= 16 \cdot 10^{80},$$

which is 16 followed by 80 zeros, giving $80 + 2 = 82$ digits. Hence (B).

**Solution 10:**   The table below shows that 0 is the most common digit, occurring 9 times whereas all other digits occur only 8 times.

|   | 1 | 2 | 3 | 4 | 5 | 6 | 7 | 8 | 9 |
|---|---|---|---|---|---|---|---|---|---|
| 1 | 2 | 3 | 4 | 5 | 6 | 7 | 8 | 9 | 0 |
| 2 | 3 | 4 | 5 | 6 | 7 | 8 | 9 | 0 | 1 |
| 3 | 4 | 5 | 6 | 7 | 8 | 9 | 0 | 1 | 2 |
| 4 | 5 | 6 | 7 | 8 | 9 | 0 | 1 | 2 | 3 |
| 5 | 6 | 7 | 8 | 9 | 0 | 1 | 2 | 3 | 4 |
| 6 | 7 | 8 | 9 | 0 | 1 | 2 | 3 | 4 | 5 |
| 7 | 8 | 9 | 0 | 1 | 2 | 3 | 4 | 5 | 6 |
| 8 | 9 | 0 | 1 | 2 | 3 | 4 | 5 | 6 | 7 |
| 9 | 0 | 1 | 2 | 3 | 4 | 5 | 6 | 7 | 8 |

**Solution 11:**   (See the related Problem 17.) Consider the simpler question: How many factors does the number $p^n$ have, where $p$ is a prime and positive integer and $n$ a non-negative integer? It's not too difficult to list these factors:

$$1, p, p^2, p^3, p^4, \ldots, p^{n-2}, p^{n-1}, p^n.$$

The above list is exhaustive of all the possibilities and we can see that the list has a total of $(n + 1)$ members. Hence, $p^n$ has $(n + 1)$ factors. Now consider a number $X$ which is given by

$$X = p_1^{n_1} \cdot p_2^{n_2}$$

where $p_1$ and $p_2$ are prime numbers and $n_1, n_2$ are positive whole numbers. Using the rule we have just discovered above, $p_1^{n_1}$ has $(n_1 + 1)$ factors and $p_2^{n_2}$ has $(n_2 + 1)$ factors. Now, for each one of the $(n_1 + 1)$ factors of $p_1^{n_1}$, there are $(n_2 + 1)$ factors of $p_2^{n_2}$. Hence, for the product $p_1^{n_1} \cdot p_2^{n_2}$, there are $(n_1 + 1)(n_2 + 1)$ factors in all. Hence, $X$ has $(n_1 + 1)(n_2 + 1)$ factors. Using the same sort of reasoning, we can conclude that the number $Y$ given by

$$Y = p_1^{n_1} \cdot p_2^{n_2} \cdot p_3^{n_3} \cdots p_m^{n_m}$$

where $p_1, p_2, p_3, \ldots, p_m$ are all prime and $n_1, n_2, n_3, \ldots, n_m$ are non-negative whole numbers, has a total of $(n_1 + 1)(n_2 + 1)(n_3 + 1) \cdots (n_m + 1)$

factors. This rule gives us a wonderfully simple recipe for calculating the number of factors any number has:

Step (1): Express the number as a product of prime factors and write down repeating factors in power form.

Step (2): To each power, add 1, then take the product of the numbers you obtain. This is the number of factors of the original number. To see how this works with the given number 576, we have

$$576 = 24^2 = 2^2 \cdot 3^2 \cdot 4^2 = 2^2 \cdot (2^2)^2 \cdot 3^2 = 2^{2+4} \cdot 3^2 = 2^6 \cdot 3^2.$$

Hence, the number of factors is $(6+1)(2+1) = 7 \times 3 = 21$, making (C) the correct choice.

**Solution 12:** $3^a$ has units digit one of $1, 3, 9, 7$ as has $7^b$. (This we have established by looking at the last digit of $3^a$ for the first few values $a$, then the same being done for $7^b$.) The units digits of all 16 possible sums are summarized in the table below.

|   | 1 | 3 | 7 | 9 |
|---|---|---|---|---|
| 1 | 2 | 4 | 8 | 0 |
| 3 | 4 | 6 | 0 | 2 |
| 7 | 8 | 0 | 4 | 6 |
| 9 | 0 | 2 | 6 | 8 |

Then from the table we have three cases in which the units digit 8 arises out of a total of 16 possible (equally likely) cases. This gives the required probability as $\frac{3}{16}$. Hence (C). Would this work for $\{1, 2, \ldots, 99\}$?

**Solution 13:** In the table below we have tabulated the units digit of $2^n$ as $n$ takes the values $1, 2, 3, 4, \ldots$

| $2^n$ | 1 | 2 | 3 | 4 | 5 | 6 | ... |
|---|---|---|---|---|---|---|---|
| Last digit of $2^n$ | 2 | 4 | 8 | 6 | 2 | 4 | ... |

and the units digit of $2^n$ is seen to follow through the sequence $2, 4, 8, 6, 2, 4, 8, 6, 2, \ldots$, cycling with a period of 4. Since $37 \equiv 1 \pmod 4$, the units digit of $2^{37}$ is 2.

The units digit of $n^3$ when $n$ has last digit 7 is 3, because $37 = 30 + 7$ and so $37^n$ has a sequence of last digits congruent to that of $7^n$.

Thus the units digit of $37^3 - 2^{27}$ is $3 - 2 = 1$, so the required fractional part is $\frac{1}{10} = 0.1$ Hence (E).

**Solution 14:** The first factor has 16 nines and thus is $10^{16} - 1$. By the theorem $10^{17} - 10 = 10(10^{16} - 1)$ is divisible by 17. Since 10 is not,

$10^{16} - 1$ must be divisible by 17. Similarly, the second factor is $10^{10} - 1$ and $10(10^{10} - 1)$ is divisible by 11, i.e. $10^{10} - 1$ is divisible by 11.

Obviously, the product must be divisible by $17 \times 11 = 187$. Hence (E).

**Solution 15:**  If the number is said to be prime it must of course be an integer, and thus the last digit of $n^3 - 1$ must be 0 or 5. The last digit of $n^3$ must therefore be 1 or 6, that is $n^3 \equiv 1 \pmod 5$. We can deduce from this that $n \equiv 1 \pmod 5$ too. (Can you do this? It is like showing that if $m^2$ is odd then $m$ is also odd, by trying the two possibilities: $m = 2q, m = 2q + 1$ and seeing which form their squares take. Thus, for the present problem, try the five possible forms $n = 5q + r, \ 0 \le r \le 4$ look at their cubes. You really need only look at the last digit of the cubes of $0, 1, 2, 3, 4$.)

Thus $n = 5k + 1, k = 0, 1, 2, 3, \ldots$ giving

$$\frac{n^3 - 1}{5} = \frac{(5k + 1)^3 - 1}{5}$$
$$= 25k^3 + 15k^2 + 3k$$
$$= k(25k^2 + 15k + 3).$$

This will be prime only for $k = 1$ (think carefully about what a prime number is). Hence $n = 6$ is the only possible solution, so that (B) is the correct choice.

**Solution 16:**  Observe that $42^{42} = (40 + 2)^{42}$, so we need consider only the last digit. (Think of the binomial expansion in Toolchest 7.) But $2^{42} = 2^{40+2} = 4(2^{40}) = 4(2^4)^{10} = 4(16^{10})$.

It is easy to check that $16^n$ has units digit 6 for any $n$. Trying a few powers will convince you, and induction will prove it: $16 \equiv 6 \pmod{10}$ and $16^k \equiv 6 \pmod{10}$, and hence $16^{k+1} \equiv 6 \cdot 16^k \pmod{10} \equiv 36 \pmod{10} \equiv 6 \pmod{10}$. Therefore the units digit we want is that of $4 \times 6$, i.e. 4. Hence (C).

We could also use a table to see the pattern in the last digit of $2^n$ although this would be a much longer way of doing it. A better alternative would be to find the last digit using congruences all the way: $42^{42} \equiv 6^{42} \cdot 7^{42} \pmod{10} \equiv 6 \cdot 9 \pmod{10}$. You fill in the gaps!

**Solution 17:**  (We will assume you have not done Problem 11 before this one, but you might want to refer to it!) Let's start with smaller numbers and see whether we can get some sort of pattern. Consider $6 = 2^1 \times 3^1$. Its divisors are $1, 2, 3$ and $3 \times 2$ and their sum is $1 + 2 + 3 + (3 \times 2)$.

Consider $18 = 2^2 \times 3^2$. The divisors are $1, \ 3, \ 3^2, \ 2, \ 2^2, \ 2 \times 3, \ 2 \times 3^2, \ 2^2 \times 3, \ 2^2 \times 3^2$, and their sum is $1 + 3 + 3^2 + 2 + (2 \times 3) + (2 \times 3^2) + 2^2 + (2^2 \times 3) + (2^2 \times 3^2)$.

The question now is 'Can we find a convenient way of getting these divisors and their sums without risking getting tangled up in the count and sum?' As can be seen from above, the building blocks of the divisors are the prime factors of the number raised to the various powers in the various terms. How can we exhaust the list of possible ways of twinning any two prime factors, each raised to an allowed power?

Considering 6 again: the product is: $(1 + 2)(2 + 3) = 1 + 3 + 2 + 3 \cdot 2$. We see that the terms on the right-hand side are the divisors of the number 6 and their sum is thus the sum of all the divisors of 6. For the number $18 = 2^2 \cdot 3^2$ the corresponding product is:

$$(1 + 2 + 2^2)(1 + 3 + 3^2)$$
$$= 1(1 + 3 + 3^2) + 2(1 + 3 + 3^2) + 2^2(1 + 3 + 3^2)$$
$$= 1 + 3 + 3^2 + 2 + 2 \cdot 3 + 2 \cdot 3^2 + 2^2 + 2^2 \cdot 3 + 2^2 \cdot 3^2.$$

Again, all the nine terms on the right-hand side are the divisors of 18 and their sum is the sum of the divisors of 18.

This is a particular case of a general result. Try to prove it using mathematical induction! If

$$Y = p_1^{n_1} \cdot p_2^{n_2} \cdot p_3^{n_3} \cdots p_m^{n_m},$$

where $p_1, p_2, p_3, \ldots, p_m$ are all prime and $n_1, n_2, n_3, \ldots, n_m$ are non-negative whole numbers, then the divisors of $Y$ are the terms of the product

$$P = (1 + p_1{}^1 + p_1{}^2 + \cdots + p_1{}^{n_1})(1 + p_2{}^1 + p_2{}^2 + \cdots + p_2{}^{n_2}) \cdots$$
$$\times (1 + p_m{}^1 + p_m{}^2 + \cdots + p_m{}^{n_m})$$

and so the sum of all the terms of this product (which is $P$) is the sum of all the divisors of $Y$.

But (since each of the bracketed series in the right-hand side of the product $P$ is a GP) the sum of these terms is

$$P = \left( \frac{p_1{}^{n_1+1} - 1}{p_1 - 1} \right) \times \left( \frac{p_2{}^{n_2+1} - 1}{p_2 - 1} \right) \times \left( \frac{p_m{}^{n_m+1} - 1}{p_m - 1} \right).$$

We have thus found a convenient method of solving the problem:

$$1800 = 2^3 \times 3^2 \times 5^2.$$

Hence, using the formula above, the sum of all the terms of this product is

$$\left( \frac{2^4 - 1}{2 - 1} \right) \times \left( \frac{3^3 - 1}{3 - 1} \right) \times \left( \frac{5^3 - 1}{5 - 1} \right)$$
$$= 15 \times 13 \times 31 = 6045. \quad \text{Hence (B).}$$

**Solution 18:** We need to express the recurring decimals as fractions:

$$0.\dot{1}0\dot{0}\% = 0.00\dot{1}0\dot{0}.$$

$$\text{Now let } x = 0.00\dot{1}0\dot{0},$$

$$\text{therefore} \quad 1000x = 1.00\dot{1}0\dot{0},$$

$$\text{and subtracting gives} \quad 999x = 1$$

$$\text{so that} \quad x = \frac{1}{999}.$$

In a similar fashion $27.\dot{2}\dot{7}$ is found to be $\frac{27}{99} = \frac{3}{11}$.

Let the number of students who study all the time be $S$ and the number of those who study only at exam time be $E$. Let the rest – the number of students who satisfy neither condition – be $R$. The total population is therefore

$$T = E + S + R$$

$$= \frac{3}{11}T + \frac{1}{999}T + R$$

This implies that $T$ is divisible by both 11 and 999. There is exactly one natural number less than 20 000 which satisfies this, and this is $11 \times 999 = 10\,989$. Hence (A).

**Solution 19:** Observe that, if $n$ is an odd natural number:

$$a^n + b^n = (a + b)(a^{n-1} - a^{n-2}b + a^{n-3}b^2 - a^{n-4}b^3 \cdots + b^{n-1}).$$

When applied to $5^{1999} + 6^{1999}$, this rule shows us that $5 + 6 = 11$ is a factor of $5^{1999} + 6^{1999}$.

What's left now is to find out whether 2, 3, 5, 7 are also factors of $5^{1999} + 6^{1999}$. Now 2 is not a factor, since $6^n$ is always even and $5^n$ always odd, so that their sum is odd. And 3 cannot be a factor either, since it clearly always divides $6^n$ but not $5^n$. (If a prime $p$ divides $a$ and $b$ then it would have to divide $a - b$.) For the same reason, $5^n + 6^n$ is never a multiple of 5 because $6^n = 2^n 3^n$ is always a non-multiple of 5 yet $5^n$ is a multiple of 5.

Checking for 7 now, the possible remainders when $5^n$ is divided by 7 occur in cycles and are shown by the table below:

| Number | $5^1$ | $5^2$ | $5^3$ | $5^4$ | $5^5$ | $5^6$ | $5^7$ | $\cdots$ |
|---|---|---|---|---|---|---|---|---|
| Remainder when divided by 7 | 5 | 4 | 6 | 2 | 3 | 1 | 5 | $\cdots$ |

Similarly, the remainders when $6^n$ is divided by 7 occur in cycles as shown by the table below.

| Number | $6^1$ | $6^2$ | $6^3$ | $6^4$ | $6^5$ | $6^6$ | $6^7$ | $\ldots$ |
|---|---|---|---|---|---|---|---|---|
| Remainder when divided by 7 | 6 | 1 | 6 | 1 | 6 | 1 | 6 | $\ldots$ |

Let us denote the set of all multiples of seven by $\mathbb{Z}_7$. The remainders when $5^n + 6^n$ is divided by 7 are also cyclical as summarised below:

| Number | $5^1+6^1$ | $5^2+6^2$ | $5^3+6^3$ | $5^4+6^4$ | $5^5+6^5$ | $5^6+6^6$ | $5^7+6^7$ | $\ldots$ |
|---|---|---|---|---|---|---|---|---|
| Rem. when div. by 7 | $4 \notin \mathbb{Z}_7$ | $5 \notin \mathbb{Z}_7$ | $5 \notin \mathbb{Z}_7$ | $3 \notin \mathbb{Z}_7$ | $2 \notin \mathbb{Z}_7$ | $3 \notin \mathbb{Z}_7$ | $4 \notin \mathbb{Z}_7$ | $\ldots$ |

and the pattern repeats itself periodically after this. We cannot therefore have a number of the form $5^n + 6^n$ which is a multiple of 7. Hence 11 is the smallest prime divisor, making (D) the correct choice. Is there any quicker way of showing that 7 does not divide $5^n + 6^n$, using congruence methods?

**Solution 20:**   We must compare the given expression with square expressions. Observe that $c^2 + 5c + 6 = (c+2)(c+3)$, and, since $c$ is a positive whole number,

$$(c+2)^2 < (c+2)(c+3) < (c+3)^2.$$

This places it between two consecutive squares, hence $c^2 + 5c + 6$ can never be a perfect square. There is no number of required form between 1 and 100: (D) is correct.

**Solution 21:**

$$Row\ 1 : 1 \ldots Y$$
$$Row\ 2 : Y + 1 \ldots 2Y$$
$$Row\ 3 : 2Y + 1 \ldots 3Y$$
$$Row\ 4 : 3Y + 1 \ldots 4Y$$
$$Row\ 5 : 4Y + 1 \ldots 5Y$$

$$\vdots \qquad \vdots$$

Now 20 is in row 3, hence 20 lies between $2Y + 1$ and $3Y$ inclusive. Expressed in symbols, we have $2Y + 1 \leq 20 \leq 3Y$. This gives $Y \leq 9$ and $7 \leq Y$. Similarly $4Y + 1 \leq 41 \leq 5Y$ giving $9 \leq Y$ and $Y \leq 10$. Hence $Y = 9$. Thus the first entries in the rows are $1, 10, 19, 28, \ldots, 100$ to give 105 in the last row, which is the $12^{th}$. Hence (E).

**Solution 22:**   A good starting point is experimentation, choosing values for $a$ and $b$ purposely so that the expression cancels to give an integer.

$a = 4$,   $b = 2$,   gives $\dfrac{a^3 + b^3}{a - b} = 6^2$.

$a = 2$,   $b = 1$   gives $\dfrac{a^3 + b^3}{a - b} = 9 = 3^2$.

$a = 8 = 2^3$,   $b = 2^2 = 2^{3-1}$   gives $\dfrac{a^3+b^3}{a-b} = \dfrac{8^3+4^3}{4} = \dfrac{512+64}{4} = 144 = 12^2$.

From these results we make a guess. It seems that if $a = 2^n$ and $b = 2^{n-1}$, we always get a perfect square. Let us verify this:
If $a = 2^n$ and $b = 2^{n-1}$ then

$$\frac{a^3 + b^3}{a - b} = \frac{2^{3n} + 2^{3n-3}}{2^n - 2^{n-1}}$$

$$= \frac{2^{3n-3}(2^3 + 1)}{2^{n-1}(2 - 1)}$$

$$= 2^{(3n-3)-(n-1)}(9)$$

$$= 9 \cdot 2^{2n-2} = \left(3 \cdot 2^{n-1}\right)^2.$$

Thus, the expression $2^{2n-2} \cdot 9$ is always an integral perfect square, for all $n \in \mathbb{Z}$. So we have infinitely many pairs, hence (D) is the correct choice.

**Solution 23:**   What *is* easy to write down, is the greatest 4-digit number in base 6: it is $5\,555_{(6)}$. (Just as $9\,999$ is the greatest 4-digit number in base 10.) To express this in base 10:

$$5\,555_{(6)} = (5 \times 6^3) + (5 \times 6^2) + (5 \times 6^1) + (5 \times 6^0)$$

$$= 1\,080 + 180 + 30 + 5 = 1\,295_{(10)}.$$

Now that we are back to the more familiar base 10, we can easily find the greatest perfect square less than 1295 – it is $35^2 = 1225$. Now we have to convert it back to base 6, as below:

| 6 | 1225 |
|---|------|
| 6 | 204 r 1 |
| 6 | 34 r 0 |
| 6 | 5 r 4 |
| 6 | 0 r 5 |

So that the largest 4-digit perfect square in base 6 is $5401_{(6)}$. Hence (D).

An alternative method is based on the observation that $55_{(6)}$, the largest 2-digit number in base 6, will give, when squared, the largest 4-digit perfect square in base 6. (Don't forget to carry the number of 6 s not 10 s when you calculate it!)

**Solution 24:**    This is a neat illustration of how a question can change the style in which it is phrased, disguising the simplicity of the basic idea. We are required to find a positive number $x$, which when multiplied by 45 gives us a perfect square. There may be many such numbers, and the question could have simply asked for the smallest of these, which will naturally give the smallest square, hence the smallest $y$, and hence the smallest sum $x + y = x + \sqrt{45x}$, which can be calculated easily. Thus, this question is just a disguised version of the following more familiar question style: 'What is the smallest nonzero whole number which when multiplied with 45 gives us a perfect square?'

$$45x = (3 \times 3 \times 5)x$$
$$= 3^2 \cdot 5^1 x.$$

For a perfect square we must have even numbers of each factor. The smallest value of $x$ to achieve this is clearly $x = 5$, and so

$$y^2 = 3^2 \cdot 5^2, \quad \text{so that } y = 3 \times 5,$$
$$\text{giving } x + y = 5 + 15 = 20. \quad \text{Hence (A)}.$$

**Solution 25:**    Any positive integer must have one of these forms: $3m + 1$, $3m + 2$, or $3m$, where $m \in \mathbb{Z}$. We proceed to test these cases on the possible answers:

Let $P_n$ be the product $n(n + 2)(n + 4)$ (i.e. answer (C)).
If $n = 3m$, $P_n$ is obviously a multiple of 3.
If $n = 3m + 1$, then $n + 2 = 3m + 1 + 2 = 3(m + 1)$, which is again a multiple of three.
If $n = 3m + 2$, then $n + 4 = 3m + 2 + 4 = 3(m + 2)$, again a multiple of three.

Hence $P_n$ is a multiple of three for all positive integer values of $n$. If you consider the other given products, such as $(n + 2)(n + 3)(n + 5)$, the above argument shows that the product under consideration is not always divisible by three. Hence (C) is the correct answer.

**Solution 26:**    The calculation is summarised by the tables below, showing how the last digits occur in cycles. For the first table, we have tabulated the last digit of $2^n$, for each value of $n$.

| Power: | 1 | 2 | 3 | 4 | 5 | 6 | 7 | 8 | 9 | 10 | 11 | 12 | 13 |
|---|---|---|---|---|---|---|---|---|---|---|---|---|---|
| Last digit: | 2 | 4 | 8 | 6 | 2 | 4 | 8 | 6 | 2 | 4 | 8 | 6 | 2 |

So powers with, say, 6 as the last digit form the sequence 4, 8, 12, etc., i.e. multiples of 4. But 1996 is a multiple of 4, so it is part of this sequence, and hence $2^{1996}$ ends in 6.

A similar table for powers of 3 is shown below.

| Power | 1 | 2 | 3 | 4 | 5 | 6 | 7 | 8 | 9 | 10 | 11 | 12 |
|---|---|---|---|---|---|---|---|---|---|---|---|---|
| Last digit | 3 | 9 | 7 | 1 | 3 | 9 | 7 | 1 | 3 | 9 | 7 | 1 |

A similar argument shows that $3^{1996}$ ends in 1, so $2^{1996} + 3^{1996}$ ends in $6 + 1 = 7$. Hence (B).

**Solution 27:**   By partially writing out the first few powers of 6 we see that the last two digits follow a cycle of order five:

6, 36, 216, ... 96, ... 76, ... 56, ... 36, ... 16, ... 96, ... 76, ... 56, ... 36, ....

Now $2007 \equiv 2 \pmod 5$, so that $6^{2007}$ has the same last two digits as $6^2 = 36$.

Similarly for powers of 7, the last two digits follow a cycle of order four:

7, 49, 343, ... 401, ... 07, ... 49, ... 43, ... 01, ...

Now $2007 \equiv 3 \pmod 4$, so that $7^{2007}$ has the same last two digits as $7^3 = 343$.

Therefore, finally, $6^{2007} + 7^{2007}$ has last two digits $36 + 43 = 79$. Hence (C).

That was done based upon pattern-spotting. Here is a somewhat more careful presentation of the congruence arithmetic involved:

$$6^5 = 11\,776 \equiv 76 \pmod{100},$$

$$6^{10} = (6^5)^2 \equiv (76)^2 \pmod{100}$$

$$\equiv 76 \pmod{100},$$

hence   $6^{5k} = (6^5)^k \equiv 76^k \pmod{100} \equiv 76 \pmod{100},$

so   $6^{2007} = 6^{2005+2} = 6^{5(401)+2} \equiv 6^2 \cdot 76 \pmod{100} \equiv 36 \pmod{100}.$

Similarly,   $7^4 = 401 \equiv 1 \pmod{100},$

hence   $7^{4k} = (7^4)^k \equiv 1^k \pmod{100} \equiv 1 \pmod{100},$

therefore   $7^{2007} = 7^{4(501)+3} \equiv 1 \cdot 7^3 \pmod{100} \equiv 43 \pmod{100}.$

Finally, $6^{2007} + 7^{2007} \equiv (36 + 43) \pmod{100} \equiv 79 \pmod{100}.$

# 5 | Trigonometry

*The traditional native-American way of measuring the height of a tree was to pace away from the base of the tree – constantly pausing to check if one could just see the top of the tree when one looked between one's legs. We now know that the mathematical basis of this method is that when one just sees the top of a tree in the manner described above, one's line of vision is approximately 45° to the horizontal. Hence a right angled isosceles triangle is roughly defined. It is not known how this method was arrived at or how it was validated.*

By the end of this topic you should be able to:

(i) Understand the definitions of *angle*, *degree* and *radian*.
(ii) Find the arc-length and area of a sector of a circle using the relevant radian formula.
(iii) Understand and use the trigonometric functions of general angles.
(iv) Understand the properties of the sine, cosine and tangent functions.

(v) Derive and use the Pythagorean set of identities.

(vi) Derive and apply the *addition, double angle, product* and *sum* formulas.

(vii) Solve trigonometric equations.

The term 'trigonometry' is a fusion of two Greek words: *trigonon* and *metria*. The former means 'triangle' and the latter, 'measurement'. Trigonometry – 'the measurement of triangles' – was originally invented by the ancients to solve problems in astronomy, where the positions of stars must be specified by angles measured by instruments like the sextant. Since then, other applications sprouted and ages of revision and restructuring led to its present form.

As an appetizer, here is a typical problem of the kind you should be able to solve when you have worked through Sections 1–8. You are invited to think about it now. You may find it hard, but by the end of Section 8, where its solution is given, it should not look difficult to you.

**Appetizer Problem:**    *If a and b are positive real numbers such that* $\cos(x + 2y) = a$ *and* $\cos(x + y) = b$, *what, in terms of a and b, is the maximum value of* $\sin(2x + 3y)$?

(A) $a\sqrt{1 - a^2} + b\sqrt{1 - b^2}$    (B) $b\sqrt{1 - a^2} + a\sqrt{1 - b^2}$
(C) $\sqrt{1 - a^2} + \sqrt{1 - b^2}$    (D) $a\sqrt{a} + b\sqrt{b}$    (E) $a\sqrt{b} + b\sqrt{a}$

## 5.1    Angles and their measurement

We define a **ray** as the straight line that originates from some point $O$ in a plane, passes through some point $P$ in the plane, and extends indefinitely – the figure shows such a ray. The arrow denotes $OP$ extended indefinitely. The point $O$ is known as the **origin** of the ray.

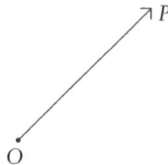

**Definition:**    An **angle** is a geometrical object formed by rotating a ray about its origin. The initial position $OP$ of the ray is called the **initial side** of the angle, and the final position $OQ$ of the ray after it has been rotated is called the **terminal side** of the angle. The origin $O$ of the ray is called the **vertex** of the angle:

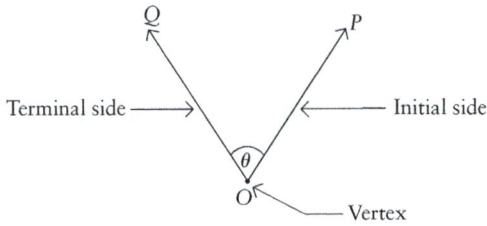

**Positive and negative angles:** Angles formed by rotating the ray in the counterclockwise direction are said to be positive angles, while angles formed by a clockwise rotation are considered negative. This assignment is by convention.

**Measurement of angles:** The measure of an angle is the amount of rotation of the ray. The two units in common use are the **degree** and the **radian**.

### The degree measure of an angle

**Historical background:** The Babylonians (and many other cultures) observed a periodicity of roughly 360 days in the occurrence of seasons. The number $360 = 6 \times 60$ was consistent with their use of a sexagesimal base – base 60 – in counting, which in turn may have been derived from the coinage system: 60 shekels equals 1 mina, based on relative values and weights of the metals.

Working with the model that the Earth was at the centre of the Universe (a model which was to be replaced with a Sun-centred model by Copernicus, Kepler and Galileo a few thousand years later), ancient Babylonians assumed that the sky processed through one complete cycle in 360 days. This may have inspired them to divide the complete revolution into 360 equal parts. This practice survives today in our method of measuring angles in degrees, minutes and seconds – base 60! Perhaps developing out of this way of angle measurement (because sundials and clocks register time by angles) the Babylonians' influence is also seen in our measurement of *time* in hours, minutes and seconds, using base 60. Another possible origin of the degree is this: given their sexagesimal counting base, what would be a natural unit angle for the Babylonians to use? The simplest regular shape is the equilateral triangle, but its angle is rather large to use as a basic unit so why not divide it into 60 units?

**Definition of a degree:** If a ray is rotated $\frac{1}{360}$ of a complete revolution, then the angle so formed has a measure of one degree, denoted by $1°$. Or, one complete revolution has a measure of $360°$.

### Some special angles

A **straight angle** is formed by rotating a ray through half of a complete revolution. A straight angle therefore has a measure of $180°$.

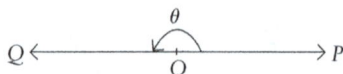

A **right angle** is formed by rotating a ray through a quarter of a complete revolution. A right angle therefore has a measure of 90°.

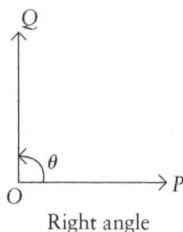

Right angle

An **acute angle** $\theta$ is one such that $0° < \theta < 90°$.

An **obtuse angle** $\theta$ is one such that $90° < \theta < 180°$.

Two angles $\theta_1$ and $\theta_2$ are said to be **complementary** if $\theta_1 + \theta_2 = 90°$. $\theta_1$ is called the complement of $\theta_2$ and *vice versa*.

Two angles $\theta_1$ and $\theta_2$ are said to be **supplementary** if $\theta_1 + \theta_2 = 180°$. $\theta_1$ is called the supplement of $\theta_2$ and *vice versa*.

**NB:** It is possible to measure angles greater than 360° (a complete revolution). For example, an angle of 600° is $1\frac{2}{3}$ revolutions counterclockwise, and is shown in the diagram.

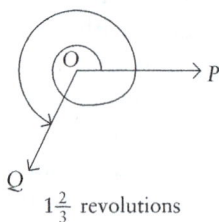

$1\frac{2}{3}$ revolutions

### The radian measure of an angle

As a prelude to our definition of the radian, let us consider a circle, radius $r$ and two of its radii, $OP$ and $OQ$:

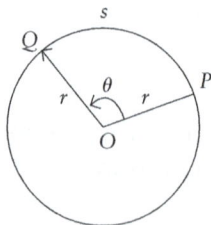

The angle $\theta$ formed by the two radii is known as a central angle, and the part of the circle lying between the points $P$ and $Q$ is called an arc of the circle, and is written arc $PQ$. In the figure above, we say that arc $PQ$ **subtends** the angle $\theta$ at the centre.

**Definition of a radian:** One **radian** is the measure of the central angle subtended by an arc of length $r$ on a circle of radius $r$. One radian is denoted by $1^c$.

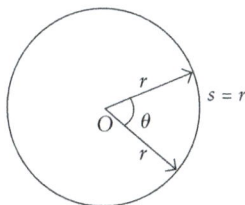

Definition of a radian

The definition above gives us a simple rule for finding the number of radians in any central angle $\theta$:

Given an arc of length $s$ on a circle of radius $r$, the radian measure of the angle subtended by the arc is $\theta = \frac{s}{r}$. For example, a circular arc of length 3 cm on a circle with radius 2 cm subtends an angle of $\frac{3}{2} = 1.5$ radians.

Some consequences of the definition of radian measure are shown below:

(i)

$$\theta = \frac{s}{r} = \frac{\pi r}{2} \div r = \frac{\pi}{2} \text{ radians}$$
$$\theta = 90° = \frac{\pi^c}{2}$$

(ii)

$$\theta = \frac{s}{r} = \frac{\pi r}{r} = \pi \text{ radians}$$
$$\theta = 180° = \pi^c$$

(iii)

$$\theta = \frac{s}{r} = \frac{2\pi r}{r} = 2\pi \text{ radians}$$
$$\theta = 360° = 2\pi^c$$

**Radian/degree conversions:**   You need only observe that $1^c = \dfrac{180°}{\pi}$, and equivalently, $1° = \dfrac{\pi}{180}^c$.

### Two applications of radian measure

The radian measure is used extensively in mathematics and physics. Here are two common applications.

#### (i) Finding the length of a circular arc

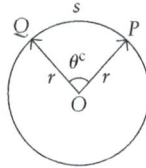

In the figure we have:

$$\theta^c = \frac{s}{r} \text{ or } s = r\theta^c$$

$$\boxed{\text{Arc-length } s = r\theta^c}$$

#### (ii) Finding the area of a sector of a circle

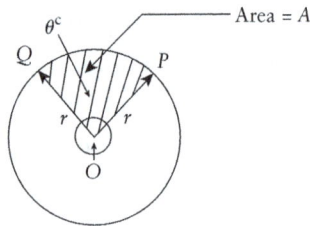

By definition, a sector is the region bounded by two radii and the intercepted arc, which is the shaded region in the figure above. Now,

$$\frac{A}{\pi r^2} = \frac{\theta^c}{2\pi}, \text{ hence we have}$$

$$\boxed{\text{Area } A = \frac{1}{2}r^2\theta^c}$$

For example, if $\theta = \frac{2\pi}{3}$ (equivalently, $120°$) and $r = 1$ cm, then the area of the sector above is

$$\frac{1}{2}(1)^2 \cdot \frac{2\pi}{3} = \frac{\pi}{3} \text{ cm}^2.$$

## 5.2  Trigonometric functions of acute angles

A **right triangle** is, by definition, a triangle that contains a $90°$ (right) angle. Since the angle sum of a triangle is known to be two right angles, it follows that the other two angles in a right triangle are both acute angles. The side opposite the right angle is called the **hypotenuse**. In the diagram below, the other two sides are given names that indicate their positions relative to the angle $\theta$:

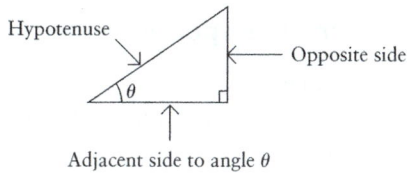

We can form six different ratios using the measures of the three sides of the triangle above, and these ratios are known as the **trigonometric functions of the acute angle** $\theta$.

## Definitions

If $\theta$ is an acute angle in a right triangle, as shown in the figure above, then:

| Function | Abbreviation | Ratio |
|---|---|---|
| (a) Sine of angle $\theta$ | $\sin\theta$ | $\dfrac{\text{opposite}}{\text{hypotenuse}}$ |
| (b) Cosecant of angle $\theta$ | $\csc\theta$ | $\dfrac{\text{hypotenuse}}{\text{opposite}}$ |
| (c) Cosine of angle $\theta$ | $\cos\theta$ | $\dfrac{\text{adjacent}}{\text{hypotenuse}}$ |
| (d) Secant of angle $\theta$ | $\sec\theta$ | $\dfrac{\text{hypotenuse}}{\text{adjacent}}$ |
| (e) Tangent of angle $\theta$ | $\tan\theta$ | $\dfrac{\text{opposite}}{\text{adjacent}}$ |
| (f) Cotangent of angle $\theta$ | $\cot\theta$ | $\dfrac{\text{adjacent}}{\text{opposite}}$ |

**NB:** We say that the cosine function $\cos\theta$ is the **cofunction** of the sine function $\sin\theta$, and *vice versa*. Also, $\cot\theta$ is the cofunction of $\tan\theta$, and *vice versa*, etc. The cofunction of an angle is the function of the *other* acute angle – the complementary angle. It is obtained by simply interchanging 'opposite' and 'adjacent' in the definition. Observe that $\cot\theta$ is the reciprocal function of $\tan\theta$, but $\cos\theta$ is *not* the reciprocal function of $\sin\theta$.

**Example:**   Given that $\theta$ is an acute angle such that $\tan\theta = \frac{1}{3}$, find the values of the remaining trigonometric functions of $\theta$.

**Solution:**   $\tan\theta = \frac{1}{3}$ leads to the figure below:

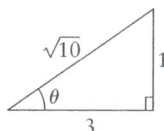

It follows that $\sin\theta = \frac{1}{\sqrt{10}}$, $\csc\theta = \sqrt{10}$, $\cos\theta = \frac{3}{\sqrt{10}}$, $\sec\theta = \frac{\sqrt{10}}{3}$, and $\cot\theta = 3$.

## Trigonometric values for some special angles

Convince yourself that the diagonal in a square with side 1 unit forms the right triangle with a $45°(\frac{\pi}{4}^c)$ angle, and that the altitude in an equilateral triangle with a side length of 2 units forms the right triangle with angles of $30°(\frac{\pi}{6}^c)$ and $60°(\frac{\pi}{3}^c)$.

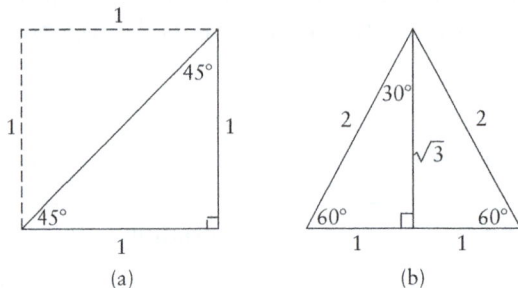

(a)    (b)

This leads us to the table below (which you don't need to memorize! – just sketch the relevant triangle when you need it):

| $\theta$ | $\sin\theta$ | $\csc\theta$ | $\cos\theta$ | $\sec\theta$ | $\tan\theta$ | $\cot\theta$ |
|---|---|---|---|---|---|---|
| $30°$ | $\frac{1}{2}$ | $2$ | $\frac{\sqrt{3}}{2}$ | $\frac{2}{\sqrt{3}}$ | $\frac{1}{\sqrt{3}}$ | $\sqrt{3}$ |
| $45°$ | $\frac{1}{\sqrt{2}}$ | $\sqrt{2}$ | $\frac{1}{\sqrt{2}}$ | $\sqrt{2}$ | $1$ | $1$ |
| $60°$ | $\frac{\sqrt{3}}{2}$ | $\frac{2}{\sqrt{3}}$ | $\frac{1}{2}$ | $2$ | $\sqrt{3}$ | $\frac{1}{\sqrt{3}}$ |

Observe that in the table above the value of a each trigonometric function is equal to the value of its cofunction at the complementary angle. For example, $\sin 30° = \frac{1}{2} = \cos 60°$.

In general, given two complementary angles $\alpha$ and $\beta$ (i.e. $\alpha + \beta = \frac{\pi}{2}^c$), the side opposite to angle $\alpha$ is adjacent to angle $\beta$, so any trigonometric function of $\alpha$ is equal to the corresponding cofunction of the complementary angle. Thus, if $\alpha + \beta = \frac{\pi}{2}$, then

$$\cos\alpha = \sin\beta, \quad \cot\alpha = \tan\beta, \quad \text{and} \quad \csc\alpha = \sec\beta.$$

Writing $\beta$ as $\frac{\pi}{2} - \alpha$ gives these trigonometric relationships:

$$\sin\alpha = \cos\left(\frac{\pi}{2} - \alpha\right)$$

$$\tan\alpha = \cot\left(\frac{\pi}{2} - \alpha\right)$$

$$\sec\alpha = \csc\left(\frac{\pi}{2} - \alpha\right)$$

## 5.3 Trigonometric functions of general angles

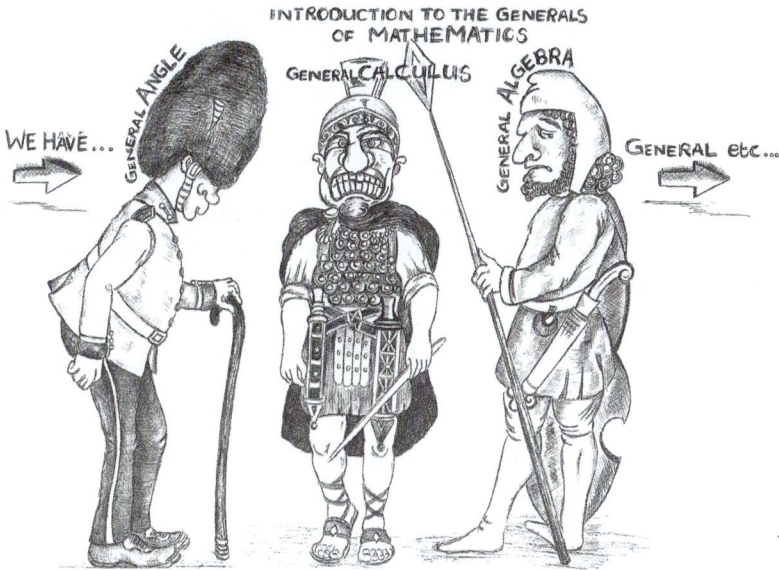

INTRODUCTION TO THE GENERALS
OF MATHEMATICS

GENERAL CALCULUS

GENERAL ALGEBRA

GENERAL ANGLE

WE HAVE...

GENERAL etc...

*It is said 'The Generals' are the reason people are generally scared of mathematics. There is General Angle, leader of the Anglo troops. Then there is General Calculus, leader of the Roman legions. We have General Algebra, leader of the Persian guard, and the list goes on ...*

Because right-triangle trigonometry is limited to the interval $0° < \theta < 90°$, we need to define trigonometric functions more generally. This allows us to solve triangles that might not contain a right angle. We shall use the Cartesian coordinate system as the frame of reference, to define the trigonometric functions. First, we need to put the general angle $\theta$ in **standard position**, by placing the vertex of the angle at the origin $(0, 0)$ and then placing the initial ray of the angle along the positive $x$-axis as shown below:

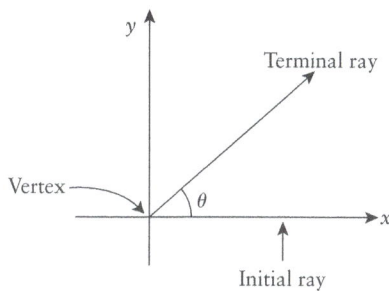

Having placed the angle in standard position, we can proceed to define the trigonometric functions of the angle $\theta$ by considering any point $(x, y)$ on the terminal ray of $\theta$, except $(0, 0)$.

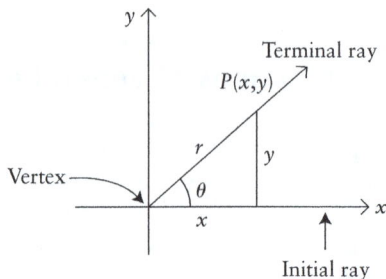

Observe that three numbers can be associated with this point, namely, the $x$-coordinate, the $y$-coordinate, and the distance $r$ from the origin, connected by Pythagoras' theorem:

$$r = \sqrt{(x-0)^2 + (y-0)^2} = \sqrt{x^2 + y^2}.$$

## Definitions

If $\theta$ is an angle in *standard position*, and if $(x, y)$ is any point on the terminal ray of $\theta$ (except $(0, 0)$), then:

| Function | Abbreviation | Ratio |
|---|---|---|
| (a) Sine of angle $\theta$ | $\sin \theta$ | $\dfrac{y}{r}$ |
| (b) Cosecant of angle $\theta$ | $\csc \theta$ | $\dfrac{r}{y}$ |
| (c) Cosine of angle $\theta$ | $\cos \theta$ | $\dfrac{x}{r}$ |
| (d) Secant of angle $\theta$ | $\sec \theta$ | $\dfrac{r}{x}$ |
| (e) Tangent of angle $\theta$ | $\tan \theta$ | $\dfrac{y}{x}$ |
| (f) Cotangent of angle $\theta$ | $\cot \theta$ | $\dfrac{x}{y}$ |

Next comes the question of actually evaluating trigonometric functions for any angle. Let us start by considering the simplest case, namely the *quadrantal angles*.

## (A) Quadrantal angles

Angles which are integral multiples of $90°$, such as $180°$, $270°$, etc. are called quadrantal angles, and any quadrantal angle can be expressed in the form: $90n°$ where $n \in \mathbb{Z}$. In radian measure, this is $n\dfrac{\pi^c}{2}$. To find the six trigonometric functions of, for example, the smallest positive quadrantal angle (90°), we proceed as follows, using the diagram on the left:

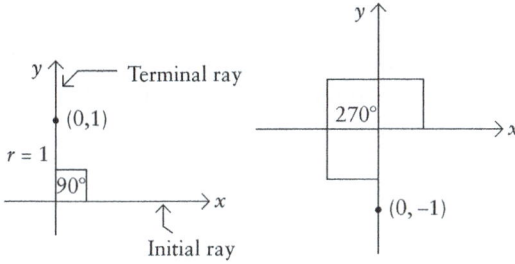

Choose the point $(0, 1)$ on the terminal ray for evaluation purposes (any other point save for $(0, 0)$ could do). Hence $x = 0$, $y = 1$, and $r = \sqrt{(1-0)^2 + (0-0)^2} = 1$.

$$\sin 90° = \frac{y}{r} = \frac{1}{1} = 1; \quad \csc 90° = 1;$$

$$\cos 90° = \frac{x}{r} = \frac{0}{1} = 0; \quad \sec 90° = \frac{1}{0}, \text{ undefined};$$

$$\tan 90° = \frac{y}{x} = \frac{1}{0}, \text{ undefined}; \quad \cot 90° = 0.$$

The trigonometric functions of other quadrantal angles can be evaluated in a similar manner. For example, to calculate them for 270°, we use the diagram on the right.

## (B) Non-quadrantal angles

The way to evaluate trigonometric functions for these angles is to use the idea of an associated *reference* acute angle.

**Definition:**   The **reference angle** of an angle $\theta$ is the positive acute angle formed by the terminal ray of $\theta$ and the horizontal axis.

For example, if $\theta = 120°$, the reference angle is 60° as shown below:

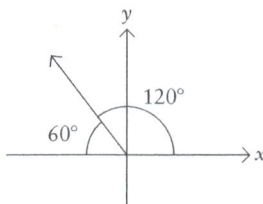

It is easy to observe that:

*The values of trigonometric functions of a given angle are equal in magnitude to the values of the corresponding trigonometric functions of the reference angle.*

It remains to determine the correct sign. This is found according to the quadrant containing the terminal ray of the angle. Using the definition of the trigonometric functions and the signs of $x$ and $y$ in the various quadrants, we construct the figure below, sometimes called the 'ASTC diagram' (for 'all-sin-tan-cos') giving the functions which are *positive* in each of the four quadrants, enumerated anticlockwise as usual:

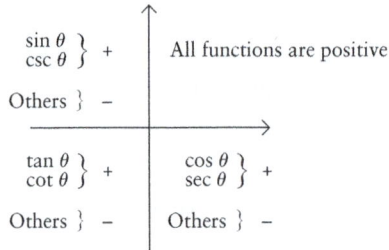

**Example:** Find $\sin 315°$.

**Solution:** Determine the reference angle:

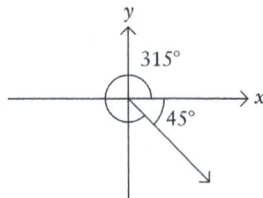

From the figure, we can see that the reference angle is $45°$ so that $|\sin 315°| = |\sin 45°|$. Now, to find the sign, use the ASTC diagram above, which tells us that $\sin 315°$ is negative, hence

$$\sin 315° = -\sin 45° = -\frac{\sqrt{2}}{2}.$$

**Trigonometric functions of real numbers**

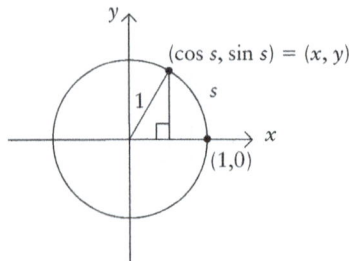

This is a slightly more general approach to defining these same functions. The figure above shows a point $(x, y)$, on the unit circle $x^2 + y^2 = 1$ at arc-length $s$ from $(1, 0)$, i.e. the distance from $(1, 0)$ to $(x, y)$ along the curve is $s$.

Let us define the cosine of $s$ to be the $x$-coordinate of the point, and the sine of $s$ to be the $y$-coordinate. Thus, $x = \cos s$, $y = \sin s$.

Observe that defining the sines and cosines in this way is consistent with all our previous definitions, where we began with an *angle*, and based everything on the sides and hypotenuse of a right triangle. Also, we observe that the domain of both functions is the set of all real numbers – this is because our length $s$ is determined by wrapping a length (given by a real number) around the unit circle.

We also define the other functions of $s$ as:

$$\tan s = \frac{y}{x} = \frac{\sin s}{\cos s} \text{ for } x \neq 0; \quad \sec s = \frac{1}{x}, \text{ and } \csc s = \frac{1}{y}.$$

Once again, these definitions are consistent with all previous definitions. Observe that the $x$ and $y$ coordinates in a unit circle vary between $-1$ and $1$ inclusive, so that we have, for any real number $s$:

$$-1 \leq \sin s \leq 1$$
$$-1 \leq \cos s \leq 1$$

We also observe that, if we wrap a length of $2\pi$ around the unit circle, we go round the circle and return to our original point; hence, for any trigonometric function $F$ and all real $s$:

$$F(s + 2\pi a) = F(s), \quad a \in \mathbb{Z}$$

Consider the figure below, which illustrates the symmetry existing between $s$ and $-s$:

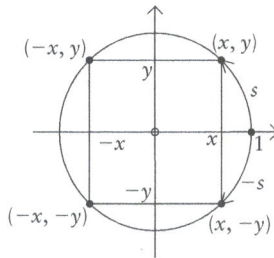

Since the arc-lengths $s$ and $-s$ have the same $x$-coordinate, it follows that $x = \cos(-s) = \cos s$. Also, since the $y$-coordinates differ only in sign, we have $-y = \sin(-s) = -\sin s$. Hence, summarizing, we have for any real number $A$:

$$\begin{array}{l} \sin(-A) = -\sin A \\ \cos(-A) = \cos A \end{array}$$

It is easy to see that $\tan(-A) = -\tan A$, $\sec(-A) = \sec A$, etc.
    For any real function $f(x)$:

(1) if $f(-x) = f(x)$, for all $x$ in the domain, then we say that $f$ is an **even** function;
(2) if $f(-x) = -f(x)$, for all $x$ in the domain, then we say that $f$ is an **odd** function.

For example, $f(x) = x^2$ is even and $f(x) = x^3$ is odd. From the above, cos and sec are even, while sin and tan, and also csc and cot, are odd. The graph of an even function is symmetrical about the $y$-axis, while for an odd function we have to perform a double reflection of the graph for $x > 0$ to get the graph for $x < 0$: first in the $y$-axis and then in the $x$-axis.

## 5.4  Graphs of sine and cosine functions

Before we consider the actual graphs, we need to understand the meaning of the terms *periodic function* and *amplitude*.
    A function $f$ is **periodic with period** $k$ if $f(x) = f(x + k)$, $\forall x \in D$, where $D$ is the domain of $f$, and $k$ is the smallest positive number for which this is true. Intuitively, the function 'keeps repeating itself', and 'if you know its values on any interval of length $k$ then you know its values everywhere'.
    The maximum value of a periodic function which is centred about the horizontal $x$-axis is called the **amplitude** of the function.

### The sine function

A stylized sketch of the graph of $y = \sin x$ is shown in the diagram.

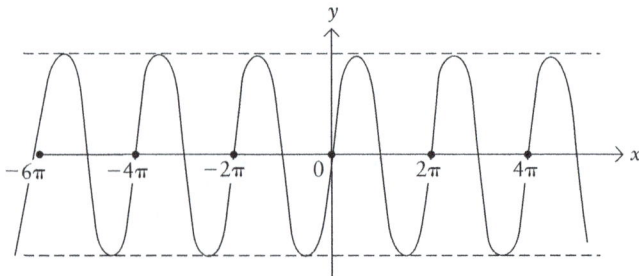

We observe the following points:

(1) The sine function is odd, hence the portion of graph for $x < 0$ can be obtained from the portion for $x > 0$ by a double reflection (in each axis).
(2) The sine function is periodic, with period $2\pi$. That is, we need only sketch its graph for $0 \leq x \leq 2\pi$, then replicate:

$$\sin x = \sin(x + 2\pi a), \quad \forall a \in \mathbb{Z}$$

(3) The amplitude of the sine function is 1.

Using the information above, and all other facts (which we shall assume familiar) concerning transformations of curves along the $x$ and $y$ axes, it is quite an easy task to sketch the graph of $y = a \sin bx$: the curve has the same basic shape as $y = \sin x$, but with these important differences:

(a) The period of $y = a \sin bx$ is $\frac{2\pi}{b}$. To see why, plot the graph of $y = \sin 2x$, where $b = 2, a = 1$. You will be 'moving twice as fast" as in $y = \sin x$, so that the period is shorter.
(b) The amplitude is $|a|$, since $|a \sin bx| = |a||\sin bx| = |a| \cdot 1 = |a|$, for $x = \frac{\pi}{2b}$.

**Example:**   Sketch the graph of $y = \sqrt{2} \sin 3x$.

**Solution:**   The period is $\frac{2\pi}{3}$ and the amplitude is $\sqrt{2}$, so the graph looks like:

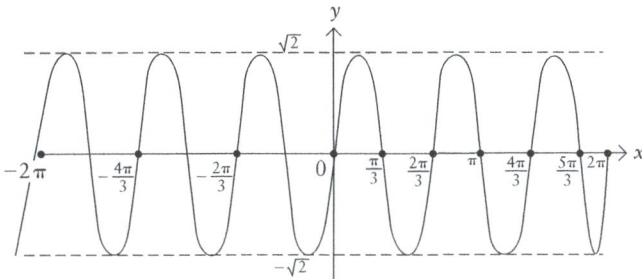

### The cosine function
A stylized sketch of the graph of $y = \cos x$ is shown in the diagram.

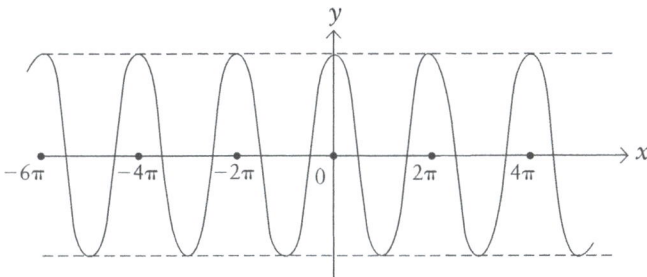

Graph of $y = \cos x$

Observe that:

(1) The graph of the cos function is even, hence the portion of graph for $x < 0$ can be obtained from the portion for $x > 0$ by a single reflection in the y-axis.
(2) The function $\cos x$ is periodic with period $2\pi$. We need only sketch its graph for $0 \leq x \leq 2\pi$, then replicate.
(3) Its amplitude is 1.
(4) The graph of $y = \cos x$ can be obtained from that of $y = \sin x$ by shifting it by an amount $\frac{\pi}{2}$ to the left along the x-axis. This should not surprise us, since $x - \frac{\pi}{2}$ and $x$ are complementary, and we have $\cos(x - \frac{\pi}{2}) = \sin x$.

## The tangent function

The graph of the tangent function is shown in the diagram.

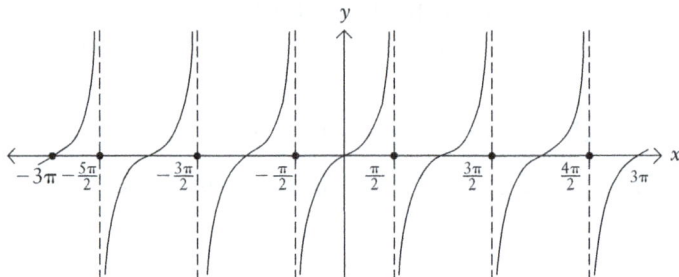

Some of the branches of the graph of $y = \tan x$

We observe the following:

(1) The tangent fuction is odd.
(2) The tangent function is periodic, with period $\pi$. That is, $\tan x = \tan(x + \pi a)$, $\forall a \in \mathbb{Z}$ and $\forall x \in D$, where $D$ is the domain of the tangent function. We need only sketch one branch of the graph, the others are replicas.
(3) The domain of the tangent function excludes all $x$ that take the form $x = \frac{\pi}{2} + k\pi$, $k \in \mathbb{Z}$. It's quite easy to see why:

$$y = \tan x = \frac{\sin x}{\cos x}; \quad \text{but } \cos x = 0, \quad \forall x : x = \frac{\pi}{2} + k\pi, \quad k \in \mathbb{Z},$$

so that $y$ is undefined for these values.
(4) Whereas the range of values taken by the sine and cosine functions is $[-1, 1]$, the range of the tangent function is the set of *all* real numbers.

## 5.5   Trigonometric identities

A trigonometric identity is, by definition, a statement that is true for all real numbers for which the expressions are defined. The identities are useful in transforming given trigonometric expressions into either simpler forms, or forms that lend themselves well to analysis.

Examples of simple identities we have already met include these:

$$\tan x = \frac{\sin x}{\cos x}$$

$$\cos(-x) = \cos x$$

$$\sin(-x) = -\sin x$$

$$\cos(x - \frac{\pi}{2}) = \sin x$$

$$\cos(x + 2\pi) = \cos x$$

### 5.5.1   The Pythagorean set of identities

Consider the unit circle that we discussed previously:

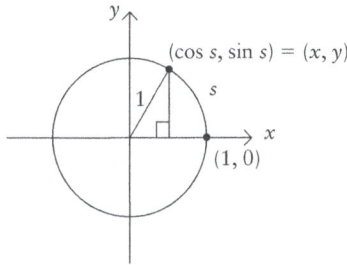

Any point $P(x, y)$ on the circle is a distance $d = 1$ from the centre O of the circle, given by Pythagoras' theorem as:

$$1 = d^2 = (x - 0)^2 + (y - 0)^2$$
$$= x^2 + y^2.$$

Hence, $(x, y)$ lies on the circle $\Longleftrightarrow x^2 + y^2 = 1$. We say that *the equation of the circle is* $x^2 + y^2 = 1$.

As we know, $x = \cos s$ and $y = \sin s$, so we have, for any $s \in \mathbb{R}$, $\cos^2 s + \sin^2 s = 1$. Hence:

$$\cos^2 A + \sin^2 A = 1 \qquad \textbf{Pythagorean Identity (i)}$$

**NB:** $\cos^n s$ and $\sin^n s$ are shorthand ways of writing $(\cos s)^n$ and $(\sin s)^n$ for all $n \in \mathbb{N}$.

Dividing (i) by $\cos^2 A$ gives:

$$\frac{\cos^2 A}{\cos^2 A} + \frac{\sin^2 A}{\cos^2 A} = \frac{1}{\cos^2 A}$$

hence $\boxed{1 + \tan^2 A = \sec^2 A}$    **Pythagorean Identity (ii)**

Dividing (ii) by $\sin^2 A$ gives:

$$\frac{\cos^2 A}{\sin^2 A} + \frac{\sin^2 A}{\sin^2 A} = \frac{1}{\sin^2 A}$$

hence $\boxed{\cot^2 A + 1 = \csc^2 A}$    **Pythagorean Identity (iii)**

**Example:**   Show that: $\sin^4 A - \cos^4 A + \cos^2 A = \sin^2 A$.

**Solution:**

$$\sin^4 A - \cos^4 A + \cos^2 A = (\sin^4 A - \cos^4 A) + \cos^2 A$$
$$= (\sin^2 A + \cos^2 A)(\sin^2 A - \cos^2 A) + \cos^2 A$$
$$= (\sin^2 A - \cos^2 A) + \cos^2 A$$
$$= \sin^2 A.$$

## 5.5.2   Addition formulas

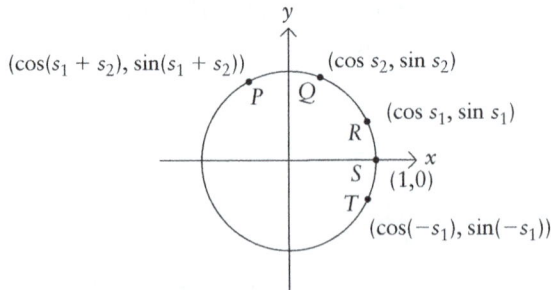

Consider the points $P, Q, R, S$ and $T$ lying on the unit circle above. Observe that the arc-lengths from $S$ to $P$ and from $T$ to $Q$ are equal. Since equal arcs in a circle subtend equal chords, it follows from Pythagoras' theorem that:

$$\sqrt{(\cos(s_1 + s_2) - 1)^2 + (\sin(s_1 + s_2) - 0)^2}$$

$$= \sqrt{(\cos s_2 - \cos s_1)^2 + (\sin s_2 - (-\sin s_1))^2},$$

$$\cos^2(s_1 + s_2) + \sin^2(s_1 + s_2) - 2\cos(s_1 + s_2)$$

$$= \cos^2 s_2 + \sin^2 s_2 - 2\cos s_1 \cos s_2 + \cos^2 s_1$$

$$+ \sin^2 s_1 + 2\sin s_1 \sin s_2,$$

hence $\quad 1 - 2\cos(s_1 + s_2) = 1 - 2\cos s_1 \cos s_2 + 2\sin s_1 \sin s_2,$

and therefore $\quad \cos(s_1 + s_2) = \cos s_1 \cos s_2 - \sin s_1 \sin s_2.$

Thus, for any $A, B \in \mathbb{R}$,

$$\boxed{\cos(A + B) = \cos A \cos B - \sin A \sin B} \qquad \textbf{Addition Formula f(i)}$$

It follows, by replacing $B$ with $-B$ and using the identities $\cos(-x) = \cos x$, $\sin(-x) = -\sin x$, that:

$$\boxed{\cos(A - B) = \cos A \cos B + \sin A \sin B} \qquad \textbf{Addition Formula f(ii)}$$

We showed earlier on that $\cos\left(\frac{\pi}{2} - \alpha\right) = \sin \alpha$. Letting $\alpha = A + B$, we have

$$\cos\left(\frac{\pi}{2} - (A + B)\right) = \sin(A + B),$$

therefore $\quad \sin(A + B) = \cos\left(\left(\frac{\pi}{2} - A\right) - B\right)$

$$= \cos\left(\frac{\pi}{2} - A\right)\cos B + \sin\left(\frac{\pi}{2} - A\right)\sin B$$

$$= \sin A \cos B + \cos A \sin B.$$

$$\boxed{\sin(A + B) = \sin A \cos B - \cos A \sin B} \qquad \textbf{Addition Formula f(iii)}$$

Next, replace $B$ with $-B$ and use the identities $\cos(-x) = \cos x, \quad \sin(-x) = -\sin x$:

$$\boxed{\sin(A - B) = \sin A \cos B - \cos A \sin B} \qquad \textbf{Addition Formula f(iv)}$$

## Addition formulas for the tangent function

$$\tan(A + B) = \frac{\sin(A + B)}{\cos(A + B)}$$

$$= \frac{\sin A \cos B + \cos A \sin B}{\cos A \cos B - \sin A \sin B}$$

$$= \frac{(\sin A \cos B + \cos A \sin B) \times \frac{1}{\cos A \cos B}}{(\cos A \cos B - \sin A \sin B) \times \frac{1}{\cos A \cos B}}$$

$$= \frac{\frac{\sin A}{\cos A} + \frac{\sin B}{\cos B}}{1 - \frac{\sin A}{\cos A} \frac{\sin B}{\cos B}}$$

$$= \frac{\tan A + \tan B}{1 - \tan A \tan B}.$$

$$\boxed{\tan(A + B) = \frac{\tan A + \tan B}{1 - \tan A \tan B}} \quad \text{Addition Formula g(i)}$$

This formula is only valid when the tan function is defined at $A$, $B$ and $A+B$, so these three must not be members of the set: $S = \{x : x = \frac{\pi}{2} + k\pi,\ k \in \mathbb{Z}\}$. (Then $\tan A \tan B \neq 1$.)

It is now easy to show (put $-B$ in place of $B$) that

$$\boxed{\tan(A - B) = \frac{\tan A - \tan B}{1 + \tan A \tan B}} \quad \text{Addition Formula g(ii)}$$

where $A$ and $B$ are real numbers such that $A$, $B$, $A - B \notin S$, so $\tan A$ $\tan B \neq -1$.

**Example:**   Find, without the help of tables, $\sin 75°$.

**Solution:**   Since $75° = 30° + 45°$, we have:

$$\sin(75°) = \sin(30° + 45°)$$

$$= \sin 30° \cos 45° + \cos 30° \sin 45°$$

$$= \frac{\sqrt{2}}{2}\left(\frac{1}{2} + \frac{\sqrt{3}}{2}\right) \quad \text{(see standard triangles on page 188)}$$

$$= \frac{\sqrt{2}}{4}(1 + \sqrt{3}).$$

## 5.5.3   Double angle formulas

These are deduced by setting $A = B$ in the addition formulas above:

from f(i) $\boxed{\cos 2A = \cos^2 A - \sin^2 A}$   **Double Angle Formula d(i)**

Now, since $\cos^2 A + \sin^2 A = 1$,

$$\cos 2A = (1 - \sin^2 A) - \sin^2 A$$
$$= 1 - 2\sin^2 A.$$

Hence $\boxed{\cos 2A = 1 - 2\sin^2 A}$  **Double Angle Formula d(ii)**

Similarly, we can show:

$\boxed{\cos 2A = 2\cos^2 A - 1}$  **Double Angle Formula d(iii)**

from f(iii) $\boxed{\sin 2A = 2\sin A \cos A}$  **Double Angle Formula d(iv)**

from g(i) $\boxed{\tan 2A = \dfrac{2\tan A}{1 - \tan^2 A}}$  **Double Angle Formula d(v)**

### 5.5.4  Product formulas

These are derived from the four addition formulas:

$$\cos(A + B) = \cos A \cos B - \sin A \sin B \qquad \text{(a)}$$
$$\cos(A - B) = \cos A \cos B + \sin A \sin B \qquad \text{(b)}$$
$$\sin(A + B) = \sin A \cos B + \cos A \sin B \qquad \text{(c)}$$
$$\sin(A - B) = \sin A \cos B - \cos A \sin B \qquad \text{(d)}$$

(a) + (b) gives:

$\boxed{\cos A \cos B = \tfrac{1}{2}(\cos(A + B) + \cos(A - B))}$  **Product Formula p(i)**

(b) − (a) gives:

$\boxed{\sin A \sin B = \tfrac{1}{2}(\cos(A - B) - \cos(A + B))}$  **Product Formula p(ii)**

(c) + (d) gives:

$\boxed{\sin A \cos B = \tfrac{1}{2}(\sin(A + B) + \sin(A - B))}$  **Product Formula p(iii)**

(c) − (d) gives:

$\boxed{\cos A \sin B = \tfrac{1}{2}(\sin(A + B) - \sin(A - B))}$  **Product Formula p(iv)**

### 5.5.5   Sum formulas

This set of formulas allows us to achieve the opposite of what the product formulas do, namely, to express the sum or difference of two cosines, or of two sines, in terms of products.

If, in formulas p(i) through p(iv) above, we multiply both sides by 2, reverse sides, and let $x = A + B$ and $y = A - B$, so that $A = \frac{x+y}{2}$, and $B = \frac{x-y}{2}$, then we have:

$$\cos x + \cos y = 2 \cos \left( \frac{x+y}{2} \right) \cos \left( \frac{x-y}{2} \right) \qquad \textbf{Sum Formula s(i)}$$

$$\cos y - \cos x = 2 \sin \left( \frac{x+y}{2} \right) \sin \left( \frac{x-y}{2} \right) \qquad \textbf{Sum Formula s(ii)}$$

$$\sin x + \sin y = 2 \sin \left( \frac{x+y}{2} \right) \cos \left( \frac{x-y}{2} \right) \qquad \textbf{Sum Formula s(iii)}$$

$$\sin x - \sin y = 2 \cos \left( \frac{x+y}{2} \right) \sin \left( \frac{x-y}{2} \right) \qquad \textbf{Sum Formula s(iv)}$$

## 5.6   Trigonometric equations

In the previous section, we defined *trigonometric identities* as equations that are true for all values of $x$ for which the expressions are defined. Now *trigonometric equations*, also known for emphasis as *conditional trigonometrical equations*, are true only for certain values of the variable $x$. 'Solving a trigonometric equation' means finding the set of all values of the variable that satisfy that equation.

**Example:**   Solve the equations:

(i) $\sin \theta = 1$, where (a) $0 \leq \theta \leq 2\pi$, (b) $\theta$ can be any real number.
(ii) $\sin \theta = -\frac{1}{2}$, where (a) $0 \leq \theta \leq 2\pi$, (b) $\theta$ can be any real number.

**Solution:**   (i)(a) $\theta = \frac{\pi}{2}$ is the only member of the set $[0, 2\pi]$ for which $\sin \frac{\pi}{2} = 1$ and is hence the only *root* (i.e. solution) of the given equation.

(i)(b) Since the period of the sine function is $2\pi$, any number of the form $\frac{\pi}{2} + 2n\pi, n \in \mathbb{Z}$ is a solution. We say that the **general solution** of the

*Okay, let me put it like this. To escape purgatory you are only required to correctly solve a single variable linear equation. On the other hand, to get into heaven you are required to correctly solve a nonlinear multivariable trigonometric equation*

above equation is

$$\frac{\pi}{2} + 2n\pi, \ n \in \mathbb{Z}.$$

This describes a *set of solutions*.

(ii)(a)  The sine function is negative in the third and fourth quadrants:

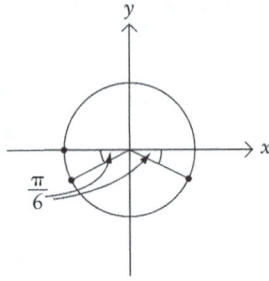

The numbers $\theta$ corresponding to $\sin\theta = -\frac{1}{2}$ are therefore

$$\pi + \frac{\pi}{6} = \frac{7\pi}{6} \quad \text{and} \quad 2\pi - \frac{\pi}{6} = \frac{11\pi}{6},$$

and hence the solutions are $\frac{7\pi}{6}$ and $\frac{11\pi}{6}$.

(ii)(b)  As before, since the period is $2\pi$, the general solutions are:

$$\frac{7\pi}{6} + 2n\pi, \ \ n \in \mathbb{Z} \ \text{ and } \ \frac{11\pi}{6} + 2n\pi, \ \ n \in \mathbb{Z}.$$

Observe that, just as there may be more than one solution $s \in [0, 2\pi]$, there may be more than one formula for the general solution to a given equation. What is the analogous general statement for the tangent function?

**Example:**  Solve the equation: $2\sin^2 x - \sin x = 1$, where $x$ is a real number.

**Solution:**

$$\begin{aligned} 0 &= 2\sin^2 x - \sin x - 1 \\ &= 2\sin^2 x + \sin x - 2\sin x - 1 \\ &= \sin x(2\sin x + 1) - (2\sin x + 1) \\ &= (\sin x - 1)(2\sin x + 1). \end{aligned}$$

The last line shows that we have found two factors whose product is zero, so that the original equation will be satisified whenever either factor is zero. We therefore treat each factor separately.

**First factor:** $\sin x - 1 = 0$, so $\sin x = 1$, hence, from the first example, we see that

$$x \in A = \left\{ x : x = \frac{\pi}{2} + 2n\pi, \ n \in \mathbb{Z} \right\}.$$

**Second factor:** $2\sin x + 1 = 0$, so $\sin x = -\frac{1}{2}$, hence $x \in B \cup C$ where

$$B = \left\{ x : x = \frac{7\pi}{6} + 2n\pi, \ n \in \mathbb{Z} \right\} \ \text{ and } \ C = \left\{ x : x = \frac{11\pi}{6} + 2n\pi, \ n \in \mathbb{Z} \right\}.$$

The general solution of our original equation is therefore the set $A \cup B \cup C$.

**Example:**   Solve the equation: $\cos 2\theta + \sin \theta = 0$ where $\theta$ is a real number.

**Solution:**   The idea is to get an equation in a single variable, so we use the identity: $\cos 2\theta = 1 - 2\sin^2 \theta$. Hence the given equation is equivalent to:

$$0 = \cos 2\theta + \sin \theta$$
$$= 1 - 2\sin^2 \theta + \sin \theta$$
$$= 2\sin^2 \theta - \sin \theta - 1.$$

The solutions are therefore as in the previous example above.

**Example:**   Solve the equation: $\sin 3x = 1$, where $x$ is a real number.

**Solution:**   Let $3x = \theta$, so the equation is simply $\sin \theta = 1$.
From the first example above, $\theta = \frac{\pi}{2} + 2n\pi$, $n \in \mathbb{Z}$, i.e. $3x = \frac{\pi}{2} + 2n\pi$.
Thus, finally:

$$x = \frac{\pi}{6} + \frac{2n\pi}{3}, \quad n \in \mathbb{Z}.$$

### The inverse circular functions

We could have written the first solution obtained in the previous example as: $\theta = \frac{\pi}{2} = \arcsin(1) = \sin^{-1}(1)$. The angle $x$, whose sine is $y$ (that is: $y = \sin x$), and such that (for there are many) $x \in [-\frac{\pi}{2}, \frac{\pi}{2}]$, is called the **inverse sine function**, and is written $x = \arcsin y$ or $x = \sin^{-1} y$. Thus, by convention, $\sin^{-1} y$ does NOT mean $\frac{1}{\sin y}$, which is equivalent to $(\sin y)^{-1}$.
   In the first example (i)(b) above, the solution could also have been written as:

$$\theta = \arcsin(1) + 2n\pi, \quad n \in \mathbb{Z}.$$

Similarly, the **inverse cosine function** and **inverse tangent function** are defined, by selecting the portions of their graphs which are strictly increasing, so represent one-to-one functions with unique inverses:

$$y = \arcsin(x) \text{ if } x = \sin y \text{ and } -\frac{\pi}{2} \leq y \leq \frac{\pi}{2}$$
$$y = \arccos(x) \text{ if } x = \cos y \text{ and } 0 \leq y \leq \pi$$
$$y = \arctan(x) \text{ if } x = \tan y \text{ and } -\frac{\pi}{2} < y < \frac{\pi}{2}.$$

**Example:**   Find the sum: $\cos^2 0° + \cos^2 2° + \cos^2 4° + \cos^2 6° + \cdots + \cos^2 358° + \cos^2 360°$.

(A) 91  (B) 181  (C) 361  (D) $\sqrt{181}$  (E) $\sqrt{91} + \sqrt{181}$

**Solution:**    Note that $\cos(90 - \theta) = \sin\theta$, $\cos(90 + \theta) = -\sin\theta$, $\sin(90 + \theta) = \cos\theta$, etc. Therefore (using $\cos^2\theta + \sin^2\theta = 1$):

(i) $\displaystyle\sum_{i=0}^{45} \cos^2 2i = \sum_{i=1}^{22} \cos^2 2i + \sum_{i=1}^{22} \sin^2 2i + \cos^2 0 + \cos^2 90 = 22 + 1 = 23.$

(ii) $\displaystyle\sum_{i=46}^{90} \cos^2 2i = \sum_{i=1}^{22} (-\cos 2i)^2 + \sum_{i=1}^{22} (-\sin 2i)^2 + \cos^2 180$

$= 22 + 1 = 23.$

Therefore $\displaystyle\sum_{i=0}^{90} \cos^2 2i = 23 + 23 = 46$   (adding (i) and (ii)).

For economy and convenience, we have once again used the summation notation, explained in Toolchest 6. Now, for the whole sum ($i = 0$ through $i = 180$), we get the value $46 + 45 = 91$. Hence (A).

Here is the solution to the appetizer problem that was posed at the beginning of the Toolchest.

**Solution of Appetizer Problem:**    Try to get the expression to be investigated in terms of the expressions for $a$ and $b$, bracketing appropriately and using the addition formulas:

$$\sin(2x + 3y) = \sin((2x + 2y) + y)$$
$$= \sin(2x + 2y)\cos y + \cos(2x + 2y)\sin y$$
$$= 2\sin(x + y)\cos(x + y)\cos y + \sin y(\cos^2(x + y)$$
$$- \sin(x + y))$$
$$= \sin(x + y)\cos(x + y)\cos y + \sin y\cos^2(x + y)$$
$$+ \sin(x + y)\cos(x + y)\cos y - \sin y\sin^2(x + y)$$
$$= \cos(x + y)[\cos y\sin(x + y) + \sin y\cos(x - y)]$$
$$+ \sin(x + y)[\cos y\cos(x + y) - \sin y\sin(x + y)]$$
$$= \cos(x + y)\sin(x + y + y) + \sin(x + y)\cos(x + y + y)$$
$$= \cos(x + y)\sin(x + 2y) + \sin(x + y)\cos(x + 2y).$$

Now, using the identity $\sin^2 A + \cos^2 A = 1$, we have

$$\cos(x + y) = b, \quad \text{hence} \quad \sin(x + y) = \pm\sqrt{1 - b^2},$$
$$\text{and} \quad \cos(x + 2y) = a, \quad \text{hence} \quad \sin(x + 2y) = \pm\sqrt{1 - a^2}.$$

Thus the maximum value of $\sin(2x+3y)$ is $b\sqrt{1 - a^2} + a\sqrt{1 - b^2}$. Hence (B).

## 5.7  Problems

**Problem 1:**   *A range of values of x for which* $\cos x < \sin x$ *is*

(A) $0° < x < 45°$   (B) $0° < x \le 45°$   (C) $45° \le x \le 90°$   (D) $45° < x \le 90°$   (E) $0° \le x \le 90°$

**Problem 2:**   *Given that* $0 < \theta < 180°$ *and that* $2 \sec \theta - \cos \theta \cot \theta \ge k$, *find the maximum value of k.*

(A) $\sin \theta$   (B) 2   (C) 1   (D) 4   (E) 3

**Problem 3:**   *If* $\cos A = \frac{3}{5}$ *and* $\pi < A < 2\pi$, *find* $\tan A - \csc A$.

(A) $-\frac{1}{2}$   (B) $\frac{1}{6}$   (C) $-\frac{1}{12}$   (D) 2   (E) $\frac{4}{5}$

**Problem 4:**   *In a right-angled triangle* $ABC$, $AC^2 = 27$, $BP^2 = 9$, *where BP is the perpendicular drawn from B to AC. Find the angle* $\hat{C}$.

(A) $\frac{\pi}{4}$   (B) $\frac{\pi}{3}$   (C) $-\frac{3\pi}{8}$   (D) $20°$   (E) $\frac{\pi}{6}$

**Problem 5:**   *If* $0 < A < \frac{\pi}{2}$ *and* $\sin A + 1 = 2\cos A$, *determine the value of* $\sin A$.

(A) 0.75   (B) $\frac{2}{3}$   (C) $-\frac{3}{5}$   (D) $\frac{1}{3}$   (E) $\frac{1}{\sqrt{6}}$

**Problem 6:**   *Solve* $\sin^2 + \cos x + 1 = 0$, *for* $0 \le x \le 2\pi$.

(A) $\frac{\pi}{2}$   (B) $\frac{\pi}{3}$   (C) 0   (D) $\pi$   (E) $\frac{4\pi}{3}$

**Problem 7:**   *Triangle ABC is such that* $AC = BC$ *and* $AB = rAC$. *Show that*
$$\cos A + \cos B + \cos C = 1 + r - \tfrac{1}{2}r^2.$$

**Problem 8:**   *Solve* $\sin 2x + \cos 2x + \sin x + \cos x + 1 = 0$, $0 \le x \le 2\pi$.

**Problem 9:**   *The sides of a triangle are* $x$, $y$ *and* $\sqrt{x^2 + xy + y^2}$. *Find its largest angle.*

**Problem 10:**   *Solve* $\cos^{10} x - \sin^{10} x = 1$, $0 \le x \le 2\pi$.

**Problem 11:**   *What is* $\cos 36°$?

(A) $\frac{2+\sqrt{5}}{8}$   (B) $\frac{1+\sqrt{2}}{5}$   (C) $\frac{1+\sqrt{5}}{4}$   (D) $\frac{5+\sqrt{2}}{10}$   (E) $\frac{\sqrt{5}}{3}$

## 5.8 Solutions

**Solution 1:**

**Method 1: Intuitive visual solution.** A stylized sketch of the graphs of $\sin x$ and $\cos x$ is given below. Now $\sin x > \cos x$ requires that the graph of $\sin x$ is always above that of $\cos x$ for the range. Since $\sin 0 = 0 < \cos 0 = 1$, the given ranges which start at zero are obviously out, so that we just have to consider (C) and (D). For $x = 45°$, $\sin x$ and $\cos x$ coincide, the common value being $\frac{\sqrt{2}}{2}$.

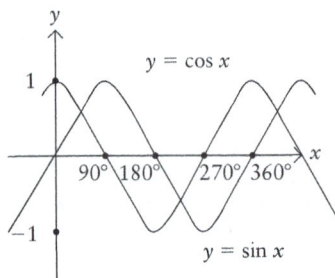

That $\sin x$ is strictly greater than $\cos x$ for values of $x$ strictly greater than $45°$ can be seen from the graphs, and at $x = 90°$, $\sin x$ is still strictly greater than $\cos x$, so that a possible range is $45° < x \leq 90°$. Hence (D).

Notice from the graphs that $\sin x$ continues to be strictly greater than $\cos x$ for values of $x$ well beyond $90°$, in fact, $\sin x > \cos x$ for the wider interval $45° < x < 225°$.

**Method 2: Analytic solution.** This method is more general; it can be used in other (even non-multiple choice) problems, for it does not rely on the graph, which may be harder to sketch for other functions. We begin by observing that, if we restrict attention to $x \in [0, \frac{\pi}{2}]$ so that both $\cos x$ and $\sin x$ are non-negative, then the equation $\cos x < \sin x$ is equivalent to $\cos^2 x < \sin^2 x$.

$$\text{Thus} \quad 1 - \sin^2 x < \sin^2 x,$$

$$\text{i.e.} \quad 1 < 2y^2, \quad \text{where } y = \sin x \leq 1,$$

$$\text{i.e.} \quad \frac{1}{2} < y^2,$$

$$\text{i.e.} \quad \frac{1}{\sqrt{2}} < y \leq 1,$$

$$\text{i.e.} \quad \frac{\pi}{4} < x \leq \frac{\pi}{2}.$$

**Solution 2:**   Try to express that LHS in terms of sin or cos only:

$$2\sec\theta - \cos\theta\cot\theta = \frac{2}{\sin\theta} - \cos\theta \cdot \frac{\cos\theta}{\sin\theta}$$

$$= \frac{2 - \cos^2\theta}{\sin\theta}$$

$$= \frac{1 + 1 - \cos^2\theta}{\sin\theta}$$

$$= \frac{1 + \sin^2\theta}{\sin\theta} = \sin\theta + \frac{1}{\sin\theta}.$$

Putting $\sin\theta = x > 0$ for $0 < \theta < 180°$, we have $(x-1)^2 \geq 0$, so $x^2 + 1 \geq 2x$. Now division by $x > 0$ gives $x + \frac{1}{x} \geq 2$. And for $x = 1$ the expression attains its minimum value 2. Thus the maximum value of $k$ is 2, hence (B).

**Solution 3:**   Consider the acute angle $\theta$ with $\cos\theta = \frac{3}{5}$, which sits in a right triangle with sides $3, 4, 5$ (draw it!). Then the angle in the required range with same cosine is in the fourth quadrant: $A = 2\pi - \theta$, and so

$$\tan A = -\frac{4}{3}, \quad \csc A = -\frac{5}{4}, \quad \text{so that} \quad \tan A - \csc A = -\frac{1}{12}. \text{ Hence (C).}$$

**Solution 4:**   First, draw the diagram! Letting $x = BC$, we have $\sin C = 3/x$, while $\cos C = x/\sqrt{27} = x/3\sqrt{3}$. Hence $\tan C = \frac{1}{\sqrt{3}}$, so $C = \frac{\pi}{6}$, from the standard triangle on page 188.

**Solution 5:**   Let $x = \sin A$, so that $x + 1 = 2\sqrt{1 - x^2}$. Since both sides are positive, this equation is equivalent to $x^2 + 2x + 1 = 4(1 - x^2)$ (squaring both sides). This gives a quadratic $5x^2 + 2x - 3 = 0$, with two roots $x = -1, \frac{3}{5}$. The positive root is applicable to the sine of an acute angle, hence (C).

**Solution 6:**   We have $0 = \sin^2 + \cos x + 1 = (1 - \cos^2 x) + \cos x + 1 = -\cos^2 + \cos x + 2$. Putting $y = \cos x$, we have a quadratic equation $0 = y^2 - y - 2 = (y + 1)(y - 2)$, with roots $y = 2$ (impossible since $\cos x \leq 1$) and $y = -1$. Therefore $x = \arccos(-1) = \pi$, because of the given range. Hence (D).

If the general solution of the equation had been required, the answer would be $x = \pi + 2k\pi, \ k \in \mathbb{Z}$.

**Solution 7:**   First draw a good diagram, observing that $\triangle ABC$ is isosceles.

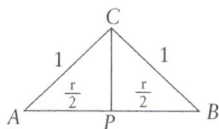

In order to evaluate the cosines, we need some right-angled triangles, so draw the perpendicular $CP$ from $C$ to $AB$; it bisects $AB$ by symmetry of isosceles triangle. We chose scale so that $AC = BC = 1$ and so $AP = PB = \frac{1}{2}r$. Now $\cos A = \cos B = \frac{1}{2}r$, and (letting angle $\hat{A}CP = C_1$)

$$\cos C = \cos 2C_1 = 1 - 2\sin^2 C_1$$

$$= 1 - 2\left(\frac{1}{2}r\right)^2 = 1 - \frac{1}{2}r^2.$$

The answer is immediate.

**Solution 8:**    A reasonable strategy is to use the double angle formulas to simplify, then replace $1 - \sin^2 x$ with $\cos^2 x$, and look for a factorization:

$$0 = \sin 2x + \cos 2x + \sin x + \cos x + 1$$

$$= 2\cos x \sin x + \cos^2 x - \sin^2 x + \sin x + \cos x + 1$$

$$= 2\cos x \sin x + 2\cos^2 x + \sin x + \cos x$$

$$= 2\cos x(\sin x + \cos x) + (\sin x + \cos x)$$

$$= (2\cos x + 1)(\sin x + \cos x).$$

Therefore    $\cos x = -\dfrac{1}{2}$, or $\sin x = -\cos x$,

hence    $x = \dfrac{2\pi}{3}$ or $x = \dfrac{4\pi}{3}$, or $x = \dfrac{3\pi}{4}$, or $x = \dfrac{7\pi}{4}$.

**Solution 9:**    Since $x, y$ are positive, we have $x$, $y < \sqrt{x^2 + xy + y^2}$, by taking the root of each side in the inequality $x^2 < x^2 + xy + y^2$, and similarly for $y$. Now the largest angle must be opposite the (largest) side of length $\sqrt{x^2 + xy + y^2}$. Applying the cosine rule:

$$\cos A = \frac{x^2 + y^2 - (x^2 + xy + y^2)}{2xy}$$

$$= \frac{-xy}{2xy} = -\frac{1}{2}.$$

Hence $A = 2\pi/3$.

**Solution 10:**    From the given range we know that $\cos^{10} x \leq 1$ and of course $1 + \sin^{10} x \geq 1$. The given equation is equivalent to $\cos^{10} x = 1 + \sin^{10} x$, and since LHS$\leq 1$ while RHS$\geq 1$, the only possible solution is when both sides are equal to 1. This gives    $\cos^{10} x = 1$ and $\sin^{10} x = 0$, implying that $\cos x = \pm 1$ and $\sin x = 0$. The solutions within the given range are $x = 0$, $x = \pi$, $x = 2\pi$.
NB. The more general equation    $\cos^{2n} x - \sin^{2n} x = 1$, for $n \geq 1$, $n \in \mathbb{Z}$, has exactly the same solutions. For odd powers the solutions are $0$, $2\pi$, $\frac{3\pi}{2}$.

**Solution 11:**   Start by observing that $36° = \frac{2}{5}90°$. Let $\theta = 18°$, so that $2\theta = 36°$ and $3\theta = 54°$, with $54° + 36° = 90°$, and recall that the cosine of an angle equals the sine of its complement (complementary angles add up to 90°). Thus

$$\cos 3\theta = \sin 2\theta.$$

In fact, this equation is satisfied by all angles $\theta = 18° + 360°n$, $n \in \mathbb{Z}$. Using the expansions for multiple angles, we see that

$$\sin 2\theta = 2 \sin \theta \cos \theta$$
$$= 2c\sqrt{1 - c^2} \quad \text{where } c = \cos \theta$$

and $\qquad \cos 3\theta = 4 \cos^3 \theta - 3 \cos \theta$
$$= 4c^3 - 3c.$$

So, by letting $2c\sqrt{1 - c^2} = 4c^3 - 3c$, we can find $\cos 18°$ :

$$2c\sqrt{1 - c^2} = 4c^3 - 3c,$$
$$\text{therefore} \quad 2\sqrt{1 - c^2} = 4c^2 - 3 \quad (\cos 18° > 0),$$
$$\text{hence} \quad 4 - 4c^2 = 16c^4 - 24c^2 + 9,$$
$$\text{and so} \quad 16c^4 - 20c^2 + 5 = 0.$$

Putting $c^2 = y$, so that $16y^2 - 20y + 5 = 0$, we have

$$c^2 = y = \frac{20 \pm \sqrt{80}}{32}$$
$$= \frac{4 \times 5 \pm 4 \times \sqrt{5}}{4 \times 8}$$
$$= \frac{5 + \sqrt{5}}{8}.$$

We discard the smaller root – why? Now we use the double angle formula $\cos 2\theta = 2 \cos^2 \theta - 1$:

$$\cos 36 = 2 \cos^2 18 - 1 = 2c^2 - 1 = 2 \left( \frac{5 + \sqrt{5}}{8} \right) - 1 = \frac{1 + \sqrt{5}}{4} = \frac{\phi}{2}.$$

# 6 Sequences and series

"Johann Carl Friedrich Gauss, mathematician and physicist, citizen of the Kingdom of Hanover, you are hereby sentenced to calculate the sum on the board all the way to infinity...we hope you will find a short-cut as you sum to infinity. You have quite a reputation of doing that"

By the end of this topic you should be able to:

(i) Use the summation notation.
(ii) Find the $n$th term of an Arithmetic Progression.
(iii) Find the sum of the first $n$ terms of an Arithmetic Progression.
(iv) Find the $n$th term of a Geometric Progression.
(v) Find the sum of the first $n$ terms of a Geometric Progression.
(vi) Find the sum to infinity of a convergent Geometric Progression.
(vii) Establish (using the method of differences) the following standard results, and use them to find sums of other series as well as the values of some products:

$$(a) \; \sum_{i=1}^{n} i = \frac{n(n+1)}{2} \quad (b) \; \sum_{i=1}^{n} i^2 = \frac{n(n+1)(2n+1)}{6}.$$

Here, as usual, is an appetizer problem of the kind you should be able to solve when you have worked through this Toolchest. You are invited to try it as soon as you wish. You may find it hard for now, but by the end of Section 6, where its solution is given, it should look much easier to you.

**Appetizer Problem:**  *Find the value of the product*

$$\left(1 - \frac{1}{2^2}\right)\left(1 - \frac{1}{3^2}\right)\left(1 - \frac{1}{4^2}\right)\left(1 - \frac{1}{5^2}\right)\cdots\left(1 - \frac{1}{1999^2}\right)$$

(A) $\frac{997}{1999}$  (B) $\frac{1001}{1999}$  (C) $\frac{999}{1999}$  (D) $\frac{1000}{1999}$  (E) $\frac{1002}{1999}$

## 6.1  General sequences

Consider this very simple sequence, which we could call the ancestor of all sequences:

$$1, 2, 3, 4, \ldots$$

It is quite easy to predict what the next term is going to be: it is obtained by adding 1, each time. Furthermore, we can find a formula for the $n$th term of the sequence. It's quite clear that the $n$th term is just $n$. Below is a list of some sequences and some corresponding formulas for the $n$th terms:

| Sequence | $n$th term |
|---|---|
| 1, 3, 5, ... | $2n - 1$ |
| 13, 8, 3, ... | $18 - 5n$ |
| 1, 4, 9, 16, ... | $n^2$ |
| $\frac{1}{2}, \frac{1}{4}, \frac{1}{8}, \ldots$ | $\frac{1}{2^n}$ |
| $\frac{1}{2}, \frac{5}{4}, \frac{7}{8}, \ldots$ | $1 + \left(-\frac{1}{2}\right)^n$ |

Note that $n \in \mathbb{N}$, and $n = 1$ corresponds to the first term of the sequence, $n = 2$ the second and so on.

## 6.2  The summation notation

Consider the sum

$$1 + 2^2 + 3^2 + 4^2 + 5^2 + 6^2 + \cdots + n^2. \tag{6.1}$$

This is the sum of the squares of the first $n$ natural numbers. Mathematicians have invented a shorthand way of writing down such sums, referred to as 'summation notation'. We represent (6.1) by

$$\sum_{i=1}^{n} i^2,$$

read as: 'the sum of terms like $i^2$, where $i$ takes values from 1 up to $n$, inclusive'. In order to represent a sum using the summation notation, we need to find a *general formula* for the terms of that sum, and we need to be careful about where the 'dummy variable' $i$ starts and stops. (It is called the **dummy** variable to distinguish it from the genuine variable $n$ because it does not appear as a variable in the whole sum, which depends on $n$ alone.) We can represent the respective sums $1 - 2 + 4 - 8 + \cdots + (-2)^n$ and $1! + 2! + 3! + \cdots + n!$ by:

$$\sum_{i=0}^{n}(-2)^i, \text{ and } \sum_{i=1}^{n} i!.$$

All the sums above are examples of **finite sums** in which we add the terms up to a certain term (say the $n$th). On the other hand, there are **infinite sums**, which we will encounter in Section 5, when we sum convergent geometric series 'to infinity'. Here are two examples of infinite series:

$$1 + \frac{1}{2} + \frac{1}{2^2} + \frac{1}{2^3} + \cdots = 2, \quad 1 + 2 + 2^2 + 2^3 + \cdots = +\infty.$$

The first is the infinite geometric series with common ratio $\frac{1}{2}$, which converges and has finite sum 2 (see Section 5). The second is the infinite geometric series with common ratio 2, which diverges – i.e. does not converge. We use the symbol $\infty$ (read 'infinity') in two different ways: to represent an infinite sum of a divergent series, as above, and also to describe infinite series of terms which do converge in the sense described in Section 5. The notation is used like this:

$$\sum_{i=0}^{\infty}\left(\frac{1}{2}\right)^i = 2, \quad \sum_{i=0}^{\infty} 2^i = +\infty$$

which are to be read as 'the sum of all terms like $\left(\frac{1}{2}\right)^i$, from $i$ equals zero to infinity, is 2', and 'the sum of all terms like $2^i$, from $i$ equals zero to infinity, is plus infinity'.

**Example:** Find

$$\sum_{i=1}^{4} \frac{(2n)!}{2^n}.$$

**Solution:**

$$\sum_{i=1}^{4} \frac{(2n)!}{2^n} = \frac{2!}{2} + \frac{4!}{4} + \frac{6!}{8} + \frac{8!}{16}$$
$$= 1 + 6 + 90 + 2520 = 2617.$$

## 6.3  Arithmetic Progressions

To taste the unlikely looking sort of problem the ideas of this section will help you to solve, here is an appetizer, whose solution appears at the end of the section.

**Appetizer Problem:**  *The value of* $1996^2 - 1995^2 + 1994^2 - 1993^2 + \cdots + 2^2 - 1^2$ *is:*

(A) 998001   (B) 998000   (C) 998002   (D) 999000   (E) 999800

Let us go back to the sequence of the natural numbers that we started off with, namely $1, 2, 3, 4, \ldots$
Consider the question of finding the sum of the first 100 members of this sequence

$$S_{100} = 1 + 2 + 3 + \cdots + 98 + 99 + 100.$$

Of course, we could do it the 'rhinoceros' way and find the sum by actually adding 1 and 2 to get 3, then adding 3 and 3 to get 6 and so on up to 100.

Clearly, this is time-consuming, and not the most desirable way of doing it. Can we find a sort of 'penny-in-the-slot-machine' that will require a minimal amount of labour to give us the desired sum?

At 9 years of age Karl Friedrich Gauss is said to have shocked his teacher (who had given the class this problem to keep them quiet) when he devised and used what is now known to many mathematicians as the Gauss method to find $S_{100}$. Born on April 23, 1777 at Brunswick, to a bricklayer father, this all-time great mathematician (some regard him as the greatest mathematician that ever lived) made many major contributions that have influenced virtually all areas of science. Much of what we know today about the invaluable Normal (Gaussian) distribution is due to him. This is how the nine-year-old boy found the sum of those hundred numbers:

$$S_{100} = 1 + 2 + 3 + \cdots + 99 + 100$$
$$S_{100} = 100 + 99 + 98 + \cdots + 2 + 1.$$

Adding the two equations together:

$$2S_{100} = 101 + 101 + 101 + \cdots + 101 + 101 \quad (100 \text{ times})$$
$$= 100 \times 101,$$

therefore   $S_{100} = 50 \times 101$
$$= 5050.$$

Just stop here a moment, and convince yourself that the method is simple yet extremely handy, by doing a similar problem, like finding the sum $S_{10}$ of the first ten natural numbers. Then do a 'rhinoceros' check, and you should discover that your two answers are the same. And you may see an interesting similarity between $11 \times 5$ and $101 \times 50$ that suggests to you a formula ...

But let us consider this more general sequence, where we get successive terms by adding not 1 but $d$, each time:

$$a, \ a + d, \ a + 2d, \ a + 3d, \ldots$$

where the first term is $a$ plus no $d^s$, the second term is $a$ plus one $d$, and so clearly the $n$th term is $a + (n-1)d$. If we add the terms of this sequence up to the $n$th term, we get the series:

$$S_n = (a) + (a + d) + (a + 2d) + \cdots + \big(a + (n-1)d\big).$$

Such a series, with a **common difference** between successive terms, is called an **Arithmetic Progression (AP)**. Here the common difference is $d$, and $a$ is the first term.

Can we find a formula for the sum $S_n$ of the first $n$ terms of an AP? Let the $n$th term of the AP be $t_n$. It is clearly arrived at by adding to the first term $a$ a total of $n-1$ common differences $d$, so that $t_n = a + (n-1)d$.

$$S_n = a + (a+d) + (a+2d) + \cdots + (t_n - 2d) + (t_n - d) + (t_n);$$

$$\text{also} \quad S_n = (t_n) + (t_n - d) + (t_n - 2d) + \cdots + (a + 2d) + (a + d) + (a).$$

Adding: $\quad 2S_n = n(a + t_n),$

therefore $\quad S_n = \dfrac{n}{2}(a + t_n).$

Since $t_n = a + (n-1)d$, we have

$$S_n = \frac{n}{2}(a + a + (n-1)d)$$

$$= \frac{n}{2}(2a + (n-1)d).$$

and we have the formula for the sum to $n$ terms of an arithmetic series:

$$S_n = \sum_{i=1}^{n} a + (i-1)d = \frac{n}{2}\big(2a + (n-1)d\big)$$

Returning to the basic sequence we began with, we see that: $1 + 2 + 3 + \cdots + (n-1) + n$, the sum of the first $n$ natural numbers, is an AP with $a = d = 1$. Hence

$$S_n = \frac{n}{2}(2a + (n-1)d) \quad \text{gives}$$

$$= \frac{n}{2}(2 + (n-1))$$

$$= \frac{n(n+1)}{2}.$$

$$\sum_{i=1}^{n} i = 1 + 2 + 3 + \cdots + (n-1) + n = \frac{n(n+1)}{2}$$

We could, of course, derive this result by a direct use of the Gauss method. We could also prove it by the *Principle of Mathematical Induction* (see Toolchest 2). And, conversely, if we *assume* this result about the sum of the first $n$ natural numbers, then we have another way of deriving the formula for the sum of an arithmetic series. For

$$S_n = (a) + (a+d) + (a+2d) + \cdots + (a + (n-1)d)$$

$$= na + d(1 + 2 + \cdots n - 1) = na + \frac{1}{2}n(n-1)d.$$

Here is the solution to the problem you met at the start of this section.

**Solution of Appetizer Problem:**    The sum can be written as

$$(1996^2 - 1995^2) + (1994^2 - 1993^2) + \cdots + (2^2 - 1^2)$$

and we observe that we have 998 pairs.

So the sum is written in a more compact way as:

$$\sum_{n=1}^{998} \left( (2n)^2 - (2n-1)^2 \right) = \sum_{n=1}^{998} 4n^2 - (4n^2 - 4n + 1)$$

$$= \sum_{n=1}^{998} 4n - 1$$

$$= 4\sum_{n=1}^{998} n - \sum_{n=1}^{998} 1$$

$$= \frac{4(998)(998+1)}{2} - 998$$

$$= 998000. \text{ Hence (B).}$$

## 6.4   Geometric Progressions

If, for the sequence 2, 4, 8, 16, ..., $2^n$, we decide to find the sum of these first $n$ terms, we get a series:

$$S_n = 2 + 4 + 8 + 16 + \cdots + 2^n$$

$$= 2 + 2 \cdot 2 + 2 \cdot 2^2 + \cdots + 2 \cdot 2^{n-1}.$$

Any series that takes the form

$$S_n = a + ar + ar^2 + \cdots + ar^{n-1},$$

where there is a **common ratio** between successive terms, is called a **Geometric Progression (GP)**; here $r$ is the common ratio, and $a$ is the **first term**. For example, in the problem we started off with, we have $a = 2$ and $r = 2$.

Now, consider what happens when we multiply that equation for the sum by the common ratio:

$$S_n = a + (ar) + (ar^2) + \cdots + (ar^{n-1})$$
$$\text{so } rS_n = \quad + (ar) + (ar^2) + \cdots + (ar^{n-1}) + (ar^n).$$

Subtracting the first equation from the second gives:

$$rS_n - S_n = ar^n - a,$$
$$\text{therefore} \quad S_n(r-1) = a(r^n - 1),$$
$$\text{hence} \quad S_n = a\left(\frac{r^n - 1}{r - 1}\right), \quad \text{provided } r \neq 1.$$

Thus we have the formula for the sum of the first $n$ terms of a geometric series when $r \neq 1$:

$$S_n = \sum_{i=1}^{n} ar^{i-1} = a\left(\frac{r^n - 1}{r - 1}\right)$$

Returning to the series we started with:

$$2 + 4 + 8 + \cdots + 2 \cdot 2^{n-1} = \frac{2(2^n - 1)}{2 - 1}$$
$$= 2(2^n - 1).$$

**Example:** The sum of the first $n$ terms of a series is given by $3^n - 1$, what is the first term of this series? Is it a GP and if so what is the common ratio?

**Solution:**

$$S_n = 3^n - 1,$$
$$\text{therefore} \quad S_{n-1} = 3^{n-1} - 1.$$

Now, the $n$th term is given by

$$U_n = S_n - S_{n-1}$$
$$= 3^n - 3^{n-1}$$
$$= 3^{n-1}(3 - 1)$$
$$= 2(3)^{n-1},$$
$$\text{therefore} \quad U_1 = 2.$$

Clearly, this is a GP with common ratio 3.

**Example:** Find a formula for the sum of the first $n$ terms of the series:

$$\frac{3}{2} + \frac{21}{4} + \frac{63}{8} + \frac{177}{16} + \cdots$$

**Solution:**    First, we look for *pattern* in the development of the series.

$$\left(\frac{3}{2}\right) + \left(\frac{21}{4}\right) + \left(\frac{63}{8}\right) + \left(\frac{177}{16}\right) + \cdots$$

$$= \left(1 + \frac{1}{2}\right) + \left(5 + \frac{1}{4}\right) + \left(8 - \frac{1}{8}\right) + \left(11 + \frac{1}{16}\right) + \cdots$$

$$= \left(2 - \frac{1}{2}\right) + \left(5 + \frac{1}{4}\right) + \left(8 - \frac{1}{8}\right) + \left(11 + \frac{1}{16}\right) + \cdots$$

$$= \left(2 + \left(-\frac{1}{2}\right)\right) + \left(5 + \left(-\frac{1}{2}\right)^2\right) + \left(8 + \left(-\frac{1}{2}\right)^3\right)\left(11 + \left(-\frac{1}{2}\right)^4\right) + \cdots$$

$$= (2 + 5 + 8 + 11 + \cdots)$$

$$+ \left(\left(-\frac{1}{2}\right)^1 + \left(-\frac{1}{2}\right)^2 + \left(-\frac{1}{2}\right)^3 + \left(-\frac{1}{2}\right)^4 + \cdots\right)$$

$$= \sum_{i=1}^{n}(3i - 1) + \sum_{i=1}^{n}\left(-\frac{1}{2}\right)^i \quad \text{(the sum of an AP and a GP)}$$

$$= 3\sum_{i=1}^{n}i - n - \frac{1}{2}\left(\frac{\left(-\frac{1}{2}\right)^n - 1}{-\frac{1}{2} - 1}\right) = \frac{3n(n+1)}{2} - n + \frac{1}{3}\left(\left(-\frac{1}{2}\right)^n - 1\right)$$

$$= \frac{3n^2 + n}{2} + \frac{1}{3}\left(\left(-\frac{1}{2}\right)^n - 1\right) = \frac{n(3n+1)}{2} + \frac{1}{3}\left(\left(-\frac{1}{2}\right)^n - 1\right).$$

## 6.5   Sum to infinity of a Geometric Progression

Here is a simple but typical problem, which not only asks for a finite sum, but also for something which is naturally called a 'sum to infinity'. The solution is at the end of the section.

**Appetizer Problem:**    *Find the sum of the first n terms of the sequence*

$$0.1, \ 0.11, \ 0.111, \ 0.1111, \ 0.11111, \ldots$$

*Also find the rational number which has infinite repeated decimal expansion* 0.11111 . . . .

An interesting case of the GP is when $|r| < 1$. As $n$ gets larger and larger $r^n$ gets smaller and smaller, i.e. gets indefinitely close to zero.

We write $\lim_{n\to\infty} r^n = 0$, and we say 'the limit of $r^n$, as $n$ approaches infinity, is zero'.

Now we define the sum of an infinite series to be the limiting value (when it exists), as $n$ approaches infinity, of the sum to $n$ terms. Then it follows, from the formula for the sum to $n$ terms of a Geometric Progression, that:

$$\lim_{n \to \infty} S_n = \lim_{n \to \infty} a\left(\frac{1 - r^n}{1 - r}\right) = \lim_{n \to \infty} = \frac{a}{1 - r}(1 - r^n) = \frac{a}{1 - r},$$

and we have what is called the sum to infinity of the GP

$$S_\infty = \sum_{i=1}^{\infty} ar^{i-1} = \frac{a}{1 - r}$$

An infinite GP for which $|r| < 1$ is called a **convergent** GP.

*Your majesty, do not panic . . . we have worked out a way. According to the great Zeno of antiquity, it is quite safe for you to jump to the ground. The plan is that you will first fall through half the distance, then half the remainder, and so on, until you are close enough to touch the ground, and then it will all be okay. I think the plan sounds good.*

**Solution of Appetizer Problem:**    First we need to find a formula for the $n$th term of the sequence.

Observe that the first term of the sequence is $0.1 = \frac{1}{10}$;
the second term is $0.11 = \frac{1}{10} + \frac{1}{100}$;
the third term is $0.111 = \frac{1}{10} + \frac{1}{100} + \frac{1}{1000}$.

Hence the $n$th term is

$$\sum_{i=1}^{n}\left(\frac{1}{10}\right)^i = \frac{\frac{1}{10}\left((\frac{1}{10})^n - 1\right)}{-\frac{9}{10}}$$

$$= \frac{1}{9}\left(1 - \left(\frac{1}{10}\right)^n\right).$$

Therefore the sum of the first $n$ terms of the sequence is given by

$$\frac{1}{9}\sum_{i=1}^{n}\left(1 - \left(\frac{1}{10}\right)^i\right) = \frac{n}{9} - \frac{1}{9}\sum_{i=1}^{n}\left(\frac{1}{10}\right)^i$$

$$= \frac{n}{9} - \frac{1}{9}\left[\frac{1}{9}\left(1 - \left(\frac{1}{10}\right)^n\right)\right]$$

$$= \frac{n}{9} - \frac{1}{81}\left(1 - \left(\frac{1}{10}\right)^n\right).$$

Now, from what we did above, the $n$th term is the sum to $n$ terms of a GP with first term $\frac{1}{10}$ and common ratio $\frac{1}{10}$. The infinite decimal expansion represents the sum to infinity of this GP, which exists because $\frac{1}{10} < 1$, and is given by

$$S_\infty = \frac{a}{1-r} = \frac{1}{10}\cdot\frac{1}{1-\frac{1}{10}} = \frac{1}{10}\cdot\frac{10}{9} = \frac{1}{9}.$$

## 6.6   Formulas for sums of squares and cubes

We already have a formula for the sum of the first $n$ natural numbers, because they form an AP. The two new formulas of this section will help you find much more interesting sums, as in the appetizer problem, whose solution will be found near the end of the section.

**Appetizer Problem:**    *Find the sum of the first n-terms of the series:*

$$(1^2) + (1^2 + 2^2) + (1^2 + 2^2 + 3^2) + (1^2 + 2^2 + 3^2 + 4^2) + \cdots$$

Our first task of this section is to find the sum of the first $n$ square numbers, $\sum_{i=1}^{n} i^2$, and the method we will use is commonly known as the *method of differences*.

This hinges upon the observation that, because of successive cancellations:

$$\sum_{i=1}^{n} \left( f(i+1) - f(i) \right) = f(n+1) - f(1). \tag{6.2}$$

You should write out the first few and last few terms of this summation to convince yourself of this result. We observe that:

$$(i+1)^3 - i^3 = 3i^2 + 3i + 1.$$

Now, if we take $f(i+1) = (i+1)^3$, and use (6.2), we obtain

$$f(n+1) - f(1) = 3\sum_{i=1}^{n} i^2 + 3\sum_{i=1}^{n} i + \sum_{i=1}^{n} 1,$$

that is   $(n+1)^3 - 1 = 3\sum_{i=1}^{n} i^2 + \dfrac{3n(n+1)}{2} + n,$

therefore   $3\sum_{i=1}^{n} i^2 = (n+1)^3 - 1 - \dfrac{3n(n+1)}{2} - n$

$$= \dfrac{2(n+1)^3 - 2 - 3n^2 - 3n - 2n}{2}$$

$$= \dfrac{2n^3 + 3n^2 + n}{2}$$

$$= \dfrac{n(2n(n+1) + 1(n+1))}{2}$$

$$= \dfrac{n(2n+1)(n+1)}{2},$$

therefore   $\sum_{i=1}^{n} i^2 = \dfrac{n(2n+1)(n+1)}{6},$

and we have the formula for sum of the squares of the first $n$ natural numbers:

$$\boxed{\; 1^2 + 2^2 + 3^2 + \cdots + n^2 = \dfrac{n(2n+1)(n+1)}{6} \;}$$

For example

$$1^2 + 2^2 + 3^2 = \dfrac{3(2 \cdot 3 + 1)(3+1)}{6} = 14.$$

By a similar argument (take $f(i+1) = (i+1)^4$ and $(i+1)^4 - i^4 = 4i^3 + 6i^2 + 4i + 1$) we can also prove the formula for the sum of the cubes:

$$\sum_{i=1}^{n} i^3 = \left(\frac{n(n+1)}{2}\right)^2$$

You can also easily prove these results using the *Principle of Mathematical Induction* (see Toolchest 2).

Here is the solution of the appetizer problem posed at the beginning of this section.

**Solution of Appetizer Problem:**    Observe that the $n$th term is:

$$\sum_{i=1}^{n} i^2 = 1^2 + 2^2 + 3^2 + \cdots + n^2$$

$$= \frac{n(n+1)(2n+1)}{6}$$

$$= \frac{2n^3 + 3n^2 + n}{6}$$

$$= \frac{n^3}{3} + \frac{n^2}{2} + \frac{n}{6}.$$

Therefore, the sum of the first $n$ terms is:

$$\sum_{i=1}^{n} \left(\frac{i^3}{3} + \frac{i^2}{2} + \frac{i}{6}\right) = \frac{1}{3}\sum_{i=1}^{n} i^3 + \frac{1}{2}\sum_{i=1}^{n} i^2 + \frac{1}{6}\sum_{i=1}^{n} i$$

$$= \frac{1}{3}\frac{n^2(n+1)^2}{4} + \frac{1}{2}\frac{n(n+1)(2n+1)}{6} + \frac{1}{6}\frac{n(n+1)}{2}$$

$$= \frac{n(n+1)[n(n+1) + 2n + 1 + 1]}{12}$$

$$= \frac{n(n+1)(n+1)(n+2)}{12}$$

$$= \frac{n(n+1)^2(n+2)}{12}.$$

Here is the solution to the problem given at the beginning of this Toolchest.

**Solution of Appetizer Problem:**    Let's evaluate the first few terms and see whether we can spot a pattern. Let $S_n$ denote the product of the first $n$ terms, so that

$$S_1 = 1 - \frac{1}{2^2} = \frac{3}{4};$$

$$S_2 = \left(1 - \frac{1}{2^2}\right)\left(1 - \frac{1}{3^2}\right) = \frac{3}{4} \times \frac{8}{9} = \frac{6}{9};$$

$$S_3 = \left(1 - \frac{1}{2^2}\right)\left(1 - \frac{1}{3^2}\right)\left(1 - \frac{1}{4^2}\right) = \frac{6}{9} \times \frac{15}{16} = \frac{10}{16};$$

$$S_4 = \left(1 - \frac{1}{2^2}\right)\left(1 - \frac{1}{3^2}\right)\left(1 - \frac{1}{4^2}\right)\left(1 - \frac{1}{5^2}\right) = \frac{10}{16} \times \frac{24}{25} = \frac{15}{25}.$$

Observe that

$$S_1 = \frac{1+2}{(1+1)^2}$$

$$S_2 = \frac{1+2+3}{(1+2)^2}$$

$$S_3 = \frac{1+2+3+4}{(1+3)^2}$$

$$S_4 = \frac{1+2+3+4+5}{(1+4)^2}.$$

Hence we guess $\quad S_n = \dfrac{1+2+3+4+\cdots+n+(n+1)}{(n+1)^2}$

$$= \frac{(n+1)(n+2)}{2(n+1)^2} \quad \left(\text{since } \sum_{r=1}^{n} r = \frac{n(n+1)}{2}\right)$$

$$= \frac{n+2}{2(n+1)}.$$

Up to this point, we have been using patterns to guess the correct expression for $S_n$. We now use the method of induction to prove that $S_n = \dfrac{n+2}{2(n+1)}$.

Let $P(n)$ denote the assertion:

$$\left(1 - \frac{1}{2^2}\right)\left(1 - \frac{1}{3^2}\right)\left(1 - \frac{1}{4^2}\right)\left(1 - \frac{1}{5^2}\right)\cdots\left(1 - \frac{1}{(n+1)^2}\right) = \frac{n+2}{2(n+1)}.$$

Then LHS of $P(1)$ is $1 - \frac{1}{2^2} = \frac{3}{4} = \frac{1+2}{2(1+1)} = \frac{3}{4}$, which is the RHS of $P(1)$.
Therefore $P(1)$ is true.

Suppose $P(k)$ is true for some $k \in \mathbb{N}$, that is:

$$S_k = \left(1 - \frac{1}{2^2}\right)\left(1 - \frac{1}{3^2}\right)\left(1 - \frac{1}{4^2}\right)\left(1 - \frac{1}{5^2}\right)\cdots\left(1 - \frac{1}{(k+1)^2}\right) = \frac{k+2}{2(k+1)}.$$

Then

$$S_{k+1} = \left(1 - \frac{1}{2^2}\right)\left(1 - \frac{1}{3^2}\right)\left(1 - \frac{1}{4^2}\right)\left(1 - \frac{1}{5^2}\right)\cdots$$
$$\times \left(1 - \frac{1}{(k+1)^2}\right)\left(1 - \frac{1}{(k+2)^2}\right)$$
$$= \left(\frac{k+2}{2(k+1)}\right)\left(1 - \frac{1}{(k+2)^2}\right)$$
$$= \frac{k+2}{2(k+1)}\left(\frac{(k+2)^2 - 1}{(k+2)^2}\right)$$
$$= \frac{(k+2)(k^2 + 4k + 3)}{2(k+1)(k+2)^2}$$
$$= \frac{k^2 + 4k + 3}{2(k+1)(k+2)} = \frac{(k+1)(k+3)}{2(k+1)(k+2)} = \frac{k+3}{2(k+2)} = \frac{(k+1)+2}{2((k+1)+1)}.$$

We have deduced the assertion $P(k+1)$ from $P(k)$. But $P(1)$ is true, hence $P(2)$ is true, $P(3)$ is true, and so on for all $n \in \mathbb{N}$, by the PMI. That is,

$$S_n = \frac{n+2}{2(n+1)} \ \forall\, n \in \mathbb{N}.$$

Therefore $S_{1998} = \frac{1000}{1999}$. Hence (D).

## 6.7  Problems

**Problem 1:**    *What is the value of* $1 - 2 + 3 - 4 + \cdots + 99 - 100$?

(A) 1   (B) $-1$   (C) 50   (D) $-50$   (E) 0

**Problem 2:**    *The sum of the first n terms of a sequence is* $n(n+1)(n+2)$. *The tenth term of the sequence is:*

(A) 114   (B) 330   (C) 396 (D) 600   (E) 1 320

**Problem 3:**    *Starting at 777 and counting backwards by 7's, a student counts* 777, 770, 763, ... *A number counted will be*

(A) 45   (B) 44   (C) 43   (D) 42   (E) 41

**Problem 4:**    *In a sequence of six numbers, the first number is 4 and the last number is 47. Each number after the second equals the sum of the previous two numbers. If S is the sum of the numbers in the sequence, then S lies in the interval*

(A) 51 to 90   (B) 91 to 100   (C) 101 to 110   (D) 111 to 120   (E) 121 to 160

**Problem 5:**   *The Fibonacci sequence: 1, 1, 2, 3, 5, 8, 13, . . . is built using the rules*

(i) $U_{n+1} = U_n + U_{n-1}$
(ii) $U_1 = U_2 = 1$

*where $U_1$ is the first term, $U_2$ the second term, $U_3$ the third term and so on. What value does the ratio $\frac{U_{n+1}}{U_n}$ approach as n gets large?*

(A) $\frac{1-\sqrt{5}}{2}$   (B) $\frac{1+\sqrt{5}}{2}$   (C) $\frac{1+\sqrt{3}}{2}$   (D) $\frac{1-\sqrt{3}}{2}$   (E) $\frac{1+\sqrt{2}}{2}$

**Problem 6:**   *A sequence is defined by the rule $U_{n+1} = \sqrt{3U_n}$ and $U_1 = 1$, where $U_1$ is the first item in the sequence, $U_2$ the second, and so on. Which of the following is the smallest whole number greater than $U_n$ for all values of n?*

(A) 2   (B) 3   (C) 4   (D) 1   (E) 5

**Problem 7:**   *The infinite geometric progression $1 + r + r^2 + r^3 + r^4 + r^5 + \cdots$ has the sum S. What is the sum of the infinite geometric progression*

$$1 + r^2 + r^4 + r^6 + r^8 + r^{10} + \cdots?$$

(A) $S - r$   (B) $\frac{S}{r}$   (C) $\frac{S}{1-r}$   (D) $S - r^2$   (E) $\frac{S}{1-r^2}$

**Problem 8:**   *a, b, c, d and e are natural numbers in an Arithmetic Progression (in that order) with common difference one, $a + b + c + d + e = f^3$ and $b + c + d = g^2$ with $f, g \in \mathbb{N}$. What is the least possible value of c?*

(A) 100   (B) 675   (C) 64   (D) 246   (E) 75

**Problem 9:**   *For a geometric sequence of positive terms to have any term equal to the sum of the next two terms, which of the following numbers must the common ratio be? Does the series consisting of these terms converge?*

(A) $\frac{\sqrt{5}-1}{2}$   (B) $\frac{\sqrt{5}}{2}$   (C) 1   (D) $\frac{1-\sqrt{5}}{2}$   (E) $\frac{2}{\sqrt{5}}$

**Problem 10:**   *The smallest positive integer x for which the sum $x + 2x + 3x + 4x + \cdots + 100x$ is a perfect square, is*

(A) 100   (B) 101   (C) 202   (D) 5 050   (E) None of these

**Problem 11:**   *I write the array of numbers with the rows as numbered below.*

| | | | 1 | | | | Row 1 |
|---|---|---|---|---|---|---|---|
| | | 1 | | 1 | | | Row 2 |
| | 1 | | 2 | | 1 | | Row 3 |
| | 1 | 3 | | 3 | 1 | | Row 4 |
| 1 | | 4 | 6 | 4 | | 1 | Row 5 |
| 1 | 5 | 10 | | 10 | 5 | 1 | Row 6 |

*The sum of all the numbers in the first 50 rows is:*

(A) $2^{49}$   (B) $2^{50} - 1$   (C) $2^{50}$   (D) $2^{49} - 1$   (E) $2^{50} + 1$

**Problem 12:**   *Below are three statements:*

I *The square of a number in any geometric progression is equal to the product of the two numbers immediately preceding and succeeding that number in the same sequence.*

II *Each term after the first term in any arithmetic progression is equal to the average of the terms preceding and succeeding it.*

III *It is possible to have a geometric progression in which each term after the first is the difference: 'preceding term minus succeeding term'.*

*Which of these statements I–III are true?*

(A) I only   (B) I and II   (C) II and III   (D) I and III   (E) all of them

## 6.8  Solutions

**Solution 1:**   We can write the sum as

$$(1 - 2) + (3 - 4) + (5 - 6) + \cdots + (99 - 100)$$
$$= (-1) + (-1) + (-1) + \cdots + (-1)$$
$$= -1 \times 50 = -50. \text{ Hence (D).}$$

**Solution 2:**   Let the sum of the first $n$ terms be $S_n$, so that the $10^{th}$ term $U_{10}$ is given by:

$$U_{10} = S_{10} - S_9$$
$$= 10.11.12 - 9.10.11 = 10.11(12 - 9) = 3.10.11 = 330. \text{ Hence (B).}$$

**Solution 3:**   The common difference is $-7$, and we start with a multiple of 7, hence every multiple of 7 less than 777 will be counted. Here, 42 is the only multiple, making (D) the correct answer. A more formal way of doing this would be to observe that we are looking for the $n$th term $a + (n - 1)d$ among those above, for some $n = 1, 2, 3 \ldots$, where $a = 777, d = -7$. That is, we are looking for integers of form: $777 - 7(n - 1) = 7(112 - n)$. Clearly 42 fits the bill, for $n = 106$, so it's the 106th term, counting backwards from 777.

**Solution 4:**   Let the second number in the sequence be $x$, so that the third term $U_3$ and subsequent terms are given by:

$$U_3 = U_2 + U_1 = x + 4,$$
$$U_4 = U_3 + U_2 = x + 4 + x = 2x + 4,$$

$$U_5 = U_4 + U_3 = 2x + 4 + x + 4 = 3x + 8,$$
$$U_6 = U_5 + U_4 = 3x + 8 + 2x + 4 = 5x + 12.$$

Thus the sequence is   4, $x$, $x + 4$, $2x + 4$, $3x + 8$, $5x + 12 \ldots$
The sum to six terms is:

$$S = (4) + (x) + (x + 4) + \cdots + (5x + 12) = 12x + 32.$$

The sixth term is $U_6 = 47$, so that $5x + 12 = 47$, or $x = 7$. Hence $S = 12x + 32 = 12 \times 7 + 32 = 116$. Since $116 \in (111, 120)$, we have (D) as the correct choice. Note that this sequence is neither a GP nor an AP.

**Solution 5:**   Let this limit be $A$, i.e. $\frac{U_{n+1}}{U_n}$ approaches $A$ as $n$ gets large. Clearly $\frac{U_n}{U_{n-1}}$ also gets closer and closer to $A$ as $n$ gets larger and larger. If, in the equation $U_{n+1} = U_n + U_{n-1}$ we divide throughout by $U_n$, we get:

$$\frac{U_{n+1}}{U_n} = 1 + \frac{U_{n-1}}{U_n}. \tag{6.3}$$

Taking limits as $n$ grows indefinitely large, we have

$$\frac{U_{n+1}}{U_n} \to A \text{ and } \frac{U_n}{U_{n-1}} \to A, \text{ hence } \frac{U_{n-1}}{U_n} \to \frac{1}{A}.$$

Therefore, in (6.3) we have $A = 1 + \frac{1}{A}$ and this allows us to find $A$ from the quadratic equation:

$$A^2 = A + 1$$
$$A^2 - A - 1 = 0$$
$$A = \frac{1 + \sqrt{5}}{2},$$

ignoring the negative root.  Hence (B).

**Remark:**   The number $\frac{1+\sqrt{5}}{2}$ appears in mathematics so frequently that it is known to mathematicians by a special name: the **Golden Ratio**, and is denoted by $\phi$, spelt *phi* and pronounced 'fie'. Here are some cases where the Golden Ratio turns out to be the answer.

I. What number greater than zero is such that the sum of itself and its inverse equals 1?
   Let the number be $x > 0$. Then, $x + \frac{1}{x} = 1$, which is the same quadratic equation as above.
II. Fractions of the form

$$x_1 + \cfrac{x_2}{x_3 + \cfrac{x_4}{x_5 + \cfrac{x_6}{x_7 + \cdots}}}$$

are called **continued fractions**. The value of a given recurring fraction is defined as the limit (if it exists) of the **convergents**, which are the numbers obtained by stopping at successive '+' signs. The continued fraction in which $x_n = 1$ for all $n$ is particularly interesting. To find its value $S > 0$, observe that

$$S = 1 + \cfrac{1}{1 + \cfrac{1}{1 + \cfrac{1}{1 + \cfrac{1}{1 + \cfrac{1}{1 + \cfrac{1}{1 + \cdots}}}}}}$$

satisfies $S = 1 + \frac{1}{S}$ (convince yourself that this is correct!), which is the quadratic equation above for the Golden Ratio.

III. If you have a calculator, it should show you that $\cos 36° \approx \phi/2$ ! In fact this is exactly true: see the solution of Problem 11 in Toolchest 5. How can we explain this? It is connected with the fact that $36°$ is the angle at each tip of the Pythagorean 'pentagram', the regular pentagon star shown in the diagram below, and the ratio $d:s$ of diagonal to side in this figure is the Golden Ratio. Try to demonstrate this for yourself by finding two similar isosceles triangles in there to give you $d/s = s/(d - s) = 1/(d/s - 1)$.

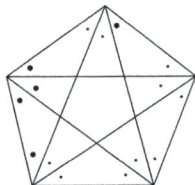

IV. What is the value of the 'recurring square root' below? The value is greater than zero.

$$\sqrt{1 + \sqrt{1 + \sqrt{1 + \sqrt{1 + \sqrt{1 + \cdots}}}}} \ ?$$

If we let $x > 0$ be the value, then

$$x^2 = 1 + x,$$

thus     $x^2 - x - 1 = 0,$

therefore     $x = \dfrac{1 + \sqrt{5}}{2} = \phi,$   the Golden Ratio again.

We could go on and on; by this time you should be able to invent your own situations in which the Golden Ratio turns out to be the answer!

**Solution 6:**    By putting $n = 1$ in $U_{n+1} = \sqrt{3U_n}$ we deduce that $U_2 = \sqrt{3U_1} = \sqrt{3} = 3^{\frac{1}{2}}$, and

$$U_3 = \sqrt{3U_2} = \sqrt{3} \cdot \sqrt{U_2} = 3^{\frac{1}{2}} \cdot \left(3^{\frac{1}{2}}\right)^{\frac{1}{2}} = 3^{\frac{1}{2}+\frac{1}{4}},$$

$$U_4 = \sqrt{3U_3} = 3^{\frac{1}{2}} \cdot \left(3^{\frac{1}{2}+\frac{1}{4}}\right)^{\frac{1}{2}} = 3^{\frac{1}{2}+\frac{1}{4}+\frac{1}{8}}.$$

Aha! – a pattern is now evident:

$$U_n = 3^{(\frac{1}{2})+(\frac{1}{2})^2+(\frac{1}{2})^3+\cdots+(\frac{1}{2})^{n-1}}$$

and we see that

$$\left(\frac{1}{2}\right)+\left(\frac{1}{2}\right)^2+\left(\frac{1}{2}\right)^3+\cdots+\left(\frac{1}{2}\right)^{n-1}$$

is a GP with a sum to infinity given by

$$S_\infty = \frac{a}{1-r} = \frac{(\frac{1}{2})}{1-(\frac{1}{2})} = 1.$$

Hence, as we increase the number of terms in the power $\left(\frac{1}{2}\right)+\left(\frac{1}{2}\right)^2+\left(\frac{1}{2}\right)^3 + \cdots + \left(\frac{1}{2}\right)^{n-1}$, we progressively get closer (but never equal) to 1. Hence, the $n$th term gets closer and closer in value to $3^1 = 3$, without ever arriving at 3. This means that the smallest whole number required is 3, making (B) the correct answer.

**Solution 7:**    What we are referring to as the sum is obviously the sum to infinity, and clearly

$$S = \frac{1}{1-r}.$$

Therefore    $1 + r^2 + r^4 + \cdots = \dfrac{1}{1-r^2}$

$$= \left(\frac{1}{1-r}\right)\left(\frac{1}{1+r}\right)$$

$$= \frac{S}{1+r}. \quad \text{Hence (D)}.$$

For this to be valid, we require that $r \neq -1$ as well as $r \neq 1$.

**Solution 8:**    From the given information we have:

$$c - 2 + c - 1 + c + c + 1 + c + 2 = f^3 \quad \text{and} \quad c - 1 + c + c + 1 = g^2.$$

Hence $5c = f^3$ and $3c = g^2$, i.e. 5 times $c$ is a perfect cube (in which the prime factors must occur in triples) and 3 times $c$ is a perfect square (in which the prime factors must occur in pairs). Therefore the smallest value of $c$ is $5^2 \cdot 3^3 = 675$. Hence (B).

**Solution 9:**    Let the sequence be

$$a, ad, ad^2, ad^3, \ldots, ad^n, ad^{n+1}, ad^{n+2}, \ldots$$

We assume $a \neq 0$, otherwise it would be a trivial case. Taking an arbitrary term $ad^n$, it is required that $ad^n = ad^{n+1} + ad^{n+2}$ which is equivalent to

$$d^n = d^n \cdot d + d^n \cdot d^2 \quad \text{(since } a \neq 0\text{)}$$

$$\text{that is} \quad 1 = d + d^2 \quad \text{(since } d \neq 0\text{)}$$

$$\text{that is} \quad d^2 + d - 1 = 0.$$

Thus, by the quadratic formula, $d = \frac{-1 \pm \sqrt{5}}{2}$ and since the terms are positive, $d > 0$ so that $d = \frac{-1+\sqrt{5}}{2}$. Hence (A). Since $0 < \frac{-1+\sqrt{5}}{2} < 1$, the series $\sum_{n=0}^{\infty} ad^n$ converges to the sum $\frac{a}{1-d} = \frac{2a}{3-\sqrt{5}}$.

**Solution 10:**

$$x + 2x + \cdots + 100x = x(1 + 2 + 3 + \cdots + 100)$$

$$= x\left(\frac{100}{2}\right)(101) \quad \left(\text{using } \sum_{i=1}^{n} i = \frac{n(n+1)}{2}\right)$$

$$= x(50)(101)$$

$$= x(2)(5^2)(101)$$

so that we have a perfect square for $x = 2(101) = 202$ (all the prime factors must be raised to even powers). Hence (C).

**Solution 11:**    The array in the problem should be familiar to you; it is called **Pascal's triangle** or **Halayudha's triangle** (see Toolchest 7 Section 1). The important thing to realize here is that:

$$1 = 2^0, \ 1+1 = 2 = 2^1, \ 1+2+1 = 4 = 2^2, \ 1+3+3+1 = 8 = 2^3, \text{ etc.}$$

That is, the sum of the $r$th row is $2^{r-1}$.

Note that this can be proved easily using the Binomial Theorem (see Toolchest 7):
$(1+x)^{n-1}$ with $x = 1$ gives the sum of the 'Pascal numbers' (or the 'binomial coefficients') chosen from row $n$ as $2^{n-1}$.

So, the sum of the rows up to the 50th row is:

$$2^0 + 2^1 + 2^2 + 2^3 + \cdots + 2^{49}.$$

This is a Geometric Progression with first term $2^0 = 1$, and common ratio 2.

$$\text{So, the sum } S_n = \frac{1(2^{50} - 1)}{2 - 1}$$

$$= 2^{50} - 1. \text{ Hence (B).}$$

**Solution 12:**

I Let the GP be

$$a + ar + ar^2 + \cdots + ar^{k-1} + ar^k + ar^{k+1} + \cdots$$

and let the number in question be $ar^k$. Then

$$(ar^k)^2 = a^2 r^{2k},$$

and the product of the immediate succeeding and preceding numbers is

$$ar^{k-1} \times ar^{k+1} = a^2 r^{k-1+k+1} = a^2 r^{2k}.$$

Since $ar^k$ is arbitrary, the result holds for any such triplet, so that I is true.

II Let the AP be

$$a+(a+d)+(a+2d)+\cdots+(a+(k-1)d)+(a+kd)+(a+(k+1)d)+\cdots$$

and let the term in question be $a + kd$. Then

$$\frac{1}{2}\left((a+(k-1)d)+(a+(k+1)d)\right) = \frac{1}{2}\left(2a+2kd\right)$$
$$= a + kd,$$

and the average of succeeding and preceding numbers is the term itself. Since the $k$ is an arbitrary positive integer, II is true.

III (See Problem 9.) Let the GP be

$$a + ar + ar^2 + \cdots + ar^{k-1} + ar^k + ar^{k+1} + \cdots$$

and let us see what is required of $ar^k$:

$$ar^k = ar^{k-1} - ar^{k+1},$$
$$\text{therefore} \quad r = 1 - r^2, \text{ (provided } a \neq 0)$$
$$\text{and so} \quad r^2 + r - 1 = 0 \quad \text{hence } r = \frac{-1+\sqrt{5}}{2}.$$

Conversely, it is clear that if $r$ takes this value then the required equation will hold, for all terms $ar^k$ such that $k > 0$, and any $a \neq 0$. Thus III is true.

Hence (E) is the correct answer.

# 7  Binomial Theorem

*After millennia of confinement in the bracketed prison on Mount Theoremus, the superhero Binomialus finally came to the rescue of prisoner X plus prisoner Y. When asked to sum-up their experience in prison, the result was a harrowing tale the elders unanimously agreed to call the Binomial Theorem.*

By the end of this topic, you should be able to:

(i) Use Pascal's triangle to expand binomial expressions.
(ii) Discuss some of the mathematical properties of Pascal's triangle.
(iii) Prove and use the Binomial Theorem.

As an appetizer, here is a typical problem of the kind you should be able to solve when you have worked through this Toolchest. You are invited to try it as soon as you wish. You will probably find it hard for now, but by the end of Section 3, where its solution is given, it should not look difficult to you.

**Appetizer Problem:**   *What is the coefficient of $x^{13}$ in $(1 + x^4 + x^5)^{10}$?*

(A) 120   (B) 360   (C) 240   (D) 720   (E) 480

## 7.1   Pascal's triangle

Consider the expansions of the terms below (remember that $a^0 = 1$ for all real values of $a$):

$$
\begin{aligned}
(1+x)^0 &= & & & & & x^0 \\
(1+x)^1 &= & & & & x^0 &+& x^1 \\
(1+x)^2 &= & & & x^0 &+& 2x^1 &+& x^2 \\
(1+x)^3 &= & & x^0 &+& 3x^1 &+& 3x^2 &+& x^3 \\
(1+x)^4 &= & x^0 &+& 4x^1 &+& 6x^2 &+& 4x^3 &+& x^4.
\end{aligned}
$$

If the coefficients only are written down, the triangular array above becomes:

| | | | | | | | | |
|---|---|---|---|---|---|---|---|---|
| row 1: | | | | | 1 | | | |
| row 2: | | | | 1 | | 1 | | |
| row 3: | | | 1 | | 2 | | 1 | |
| row 4: | | 1 | | 3 | | 3 | | 1 |
| row 5: | 1 | | 4 | | 6 | | 4 | | 1 |

For each non-negative integer $n$, the entries in the $n$th row are the coefficients of the powers of $x$ in the expansion of $(1+x)^{n-1}$. What we want is a simple rule that helps us find the entries in the subsequent rows without having to expand $(1 + x)^n$. Observe how to get row 4, assuming we already have rows 1, 2 and 3:

Multiplying $(1+2x+x^2) \times (1+x)$, each of the three terms is multiplied by 1 and by $x$, giving six terms in all. The first and last coefficients (of $x^0$ and $x^3$) will clearly be 1 (and similarly for all rows), while the middle coefficients 3, 3 come from $(1 \times x) + (2x \times 1) = 3x$, and $(2x \times x) + (x^2 \times 1) = 3x^2$. This corresponds to each 3 in the fourth row of the triangle being the sum of the two numbers above it: $3 = 1 + 2$, $3 = 2 + 1$. The same is true of row

3: the 2 comes from summing the two numbers (each 1) above it: $2 = 1 + 1$. We observe the relationships illustrated in the diagram below:

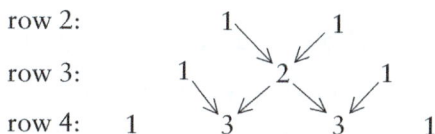

row 2:          1          1

row 3:      1       2       1

row 4:  1       3       3       1

Eureka! – we have discovered a rule for finding the entries of the subsequent rows! These entries will be the coefficients in the expansions of any sum of two numbers $(a+b)^n$ – hence they are called the **binomial coefficients**. We are, of course, treading in the footsteps of others: Blaise Pascal did it over 300 years ago, and an Indian mathematician called Halayudha, much earlier. In their honour, the triangular array of integral binomial coefficients is known to mathematicians as **Pascal's triangle**, or **Halayudha's triangle**.

**Example:**   Use Pascal's triangle to find the expansion of $(x + y)^5$.

**Solution:**   We complete the triangle up to row six to get

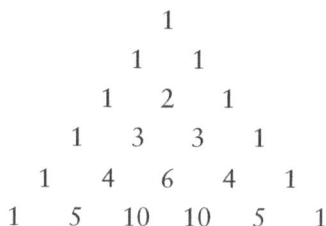

$$
\begin{array}{ccccccccccc}
 & & & & & 1 & & & & & \\
 & & & & 1 & & 1 & & & & \\
 & & & 1 & & 2 & & 1 & & & \\
 & & 1 & & 3 & & 3 & & 1 & & \\
 & 1 & & 4 & & 6 & & 4 & & 1 & \\
1 & & 5 & & 10 & & 10 & & 5 & & 1
\end{array}
$$

We now have $x + y = x\left(1 + \frac{y}{x}\right);\quad (x \neq 0)$
So $(x + y)^5 = x^5 \left(1 + \frac{y}{x}\right)^5$, and

$$
\left(1 + \frac{y}{x}\right)^5 = 1\left(\frac{y}{x}\right)^0 + 5\left(\frac{y}{x}\right)^1 + 10\left(\frac{y}{x}\right)^2 + 10\left(\frac{y}{x}\right)^3 + 5\left(\frac{y}{x}\right)^4 + 1\left(\frac{y}{x}\right)^5
$$

$$
= 1 + \frac{5y}{x} + \frac{10y^2}{x^2} + \frac{10y^3}{x^3} + \frac{5y^4}{x^4} + \frac{y^5}{x^5}.
$$

Hence

$$
(x + y)^5 = x^5 \left(1 + \frac{y}{x}\right)^5
$$

$$
= x^5 \left(1 + \frac{5y}{x} + \frac{10y^2}{x^2} + \frac{10y^3}{x^3} + \frac{5y^4}{x^4} + \frac{y^5}{x^5}\right)
$$

$$
= x^5 + 5x^4 y + 10x^3 y^2 + 10x^2 y^3 + 5xy^4 + y^5.
$$

**Note:**   Now that you have seen how the 'binomial' (two number) expansion works, with the powers of the second number $y$ rising from 0 to 5 while the powers of the first number $x$ are falling from 5 to 0, you need never again go through the lengthy routines of the previous example.

**Example:**   Expand $(3 + 2x)^4$.

**Solution:**

$$(3 + 2x)^4 = 3^4(2x)^0 + 4 \cdot 3^3 \cdot (2x)^1 + 6 \cdot 3^2 \cdot (2x)^2$$
$$+ 4 \cdot 3^1 \cdot (2x)^3 + 3^0 \cdot (2x)^4$$
$$= 81 + 216x + 216x^2 + 96x^3 + 16x^4.$$

*Hey Pascal, it looks like your triangle is going bust ... your love triangle that is!*

**Example:**   The function $f(x) = (1 + ax)^5(1 + bx)^4$, where $a$ and $b$ are positive whole numbers, is such that the coefficient of $x^2$ is 62. What is the value of $a + b$?.

(A) 3   (B) 4   (C) 5   (D) 6   (E) 9

**Solution:**   $(1 + ax)^5 = 1 + 5ax + 10ax^2 + \cdots$ (we need only go up to the term in $x^2$). $(1 + bx)^4 = 1 + 4bx + 6bx^2 + \cdots$, so that, for the full expansion of $f(x)$, the term in $x^2$ is

$$10ax^2 + 6bx^2 + 20abx^2 = x^2(10a + 6b + 20ab).$$

Hence $10a + 6b + 20ab = 62$, or $5a + 3b + 10ab = 31$. We see that $a = 1$ and $b = 2$ is the only solution, so that $a + b = 3$. Hence (A).

Now consider the question of finding the expansion of some high power like $(1 + x)^{20}$. It is not really practicable to write down Pascal's triangle up to the 21st row. So, although our generating principle for Pascal's triangle is quite a significant achievement, we need to find another more efficient method: in fact we can find a formula for the $r$th number in the $n$th row! First we need some shorthand notation.

## 7.2   A formula for the coefficients

By consideration of the number of ways of picking $k$ brackets out of $n$ brackets in the product $(x + y)^n$, we can arrive at a formula for the binomial coefficients, that is, the numbers in Pascal's triangle. We shall leave full explanation of the factorial notation and of combinations to Toolchest 8, and here content ourselves with proving inductively that the formula (wherever it came from) is true.

Let $\binom{n}{k}$ denote the coefficient of $x^k$ in the expansion of $(1 + x)^n$, or the binomial coefficient which appears at the $k$th place in the expansion of the $n$th power, for $0 \leq k \leq n$. There are two different forms of notation in use, and the formula is:

$$\binom{n}{k} = {}^nC_k = \frac{n!}{(n - k)!k!}, \quad 0 \leq n, \ 0 \leq k \leq n.$$

Note that we adopt the convention (it is forced upon us!) that $0! = 1$, and so $\binom{0}{0} = 1$ – it is the 1 at the top vertex of Pascal's triangle, or the coefficient of $x^0$ in the (trivial) expansion of $(1 + x)^0 = 1$.

In order to prove the Binomial Theorem we first establish the result about these coefficients which corresponds to that rule for finding each term in the $(n + 1)$th row of Pascal's triangle as the sum of the two above it in the $n$th row. It is, naturally, called Pascal's rule.

**Lemma (Pascal's rule):** $\binom{n}{k} + \binom{n}{k-1} = \binom{n+1}{k}$, $1 \le k \le n$, $1 \le n$.

**Proof:** We have

$$\frac{1}{k} + \frac{1}{n-k+1} = \frac{n+1}{k(n-k+1)},$$

so that multiplying each item by $\dfrac{n!}{(k-1)!(n-k)!}$ gives

$$\frac{n!}{k(k-1)!(n-k)!} + \frac{n!}{(n-k+1)(k-1)!(n-k)!}$$
$$= \frac{(n+1)!}{k(n-k+1)(k-1)!(n-k)!},$$

hence $\quad \dfrac{n!}{k!(n-k)!} + \dfrac{n!}{(n-k+1)!(k-1)!} = \dfrac{(n+1)!}{k!(n-k+1)!}$,

and so $\quad \binom{n}{k} + \binom{n}{k-1} = \binom{n+1}{k}$,

which is Pascal's rule.                                              ◇

## Theorem 1   (Binomial Theorem)

$$(x+y)^n = \sum_{k=0}^{n} \binom{n}{k} x^k y^{n-k}$$

*where $x, y \in \mathbb{R}$ and $n$ is a natural number, and* $\binom{n}{k} = \dfrac{n!}{(n-k)!k!}$.

**Proof:** We prove this by induction on $n$. For any $n \in \mathbb{N}$, let $\mathcal{P}(n)$ denote the assertion

$$(x+y)^n = \sum_{k=0}^{n} \binom{n}{k} x^k y^{n-k}, \quad \forall\, x, y \in \mathbb{R}.$$

Then

$$\text{LHS of } \mathcal{P}(1) = (x+y)^1 = x+y;$$

$$\text{RHS of } \mathcal{P}(1) = \sum_{k=0}^{1} \binom{1}{k} x^k y^{1-k} = \binom{1}{0} y + \binom{1}{1} x = y + x;$$

so that $\mathcal{P}(1)$ is true. Now assume that $\mathcal{P}(t)$ is true for some $t \in \mathbb{N}$. That is

$$(x+y)^t = \sum_{k=0}^{t} \binom{t}{k} x^k y^{t-k}.$$

Our goal is to deduce $\mathcal{P}(t+1)$, i.e. to show that the following is also true:

$$(x+y)^{t+1} = \sum_{k=0}^{t+1} \binom{t+1}{k} x^k y^{t+1-k}.$$

We now use Pascal's rule to compute the expansion of the $(t+1)$th power of $1+x$. The rule is used at the fifth line below.

$$(x+y)^{t+1} = (x+y)(x+y)^t = (x+y) \sum_{k=0}^{t} \binom{t}{k} x^k y^{t-k}$$

(by the inductive hypothesis)

$$= \sum_{k=0}^{t} \binom{t}{k} x^{k+1} y^{t-k} + \sum_{k=0}^{t} \binom{t}{k} x^k y^{t-k+1}$$

$$= \sum_{k=1}^{t+1} \binom{t}{k-1} x^k y^{t-k+1} + \sum_{k=0}^{t} \binom{t}{k} x^k y^{t-k+1}$$

$$= x^{t+1} + \sum_{k=1}^{t} \binom{t}{k-1} x^k y^{t-k+1} + \sum_{k=1}^{t} \binom{t}{k} x^k y^{t-k+1} + y^{t+1}$$

$$= x^{t+1} + \left( \sum_{k=1}^{t} \left( \binom{t}{k-1} + \binom{t}{k} \right) x^k y^{t-k+1} \right) + y^{t+1}$$

$$= x^{t+1} + \left( \sum_{k=1}^{t} \binom{t+1}{k} x^k y^{t-k+1} \right) + y^{t+1}$$

$$= \sum_{k=0}^{t+1} \binom{t+1}{k} x^k y^{t+1-k}.$$

We have proved that, if $\mathcal{P}(t)$ is true, then $\mathcal{P}(t+1)$ is also true. We have earlier shown that $\mathcal{P}(1)$ is true, so by the PMI $\mathcal{P}(n)$ is true $\forall n \in \mathbb{N}$.    ◇

**Example:**   Find the coefficient of $x^3$ in the expansion of $(2+3x)^{15}$.

**Solution:**

$$(2+3x)^{15} = \sum_{k=0}^{15} \binom{15}{k} (3x)^k 2^{15-k}.$$

For the term in $x^3$ we set $k = 3$, to get the term as $\binom{15}{3} \cdot 2^{12} \cdot 27 x^3$. Hence the coefficient is $\binom{15}{3} \cdot 2^{12} \cdot 27$, and it is better to leave it in this factor form.

**Example:** The constant term in the expansion of $\left(x + \frac{1}{x^3}\right)^{12}$ is

(A) 26   (B) 169   (C) 260   (D) 220   (E) 310

**Solution:**

$$\left(x + \frac{1}{x^3}\right)^{12} = \sum_{k=0}^{12} \binom{12}{k} x^k (x^{-3})^{12-k}$$

$$= \sum_{k=0}^{12} \binom{12}{k} x^{-36+3k+k}$$

$$= \sum_{k=0}^{12} \binom{12}{k} x^{-36+4k}.$$

To get the constant term we set $-36 + 4k = 0$, so that $k = 9$, hence the constant term is $\binom{12}{9} = \frac{12!}{9!3!} = 220$. Hence (D).

**Note:** When should you use the shorthand notation $\binom{n}{k}$ and when should you use the full formula? If we apply the Binomial Theorem directly to $(1 + x)^n$ we get

$$(1+x)^n = \sum_{k=0}^{n} \binom{n}{k} x^k$$

$$= 1 + \binom{n}{1} x + \binom{n}{2} x^2 + \cdots + x^n \tag{7.1}$$

$$= 1 + nx + \frac{n(n-1)}{2!} x^2 + \frac{n(n-1)(n-2)}{3!} x^3 + \cdots + x^n. \tag{7.2}$$

The shorthand form (7.1) is often adequate, especially when dealing with large powers of $(1+x)$ and large number of terms, while the explicit formula (7.2) may be used for the smaller powers, and where we have to find the first few terms only.

## 7.3   Some properties of Pascal's triangle

Consider the sum of the terms in each of the rows. These are summarized in the table below:

| | | | | | | | | | | |
|---|---|---|---|---|---|---|---|---|---|---|
| row 1: | | | | | 1 | | | | | $= 2^0$ |
| row 2: | | | | 1 | $+$ | 1 | | | | $= 2^1$ |
| row 3: | | | 1 | $+$ | 2 | $+$ | 1 | | | $= 2^2$ |
| row 4: | | 1 | $+$ | 3 | $+$ | 3 | $+$ | 1 | | $= 2^3$ |
| row 5: | 1 | $+$ | 4 | $+$ | 6 | $+$ | 4 | $+$ | 1 | $= 2^4$ |

From the table above, we can guess that the sum of the terms in row $n$ is $2^{n-1}$. However, we still have to prove that the sum is really is $2^{n-1}$ and one way of doing this is by using the Binomial Theorem. The entries in row $n$ are given by the coefficients in the expansion of $(1+x)^{n-1}$. Hence, the sum of the terms in row $n$ is given by $(1+x)^{n-1}$ with $x = 1$, which is $2^{n-1}$.

Next, instead of adding the terms in the rows, let us add the terms *along the diagonals* as shown below:

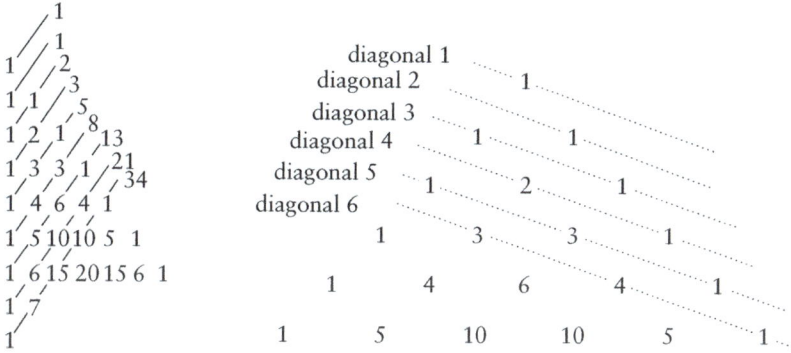

Summarizing the results by use of a table gives:

| Diagonal | Sum | |
|---|---|---|
| 1 | 1 | = 1 |
| 2 | 1 | = 1 |
| 3 | 1 + 1 | = 2 |
| 4 | 2 + 1 | = 3 |
| 5 | 1 + 3 + 1 | = 5 |
| 6 | 3 + 4 + 1 | = 8 |
| ⋮ | ⋮ | ⋮ |

We recognize the sums to be the terms of the *Fibonacci sequence* which is defined as follows: $u_1 = u_2 = 1$ and $u_n = u_{n-1} + u_{n-2}$ for all $n \geq 3$, so that the sequence is $1, 1, 2, 3, 5, 8, 13, \ldots$. (For more on these, see Toolchest 4, Problem 7 and Toolchest 6, Problem 5.) However this just remains a conjecture, unless we can prove it. Observe, from the construction of the diagonals out of the binomial coefficents, that:

If $n$ is odd, the first term (it's 1) in the $n$th diagonal $d_n$ matches the first term in the diagonal $d_{n-2}$; the second term in $d_n$ is the sum of the first unmatched terms of $d_{n-1}$ and $d_{n-2}$; the third term is the sum of the next unmatched terms of the previous two diagonals; continuing like this, we arrive at the last term (1) in $d_n$ which matches the last term in $d_{n-1}$.

If $n$ is even, the first term (it's $n/2$) in the $n$th diagonal $d_n$ is the sum of the first two terms of $d_{n-1}$ and $d_{n-2}$; the second term is the sum of the next two terms of the previous two diagonals; continuing like this, we arrive at the last term (1) in $d_n$ which matches the last term in $d_{n-1}$.

Another pattern: Consider the remainders when the sum of the entries in row $n$ is divided by $n$. As before, we draw our little 'pattern spotting' table:

| Row | Sum | Remainder on division by row number |
|---|---|---|
| 1 | 1 | 1 |
| 2 | 2 | 0 |
| 3 | 4 | 1 |
| 4 | 8 | 0 |
| 5 | 16 | 1 |
| ⋮ | ⋮ | ⋮ |

It seems that the remainder is 1 if the row number is odd, and zero if it is even. How can we explain this? After studying number theory in Toolchest 4, you should be able to explain this variation by considering the residue class of $2^{n-1} \pmod{n}$ when $n$ is even and when it is odd. Many other patterns in this triangle can be identified; try your hand at it. It is fascinating to colour in the even terms (divisible by 2), and admire the pattern – which will only become apparent if you have enough rows! Next, try colouring in all the terms which are divisible by 3.

**Solution of Appetizer Problem:**  This is actually a *trinomial* expression, but we can deal with it using the Binomial Theorem twice over, and obtain the expression for the general term.

Applying the Binomial Theorem to $\left(1 + (x^4 + x^5)\right)^{10}$, the general term in the expansion has the form:

$$\binom{10}{m}(1)^{10-m}(x^4 + x^5)^m = \binom{10}{m}(x^4 + x^5)^m, \quad 0 \le m \le 10.$$

Applying the Binomial Theorem again to $(x^4 + x^5)^m$, the general term becomes

$$\binom{10}{m}\binom{m}{s}(x^4)^s(x^5)^{m-s}, \quad 0 \le s \le m \le 10.$$

We need $4s + 5(m - s) = 5m - s = 13$ which is only possible for $m = 3$ and $s = 2$.

Thus we will have

$$\binom{10}{3}\binom{3}{2}(x^4)^2(x^5)^1 = \frac{10!3!}{3!7!2!1!}x^{13} = 360x^{13} \text{ Hence (B).}$$

## 7.4  Problems

**Problem 1:**  *A jury of 12 people has to decide whether or not a defendant is guilty. An absolute majority (i.e. seven or more) is needed. It is known* *that four will say YES and three will say NO. Of the rest four will each toss a coin. The remaining one will vote with majority. What is the probability that the defendant is found guilty?*

*Now, ladies and gentlemen, do you find the defendant guilty or not guilty?*

(A) $\frac{11}{16}$   (B) $\frac{1}{4}$   (C) $\frac{7}{12}$   (D) $\frac{6}{11}$   (E) $\frac{1}{8}$

**Problem 2:**   *What are the first two digits in the number $11^{11}$?*

(A) 11   (B) 18   (C) 26   (D) 28   (E) 25.

**Problem 3:**   *What is the coefficient of $x^3$ in $(1 + x + x^2 + x^3)^{12}$?*

(A) 54   (B) 120   (C) 364   (D) 684   (E) 340

**Problem 4:**   *If $x^2 + \dfrac{1}{x^2} = 14$ and $x > 0$, what is the value of $x^5 + \dfrac{1}{x^5}$?*

(A) 70   (B) 700   (C) 125   (D) 724   (E) 684

**Problem 5:**   *Sum to infinity the series:*

$$1 - \left(\frac{1}{6}\right) + \frac{1 \cdot 3}{2!}\left(\frac{1}{6}\right)^2 - \frac{1 \cdot 3 \cdot 5}{3!}\left(\frac{1}{6}\right)^3 + \frac{1 \cdot 3 \cdot 5 \cdot 7}{4!}\left(\frac{1}{6}\right)^4 - \cdots$$

*To start you off on this one, take the leap of faith that the Binomial Theorem holds not only for positive integral exponent n, but even for such a number as $n = 1/2$.*

## 7.5  Solutions

**Solution 1:**  Of the four tossers we need at least two to say YES for the defendant to be found guilty. For, if two tossers say YES, they will join the four that are known to say YES to form a total of six members of the jury who will say YES, against five saying NO. Those saying YES being in the majority, the remaining member will join them to make a total of seven – a majority vote.

Of course, if more than two tossers say yes, then the decision is GUILTY too.

We have a total of $2^4$ possible outcomes when four coins are tossed (or, equivalently, when a single coin is tossed four times). Observe further, from Pascal's triangle, that there is $^4C_4 = 1$ way in which four tossers say YES, there are $^4C_3 = 4$ ways in which exactly three tossers say YES, and $^4C_2 = 6$ ways in which exactly two tossers say YES. Hence the probability that the dependent is found guilty is

$$\frac{1+4+6}{2^4} = \frac{11}{16}. \quad \text{Hence (A)}.$$

**Solution 2:**  The insight is to express the power in a form allowing us to use the Binomial Theorem:

$$11^{11} = (1+10)^{11}$$

$$= 1 + \binom{11}{1}10 + \binom{11}{2}10^2 + \cdots + \binom{11}{9}10^9 + \binom{11}{10}10^{10}$$

$$+ \binom{11}{11}10^{11}$$

$$= 10^{11} + 11 \cdot 10^{10} + 55 \cdot 10^9 + 165 \cdot 10^8 + 330 \cdot 10^7$$

$$+ 462 \cdot 10^6 + 462 \cdot 10^5 + \cdots$$

This should be enough to give us the first two digits, as we will see more clearly in a short while.

Factoring out $10^{11}$ gives

$$11^{11} = 10^{11}(1 + 1.1 + 0.55 + 0.165 + 0.033 + 0.00462$$

$$+ 0.000462 + \cdots)$$

$$= 10^{11}(2.86\ldots)$$

From this we see that the first two digits are 28. Hence (D).

There is an alternative method which relies on a quick way of finding powers $a^n$ of a number $a$: find the first few terms in the series $a^2, a^4, \ldots$ (found by squaring each preceding term) and then using the fact that any number can be expressed as a sum of powers of 2.

For example, in this case, $11^2 = 121, 11^4 = 14\,641, 11^8 = 2143\ldots$ (we do not need to find the entire number since we are only looking for the first two digits of the final result). Then $11^{11} = 11 \cdot 11^2 \cdot 11^8 = 1331 \cdot (2143\cdots) = 28\ldots$

**Solution 3:**   This is a *quadrinomial* expression, but we can deal with it using the Binomial Theorem three times over, and obtain the expression for the general term, as in the Appetizer Problem.

Applying the Binomial Theorem first to $\left(1 + (x + x^2 + x^3)\right)^{12}$, the general term in the expansion has the form:

$$\binom{12}{m}(1)^{12-m}(x + x^2 + x^3)^m = \binom{12}{m}\left(x + (x^2 + x^3)\right)^m, \quad 0 \le m \le 12,$$

$$= \binom{12}{m}\binom{m}{s}x^{m-s}(x^2 + x^3)^s, \quad 0 \le s \le m \le 12,$$

$$= \binom{12}{m}\binom{m}{s}\binom{s}{t}x^{m-s}\left(x^2\right)^{s-t}\left(x^3\right)^t, \quad 0 \le t \le s \le m \le 12.$$

We need $m - s + 2(s - t) + 3t = m + s + t = 3$ which holds for only three cases: $(m, s, t) = (1, 1, 1), (2, 1, 0), (3, 0, 0)$. Therefore the coefficient of $x^3$ is

$$\binom{12}{1}\binom{1}{1}\binom{1}{1} + \binom{12}{2}\binom{2}{1}\binom{1}{0} + \binom{12}{3}\binom{3}{0}\binom{0}{0}$$

$$= \frac{12!}{1!1!}\cdot 1 \cdot 1 + \frac{12!}{2!10!}\cdot 2 \cdot 1 + \frac{12!}{3!9!}\cdot 1 \cdot 1$$

$$= 12 + 132 + 220 = 364. \quad \text{Hence (C)}.$$

**Solution 4:**

$$\left(x + \frac{1}{x}\right)^2 = x^2 + \frac{1}{x^2} + 2 = 14 + 2 = 16,$$

therefore   $x + \dfrac{1}{x} = 4$,   because $x > 0$.

We also note, for use below, that $x^2 + \frac{1}{x^2} = \left(x + \frac{1}{x}\right)^2 - 2$. Next, consider the various higher powers of $x + \frac{1}{x}$:

$$\left(x + \frac{1}{x}\right)^3 = \left(x + \frac{1}{x}\right)^2\left(x + \frac{1}{x}\right) = \left(x^2 + \frac{1}{x^2} + 2\right)\left(x + \frac{1}{x}\right)$$

$$= x^3 + \frac{1}{x^3} + 3\left(x + \frac{1}{x}\right),$$

therefore   $x^3 + \dfrac{1}{x^3} = \left(x + \dfrac{1}{x}\right)^3 - 3\left(x + \dfrac{1}{x}\right).$

Also   $\left(x + \dfrac{1}{x}\right)^4 = \left(x + \dfrac{1}{x}\right)^3 \left(x + \dfrac{1}{x}\right) = x^4 + \dfrac{1}{x^4} + 4\left(x^2 + \dfrac{1}{x^2}\right) + 6,$

therefore   $x^4 + \dfrac{1}{x^4} = \left(x + \dfrac{1}{x}\right)^4 - 4\left(x^2 + \dfrac{1}{x^2}\right) - 6$

$$= \left(x + \dfrac{1}{x}\right)^4 - 4\left(\left(x + \dfrac{1}{x}\right) - 2\right) - 6$$

$$= \left(x + \dfrac{1}{x}\right)^4 - 4\left(x + \dfrac{1}{x}\right) + 2.$$

Also,   $\left(x + \dfrac{1}{x}\right)^5 = \left(x + \dfrac{1}{x}\right)^4 \left(x + \dfrac{1}{x}\right) = x\left(x^4 + \dfrac{1}{x^4} + 4x^2 + \dfrac{4}{x^2} + 6\right)$

$$+ \dfrac{1}{x}\left(x^4 + \dfrac{1}{x^4} + 4x^2 + \dfrac{4}{x^2} + 6\right),$$

hence   $x^5 + \dfrac{1}{x^5} = \left(x + \dfrac{1}{x}\right)^5 - 5\left(x^3 + \dfrac{1}{x^3}\right) - 10\left(x + \dfrac{1}{x}\right)$

$$= \left(x + \dfrac{1}{x}\right)^5 - 5\left(\left(x + \dfrac{1}{x}\right)^3\right)$$

$$-3\left(x + \dfrac{1}{x}\right)\right) - 10\left(x + \dfrac{1}{x}\right)$$

$$= \left(x + \dfrac{1}{x}\right)^5 - 5\left(x + \dfrac{1}{x}\right)^3 + 5\left(x + \dfrac{1}{x}\right).$$

So, using $x + \dfrac{1}{x} = 4$, we have

$$x^5 + \dfrac{1}{x^5} = 4^5 - 5 \cdot 4^3 + 5 \cdot 4 = 724.$$

Hence (D). Now, you should notice that in all of that, which is instructive in its own right, we did not use the Binomial Theorem. Would its use result in a quicker solution? Try it, applied to both $\left(x + \dfrac{1}{x}\right)^5$ and $\left(x + \dfrac{1}{x}\right)^3$.

**Solution 5:**   The Binomial Theorem has a wonderful generalization, usually encountered by students as an application of calculus. The formula we have used in this toolchest to expand $(1 + x)^n$, for positive integral exponent $n$, is actually just the particular (terminating) case of a theorem which allows this expression to be expanded as an infinite series, for any real exponent $n$, with the restriction $|x| < 1$ to ensure convergence. The theorem was discovered by Isaac Newton as a young man in the 1660s, but only proved rigorously in the nineteenth century:

$$(1+x)^n = 1+nx+\frac{n(n-1)}{2!}x^2+\frac{n(n-1)(n-2)}{3!}x^3+\cdots, \quad n \in \mathbb{R}, \ |x| < 1.$$

Our series can be written as

$$1+\frac{\left(-\frac{1}{2}\right)}{1!}\left(\frac{1}{3}\right)+\frac{\left(-\frac{1}{2}\right)\left(-\frac{1}{2}-1\right)}{2!}\left(\frac{1}{3}\right)^2$$

$$+\frac{\left(-\frac{1}{2}\right)\left(-\frac{1}{2}-1\right)\left(-\frac{1}{2}-2\right)}{3!}\left(\frac{1}{3}\right)^3$$

$$+\frac{\left(-\frac{1}{2}\right)\left(-\frac{1}{2}-1\right)\left(-\frac{1}{2}-2\right)\left(-\frac{1}{2}-3\right)}{4!}\left(\frac{1}{3}\right)^4+\cdots$$

$$=\left(1+\frac{1}{3}\right)^{-\frac{1}{2}}=\left(\frac{4}{3}\right)^{-\frac{1}{2}}=\left(\frac{3}{4}\right)^{\frac{1}{2}}=\frac{\sqrt{3}}{2}.$$

# 8 Combinatorics (counting techniques)

By the end of this topic you should be able to:

(i) Describe and apply the fundamental principle of enumeration (FPE).
(ii) Use factorial notation.
(iii) Define the terms 'partition' and 'permutation'.

(iv) Use the general partitioning formula.
 (v) Use the general permutation formula.
(vi) Solve problems involving circular permutations.
(vii) Define the term 'combination'.
(viii) Use the general combinatorics formula.
(ix) Understand the idea of derangements.
(xi) Use the exclusion–inclusion principle.
(xii) Use the pigeon-hole principle (PHP).

Here is a typical problem of the kind you should be able to solve when you have worked through this Toolchest. You are invited to try it as soon as you wish. You will probably find it hard for now, but by the end of Section 3, where its solution is given, it should not look difficult to you.

**Appetizer Problem:**   *Distinct 3-digit numbers are formed using only the digits 1, 2, 3, and 4, with each digit used at most once in each number formed. What is the sum of all possible numbers so formed?*
(A)  6000    (B)  6600    (C)  6660    (D)  6666    (E)  11234

## 8.1   The fundamental principle of enumeration

Suppose a traveller is to move from town A to C via town B using the road network shown below:

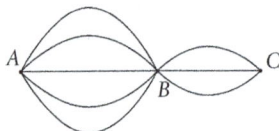

Consider the following question: 'In how many different ways can a traveller move from A to C via B?' To answer this question, we could label the roads, and draw a 'possibility table', as shown in the diagram.

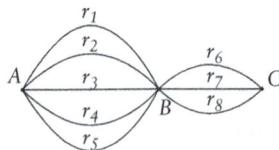

Road used from A to B

| Road used from B to C | | $r_1$ | $r_2$ | $r_3$ | $r_4$ | $r_5$ |
|---|---|---|---|---|---|---|
| | $r_6$ | $(r_1, r_6)$ | $(r_2, r_6)$ | $(r_3, r_6)$ | $(r_4, r_6)$ | $(r_5, r_6)$ |
| | $r_7$ | $(r_1, r_7)$ | $(r_2, r_7)$ | $(r_3, r_7)$ | $(r_4, r_7)$ | $(r_5, r_7)$ |
| | $r_8$ | $(r_1, r_8)$ | $(r_2, r_8)$ | $(r_3, r_8)$ | $(r_4, r_8)$ | $(r_5, r_8)$ |

We see that the table has $5 \times 3 = 15$ distinct entries. Since each entry represents a unique choice of roads from A to C via B, this is also the number of ways of moving from A to C via B, so the answer we have been looking for is 15.

It may not always be practical to draw such 'possibility tables', and, indeed, it is not necessary. There is a convenient method of counting all the possibilities without listing them. This uses a very important principle which you might describe aptly as 'The Golden Principle of Combinatorics':

### The fundamental principle of enumeration (FPE)

If there are $a_1$ ways of doing a job $A_1$, $a_2$ ways of doing a second job $A_2$ *after $A_1$ has been done*, $a_3$ ways of doing a third job $A_3$ *after $A_1$ and $A_2$ have been done*, and, in general, $a_n$ ways of doing the *n*th job $A_n$ *after $A_1, A_2, \ldots, A_{n-1}$ have been done*, then there are $a_1 \times a_2 \times a_3 \times \cdots \times a_n$ ways of doing all the jobs $A_1, A_2, \ldots, A_n$. In the problem above, the first 'job' is moving from A to B, and this can be done in five ways. The second 'job' is moving from B to C, and this can be done in three ways, so that by using the FPE we have a total of $5 \times 3$ ways of moving from A to B then to C (i.e. doing the two jobs together).

**Example:**   I have a rectangular strip of paper divided into three boxes as shown below:

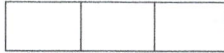

Using each of the red, blue and green crayons to colour the boxes, how many different colour patterns can I make if each crayon is used exactly once?

**Solution:**   The first (say left-hand side) box can be coloured in three ways (we can use any one of the red, blue, or green crayons). Having coloured the first box, we are left with two ways of colouring the second (say middle) box, after which we have just one way of colouring the last box. So, using the FPE, we have a total of $3 \times 2 \times 1 = 6$   ways of colouring the rectangular strip, and hence six distinct patterns.

Now, before we complacently go on to the next example, there is an important issue to raise about this solution. The problem could be accused of being ambiguous. Supposing that the strip is drawn on the page of this book as you see it, the left-hand box is clearly distinguished from the right, and there are indeed six different distinguishable colourings. But supposing the strip to be cut out, there would be no way of distinguishing left from right, and we would have three pairs of colourings which are essentially identical. The middle colour would be the distinguishing mark, and the answer to the problem as it is written above, should actually be *three* distinct patterns.

**Example:**   How many 3-digit even numbers can be made using the digits 2, 3, 4 and 5 without repeats (i.e. without using any digit more than once)?

**Solution:**   The number of such 3-digit numbers is the same as the number of ways of filling in the boxes below with the given digits but, of course, taking care to do this in such a way that we create even numbers only, that is, we make the last digit even.

| | | |
|---|---|---|

We see that there are two ways of choosing the last digit (it is either 2 or 4). Having done that, we are left with three digits from which we select one to go in the first box, after which we are left with two digits from which we select one to go in the second (middle) box.

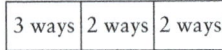

| 3 ways | 2 ways | 2 ways |
|---|---|---|

Hence we have, by the FPE, a total of $3 \times 2 \times 2 = 12$ ways of filling up all the three boxes – that is, 12 distinct 3-digit even numbers we can construct from the given set.

**Remark:** These two examples warn us that the 'first job' in applying the FPE may not necessarily refer to the 'first box' (from the left, say) – it may be better to start with the middle box or the right-hand box.

## 8.2 Factorial arithmetic

We denote by $n!$ ('factorial $n$') the product of the first $n$ consecutive natural numbers. For example:

$$4! = 1 \times 2 \times 3 \times 4 = 4 \times 3 \times 2 \times 1 = 24$$
$$n! = 1 \times 2 \times 3 \times \cdots \times n = n \times (n-1) \times (n-2) \times \cdots \times 3 \times 2 \times 1.$$

Also, by convention or definition, $0! = 1$.

*Okay sir, it looks like we have been out–numbered*
*four factorials to two – requesting exponential back–up!*

**Example:** Find alternative expressions for:

(a) $10! + 11!$   (b) $\dfrac{6! + 7!}{2!3!4!}$   (c) $n!(n^2 + 3n + 2)$.

(d) $\dfrac{(2n)! - (2n - 1)!}{2n! - (n - 1)!}$   (e) $\dfrac{(2n)!}{1 \times 3 \times 5 \times \ldots \times (2n - 1)}$.

**Solution:**

(a)
$$10! + 11! = 10! + (11 \times 10!)$$
$$= 10!(1 + 11) = 12(10!) = \frac{1}{11}12!.$$

(b)
$$\frac{6! + 7!}{2!3!4!} = \frac{6!(1 + 7)}{2 \cdot 2 \cdot 3 \cdot 4!}$$
$$= \frac{8}{12} \cdot \frac{6!}{4!} = \frac{2}{3} \cdot 5 \cdot 6 = 20.$$

(c)
$$n!(n^2 + 3n + 2) = n!(n + 2)(n + 1)$$
$$= (n + 2)(n + 1)(n!) = (n + 2)!.$$

(d)
$$\frac{(2n)! - (2n - 1)!}{2n! - (n - 1)!} = \frac{(2n - 1)!(2n - 1)}{(n - 1)!(2n - 1)}$$
$$= \frac{(2n - 1)!}{(n - 1)!}$$
$$= n(n + 1)(n + 2) \cdots (2n - 1).$$

(e)
$$(2n)! = 2n \times (2n - 1) \times \cdots \times 3 \times 2 \times 1$$
$$= ((2n - 1) \times (2n - 3) \times \cdots \times 5 \times 3 \times 1)$$
$$\times ((2n) \times (2n - 2) \times \cdots \times 4 \times 2 \times 1).$$

Factoring out 2 for each of the $n$ even terms in the second bracket, we have:

$$(2n)! = (1 \times 3 \times 5 \times \cdots \times (2n - 1)) \times 2^n \times (1 \times 2 \times 3 \times \cdots \times n)$$
$$= (1 \times 3 \times 5 \times \cdots \times (2n - 1)) \times 2^n \times n!$$
$$\frac{(2n)!}{1 \times 3 \times 5 \times \cdots \times (2n - 1)} = 2^n(n!).$$

## 8.3   Partitions and permutations of a set

### 8.3.1   Definition of terms

A **partition** is a division of a set into any finite number of non-empty and non-overlapping subsets. We can visualize the process for a set $S$ and subsets $A_1, A_2, \ldots, A_n$ as shown below. The image of a broken pane of glass helps to convey the idea that the sets (pieces of glass) are disjoint – not overlapping:

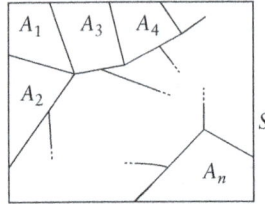

We write $S = A_1 \cup A_2 \cup \cdots \cup A_n$ and $A_i \cap A_j = \emptyset \; \forall \, i \neq j$. We say that the set $S$ is the **disjoint union** of the subsets $A_i$.

The problem of enumerating all possible partitions of a set is quite difficult. We shall give an answer in the following section.

A (linear) **permutation** of a set of objects is simply a linear ordering of them, or a seating of them in a row of boxes or chairs of the same total number $n$; mathematicians say: a one-to-one mapping of the set onto the set of positive integers $\{1, 2, 3, \ldots, n\}$. There is another type of permutation, known as **circular permutation**, which we will consider in a subsequent section. We shall concentrate first on linear permutations.

### 8.3.2   The general partitioning formula

Consider the question of the number of linear permutations of $n$ distinct objects (e.g. arrangements of $n$ different people on a bench in a park). The first position (whatever we choose that to be) on the bench can be filled in $n$ ways, after which the second is filled in $n - 1$ ways, after which $\ldots$, and so on up to the last position, which can be filled in only one way. Hence, by the FPE :

> **There are $n!$ linear permutations of n distinct objects**

Now consider this typical partitioning problem:

The Brilliant Duck in Bertrand Carroll's book 'Mathland' was indeed brilliant. She devised a quick way of counting her 14 ducklings which was basically a doubling process. Starting with two ducklings in the first row, each subsequent row contained twice as many ducklings as in the previous one. After arranging her ducklings in this manner, she would count the rows to get three of them.

In how many distinct ways could the Brilliant Duck partition her 14 ducklings into three groups described above? Let the symbol $\binom{14}{8,4,2}$ denote the number of ways of partitioning a set of 14 objects into subsets of 8, 4 and 2. The notation is suggested by the binomial coefficient $\binom{n}{k}$, which is the number of ways of choosing $k$ things from $n$, that is, the number of ways of partitioning $n$ things into subsets of $k$, $n-k$ things. We now devise a scheme comprising a finite number of stages ('jobs') which has, as outcome, a unique partition of the type that is counted by $\binom{14}{8,4,2}$.

First, convince yourself that the Brilliant Duck could order her 14 ducklings by successively carrying out these 'jobs':

Job (i) – assigning the ducklings to their rows.
Job (ii) – ordering the ducklings of the first row.
Job (iii) – ordering the ducklings of the second row.
Job (iv) – ordering the ducklings of the third row.

We wish to enumerate (i). Let $x$ be the number of ways in which job (i) can be done. We know from the previous displayed result that (ii), (iii) and (iv) can be done in 2!, 4! and 8! ways, respectively. Hence the FPE gives us a total of $x \cdot 8!4!2!$ possible ways of doing the four jobs.

Since this four-stage arranging of ducklings is equivalent to ordering them linearly (putting the rows end to end), and since the FPE tells us that we have a total of 14! possible linear permutations of the 14 ducklings, we have

$$14! = x \cdot 8!4!2! \quad \text{so that} \quad x = \binom{14}{8,4,2} = \frac{14!}{8!4!2!}.$$

Now, to generalize:

Consider the question of partitioning a set of $n$ elements into $p$ subsets, the first subset $A_1$ containing $a_1$ elements, the second subset $A_2$ containing $a_2$ elements, ..., the $p$th subset $A_p$ containing $a_p$ elements, with $a_1 + a_2 + \cdots + a_p = n$. Then an argument akin to that presented above gives the number of distinct partitions possible as:

$$\binom{n}{a_1, a_2, \dots, a_p} = \frac{n!}{a_1! a_2! \dots a_p!}$$

**Example:**   In DNA replication, one 'daughter strand' is synthesized continuously while the other is synthesized as fragments called *Okazaki* fragments, which are later joined by the appropriate enzymes to form a continuous molecule. (You don't need to understand this bit of biology to work out the question following – we just want to tell you where the strange name comes from!) How many distinct words can we make out of the letters of the word 'OKAZAKI'?

**Solution:**   Mark out seven spaces in a row for the seven letters of the word. Now the word 'OKAZAKI' has five distinct types of letters: one each of O, Z and I, and two K s, and two A s. Think of the seven letters as numbered thus: O – 1, K – 2, ... , I – 7. Now each distinct rearrangement, say ZIAAKKO, corresponds to four different orderings of the numbers: 47|35|26|1, 47|35|62|1, 47|53|26|1, and 47|53|62|1. This is because there are 2! = 2 ways of ordering the two A s and 2! = 2 ways of ordering the two K s. But we know there are 7! different orderings of the numbers. Hence there are 7!/4 distinct words.

After some thought, this problem will be seen to be equivalent to that of partitioning seven pigeon-holes into five subsets, two with two pigeon-holes, and three with one pigeon-hole each. Hence the required number is:

$$\binom{7}{2, 2, 1, 1, 1} = \frac{7!}{2!2!1!1!1!} = 1260.$$

### 8.3.3   The general permutation formula

If, from a set of $n$ objects, we select $r$ and order them, then we have what are known as *permutations of $n$ objects taken $r$ at a time*. The number of such permutations for a given $n$ and $r$ is denoted by $^nP_r$. We seek a formula for $^nP_r$.

### Derivation of the formula

Consider the possible arrangements on a bench of $n$ people taken $r$ at a time. Mark out $r$ distinct boxes on a bench (linear!) so that each box defines space large enough for someone to sit on. The first position can be occupied in $n$ ways, and (after it is occupied) the second in $(n-1)$ ways, then the third in $(n-2)$ ways, ..., and the last in $(n-r+1)$ ways. Hence by the FPE we have:

$$
\begin{aligned}
{}^nP_r &= n \times (n-1) \times \cdots \times (n-r+1) \\
&= \frac{n \times (n-1) \times \cdots \times (n-r+1) \times (n-r) \times \cdots \times 3 \times 2 \times 1}{(n-r) \times (n-r-1) \times \cdots \times 3 \times 2 \times 1} \\
&= \frac{n!}{(n-r)!}.
\end{aligned}
$$

That is, the number of permutations of $n$ things taken $r$ at a time is:

$$
\boxed{{}^nP_r = \frac{n!}{(n-r)!}}
$$

One could also observe that

$$
\begin{aligned}
{}^nP_r &= \binom{n}{1,1,1,\ldots,1,n-r} \\
&= \frac{n!}{1!1!\ldots 1!(n-r)!} = \frac{n!}{(n-r)!}.
\end{aligned}
$$

### 8.3.4  Circular permutations

King Arthur of the Round Table plans to hold a meeting with five of his top knights. How many seating arrangements (distinct) are there?

To illustrate the wide variety of situations that are essentially the same (*isomorphic*) problems, we may think also of a water-hole in the African bush, and the possible drinking positions of a male lion and five females. Now, consider a possible seating (or drinking) arrangement. If everyone (including the king!) stands up and, instead of sitting on the chair they had been originally sitting on, sit down on the chair to the immediate left, then, although each person's position has changed, the *two people he is sitting next to on either side have not changed*. In this case we do not regard these two seating arrangements as distinct, i.e. we consider two arrangements $A_1$ and $A_2$ to be distinct if and only if there is at least one person in $A_1$ for which the two people flanking him are not the same as those flanking him in $A_2$. It is clear that equivalent arrangements are *rotationally equivalent*, so we can fix one person, say King Arthur himself. By the FPE, the remaining five knights can take up the remaining five positions in $5 \times 4 \times 3 \times 2 \times 1 = 120$ ways. So in general, the number of ways in which $n$ people can sit round a circular table (or $n$ lions can drink around a water-hole) is $(n-1)!$.

> There are (n − 1)! circular permutations of n distinct objects

**Example:** A traditional Zimbabwean necklace, or *chuma*, has seven differently coloured beads threaded to a string made from the mane of a goat. How many different necklaces can one buy? How many different colour patterns can one wear by re-threading the beads?

**Solution:**   The number of ways of arranging $n$ objects round a circular ring is $(n-1)!$. Therefore the number of different colour patterns one can wear is $(n-1)!$, for $n \geq 1$. This is also the number of different necklaces for $n = 1, 2$. But for $n \geq 3$, because one may put one's head through the same necklace in two different directions (we say that there are two distinct *orientations* of the circular arrangement), the total number of different necklaces one can buy is $\frac{1}{2}(n-1)!$. Check these claims when $n = 1$, $n = 2$, $n = 3, n = 4$.

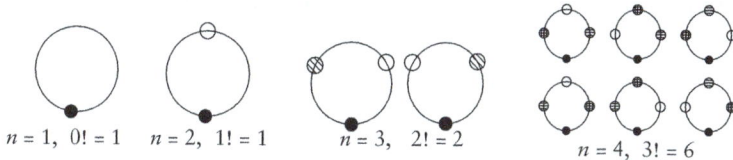

$n = 1, \ 0! = 1$    $n = 2, \ 1! = 1$      $n = 3, \ 2! = 2$

$n = 4, \ 3! = 6$

Therefore, with $n = 7$, we can buy $\frac{1}{2} \cdot 6! = 360$ different necklaces, and can wear 720 different colour patterns.

**Example:**   Two pairs of lovers are sitting on a bench in a park. No member of any pair is prepared to be separated from the other member. In how many ways can the lovers be arranged on a bench? And in how many ways around one of those circular seats encircling the trunk of a tree?

**Solution:**   Deal with the linear bench first. Tie up the two members of each 'love-pair' so that you end up with two items that can be arranged in 2! ways. Also, within each of the 'tied up pairs', the members can be arranged in 2! ways so that by the FPE we have a total of $2! \times 2! \times 2! = 8$ ways of arranging the lovers on the bench.

Now, for the circular arrangement, the two love-pairs can be arranged in $1! = 1$ way, and each pair can be arranged in 2! ways, so the total number of ways is $1! \times 2! \times 2! = 4$.

To check whether you have fully understood, do the same problem for three couples, and verify that there are 48 linear arrangements and 16 circular ones.

**Example:**   How many distinct even numbers can you form with the digits 0, 1, 2, 3, (a) if repeats are allowed, (b) if repeats are not allowed?

**Solution:**   We note that, in both cases, the numbers can be of 1, 2, 3, or 4 digits; that zero cannot be the first digit from the left (unless the number is zero); and that for evenness the units digit must be 0 or 2.

(a) There are 2 one-digit numbers, $3 \times 2 = 6$ two-digit numbers, $3 \times 4 \times 2 = 24$ three-digit numbers, $3 \times 4 \times 4 \times 2 = 96$ four-digit numbers, giving 128 numbers altogether.

(b) We have:

*One-digit even numbers*: There are two, 0 and 2.

*Two-digit even numbers:* If the units digit is 0 there are three, and if the units digit is 2 there are two numbers, giving five.

*Three-digit even numbers:* For each of the first three above, we can extend it in two ways, giving six. For each of the other two numbers there is only one extension. And then there are two more numbers with the tens digit zero. Altogether ten.

*Four-digit even numbers:* For each of the first six above, we can extend it in only one way. The next two have no extension. For each of the last two there is one extension. And then there are two numbers with the hundreds digit zero. Altogether ten again.

The total number of even numbers we can make is 27.

Is there a more elegant way? What about starting with the total number $4! = 24$ of four-digit permutations, of which half (12) must end in 0 or 2. Of the six ending in 2, two must start with 0, leaving 10 four-digit numbers. Of the 12, chopping off the first digit will leave a valid 3-digit number precisely when the third digit is not 0, and all valid 3-digit numbers arise uniquely in this way. The six ending in zero cannot have third digit zero, and of the six ending in 2, two will have third digit zero. That leaves 10 3-digit numbers. The rest is easy.

For interest, let us think about how many **odd** numbers we could have made in case (b). Because of the conditions on the 0, it is clear there is no symmetry between odd and even situations, so the total will not be the same. Perhaps the best way is to find **all** possible numbers, and subtract the even ones.

*One-digit numbers:* There are four.

*Two-digit numbers:* The simplest way is to take all $3 \times 4$ possibilities and subtract the three with zero tens digit, giving $12 - 3 = 9$.

*Three-digit numbers:* Similarly, we have $(2 \times 3 \times 4) - 6 = 18$.

*Four-digit numbers:* We have $(2 \times 3 \times 4) - 6 = 18$.

Hence the total number of numbers we can form is $4 + 9 + 18 + 18 = 49$. Therefore the total of odd numbers will be $49 - 27 = 22$.

**Example:**   Ten lines are drawn in a plane. What is the maximum number of parts into which the plane can be divided by the 10 straight lines?
(A) 56   (B) 45   (C) 10   (D) 9   (E) 55

**Solution:**   Let us consider the more general problem of finding the maximum number of parts into which a plane can be divided by $n$ straight lines. Naturally, the first step is to consider small values of $n$ and see whether any useful observations can be made. The diagrams below show some of the ways the lines can cut the plane in a manner that maximizes the number of parts into which the plane can be divided (the plane is that of the paper).

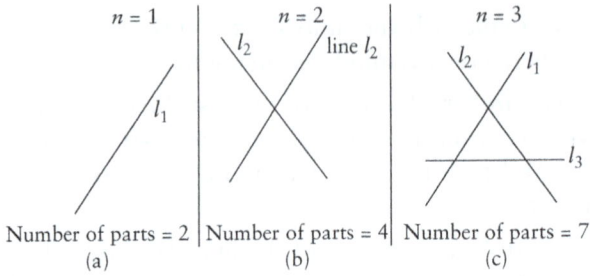

Number of parts = 2 | Number of parts = 4 | Number of parts = 7
(a)          (b)          (c)

What lessons can we derive from Figures (a), (b) and (c) concerning this partition process? Can a formula that gives the maximum number of parts for a given $n$ be found? From Figures (b) and (c), a crucial observation can be made, namely that the number of parts will be maximum when:

1. no two lines are parallel, and
2. no more than two lines pass through one point.

Try to convince yourself that this is necessary and sufficient for the maximization of the number of parts $(\overline{n})$ into which a plane can be divided by a given number of lines $n$. We will assume that the conditions above are satisfied. This means that if we draw one additional $(n+1)$th line, this line will cross each of the first $n$ lines at $n$ distinct points and as a result, pass through $n+1$ of the available parts, dividing each one of them into two parts, thereby creating an additional $n+1$ parts. This leads us to the equation:

$$\overline{n+1} = \overline{n} + (n+1). \tag{8.1}$$

What we want is a formula that gives $\overline{n}$ as a function of $n$ only. Using (8.1) repeatedly gives us:

$$\overline{1} = 1 + 1$$
$$\overline{2} = \overline{1} + 2$$
$$\overline{3} = \overline{2} + 3$$
$$\vdots$$
$$\overline{n} = \overline{n-1} + n.$$

Adding these equations gives:

$$\overline{n} = 1 + (1 + 2 + 3 + \cdots + n)$$
$$= 1 + \frac{n(n+1)}{2}.$$

Hence, the maximum number of parts into which a plane can be divided by $n$ lines is $\overline{n} = 1 + \frac{n(n+1)}{2}$ so that in the given problem, $\overline{10} = 1 + \frac{10(11)}{2} = 56$. Hence (A).

Here is the solution to the Appetizer Problem posed at the start of this Toolchest.

**Solution of Appetizer Problem 1:**  The total number of arrangements is $4 \times 3 \times 2 = 24$, with each of the digits $1, 2, 3, 4$ occurring six times as the units digit, and similarly for the tens and the hundreds digits. Thus, stacking all possible numbers for addition, the sum of each column will be $6(1+2+3+4) = 60$. So the overall sum will be $60 + 600 + 6000 = 6660$. Hence (C).

## 8.4  Combinations

As usual we start by posing a typical problem of the kind you should be able to solve when you have worked through this section. You are invited to try it as soon as you wish. You may not find it easy now, but by the end of the section, where its solution is given, it should not appear difficult to you.

**Appetizer Problem:**  *An ordered triplet of numbers $(x, y, z)$ is called a 3-tuple. How many 3-tuples $(x, y, z)$ are there such that $x$, $y$ and $z$ are whole numbers greater than or equal to 1, satisfying the equation $x + y + z = 6$?*
(A) 10   (B) 20   (C) 12   (D) 24   (E) 18

Consider the number of subsets of size $r$ that can be chosen from a collection of $n$ objects. We denote this number by ${}^{n}C_{r}$ or $\binom{n}{r}$. From the previous section, we know that the number of ways of arranging $r$ objects chosen from a collection of $n$ ($n \geq r$) is ${}^{n}P_{r}$. However, because each subset has $r!$ permutations (linear) of its members, ${}^{n}P_{r}$ is $r!$ times the number of ways of choosing $r$ objects from the $n$ objects. That is

$$ {}^{n}P_{r} = r! \; {}^{n}C_{r} $$

This observation gives us a formula for ${}^{n}C_{r}$:

$$ {}^{n}C_{r} = \frac{{}^{n}P_{r}}{r!} = \frac{n!}{(n-r)! \, r!}. $$

Alternatively, we could observe that forming a combination of $r$ objects chosen from $n$ is a partition of the given $n$-membered set into two subsets – the desired set of $r$ objects plus the remaining set of $(n-r)$ objects. Hence, by the partitioning formula, we get (as before):

$$ {}^{n}C_{r} = \binom{n}{r, n-r} = \frac{n!}{r!(n-r)!}. $$

These numbers $^nC_r$ are called the **binomial coefficients**, and feature as the entries at place $r$ in row $n$ of 'Pascal's triangle'. (See Section 3 of Toolchest 7.)

$$^nC_r = \frac{n!}{(n-r)!r!}$$

**Example:**  If each of a group of $n$ teenagers at a party shakes hands with each of the remaining $n - 1$ exactly once, how many handshakes will there be?

**Solution:**  Observe that the number of different handshakes equals the number of unordered pairs (i.e. groups of two people) we can choose from the $n$ people. This is

$$
\begin{aligned}
^nC_2 &= \frac{n!}{(n-2)!2!} \\
&= \frac{n(n-1)(n-2)!}{(n-2)!2!} \\
&= \frac{n(n-1)}{2}.
\end{aligned}
$$

**Note:** This problem appears again as the Problem 2 at the end of this Toolchest, where its solution is approached from different points of view, to illustrate some basic principles of problem-solving.

**Example:**  A committee comprising two women and three men is to be chosen from a group of five men and four ladies. In how many ways can this be done?

**Solution:**  The three men can be chosen in $^5C_3$ ways and the two women in $^4C_2$ ways so that by the FPE the committee can be chosen in $^5C_3 \times {}^4C_2 = 60$ ways.

**Example:**  How many three-element subsets drawn from the set $S = \{1, 2, 3, \ldots, 29, 30\}$ are such that the sum of the three elements is a multiple of 3?
(A) 1000   (B) 120   (C) 1120   (D) 30   (E) 10

**Solution:**  Observe (see Toolchest 4) that each of the 30 integers in $S$ fall into exactly one of the following (residue) classes, with 10 in each class:

(i) $T_0$, the set of those leaving no remainder when divided by 3, i.e. the set of multiples of 3.
(ii) $T_1$, the set of those leaving remainder 1 when divided by 3.
(iii) $T_2$, the set of those leaving remainder 2 when divided by 3.

Consider a general three element subset $\{x, y, z\}$ extracted from $S$. Because a multiple of 3 can be formed out of the three remainders in exactly four different ways: $0+0+0$, $1+1+1$, $2+2+2$ and $0+1+2$, we know that $x+y+z$ is a multiple of 3 if and only if one of the following holds:

1. all of $x, y, z$ lie in the same residue class $T_i$;
2. exactly one of $x, y, z$ lies in each of the three residue classes.

Case (1) gives $3 \times {}^{10}C_3$ such subsets.
Case (2) gives, by the FPE, $10 \times 10 \times 10 = 10^3$ such subsets.

Hence we have $10^3 + 3 \times {}^{10}C_3 = 1360$ such three element subsets. Hence (C).

**Example:**  Find the number of all subsets of an $n$-element set.

**Solution:**  In constructing any particular subset of the set $S$, there are exactly two ways of dealing with each of the $n$ elements: you either include it or exclude it. Hence, by the FPE, there are:  $2 \times 2 \times 2 \times \cdots \times 2$ (up to $n$ factors) ways of dealing with all of the $n$ elements, and thus of selecting or characterizing subsets of $S$. This gives $2^n$ as the number of subsets. Note that the empty set is included. Mathematicians say that the set of subsets is mapped one-to-one onto the set of all functions from $S$ to the *two-point set* $\{0, 1\}$. Think of assigning 1 to designate 'in' and assigning 0 to designate 'not in'. Thus each subset $A \subset S$ is said to have a 'characteristic function' $f_A$: that function mapping each of its members to 1 and all other elements of the set $S$ to 0.

Another approach (see Section 3 of Toolchest 7) is instructive. We know that ${}^nC_r$ is the number of subsets of size $r$, so that the total number will be ${}^nC_0 + {}^n C_1 + \cdots + {}^n C_{n-1} + {}^n C_n$, which is the sum of the numbers in the $(n+1)$th row of Pascal's triangle. (Note that there is one subset of size $n$, the whole set, and one subset of size zero, namely the empty set.) The sum of these 'binomial coefficients' is $(1+1)^n = 2^n$.

Here is the solution to the Appetizer Problem posed at the beginning of this section.

**Solution of Appetizer Problem:**  A very efficient counting technique is to represent the number 6 by six markers with five spaces in between them, as below:

$$1 \quad 1 \quad 1 \quad 1 \quad 1 \quad 1$$

Now we can use two asterisk signs to represent a particular ordered triplet $(x, y, z)$ like this:

$$1 \quad * \quad 1 \quad 1 \quad 1 \quad * \quad 1 \quad 1$$

by which we mean that $x = 1$, $y = 3$, and $z = 2$ with $1+3+2 = 6$. Because the asterisks are put in the spaces between, we are assured that the smallest value any of $x, y$, or $z$ can take is 1. Thus the problem is equivalent to that of finding *the number of ways in which two asterisks can fill up five spaces.* This can be done in

$$^5C_1 \times {}^4C_1 = 20 \text{ ways.}$$

Hence, there are 20 3-tuples, making (B) the correct answer.

In the example on page 274 we use the same counting technique.

## 8.5 Derangements

As an appetizer, here is a typical (and very appealing) problem about the derangements of this section and the exclusion–inclusion principle in the following section. You should be able to solve such problems when you have worked through Sections 5 and 6. You are invited to try this one as soon as you wish. You may find it hard for now, but by the end of Section 6, where its solution is given, it should not look too difficult to you – well, the first part anyway!

**Appetizer Problem:**   *In how many visually distinct ways can three married couples be placed in six chairs around a circular table with sexes alternating and no spouses seated next to each other?*

(A) 12   (B) 14   (C) 20   (D) 28   (E) 32

*Can you derive a formula for the number of such seating arrangements for n couples in 2n chairs?*

**Note:** To clarify the meaning of visually distinct, note that if everybody stands up and moves to occupy the chair to their right (say), then the resulting arrangement is *rotationally equivalent* in the sense of Section 8.3.4, but is regarded as visually *distinct* from the previous. If you (the observer) were blindfolded and started from a random chair (you know not which) to list the names in a clockwise order around the table, you would not be able to distinguish these two arrangements. Another way to get visually distinct arrangements is to number the chairs.

What is a **derangement**? Let us define the situation in which item $A_1$ is matched with item $a_1$, $A_2$ to $a_2$, ..., $A_n$ to $a_n$, as a 'normal state'. For example, a state in which Mrs Moyo is attached to Mr Moyo, Mrs Engelbrecht to Mr Engelbrecht and Mrs Vancouver to Mr Vancouver, may be defined as a 'normal state.' We are interested in the number of states where the assignments are 'totally muddled up', i.e. when none of the $n$ assignments described above is the same as in the normal state. For example, think of the situation in which Mrs Moyo is assigned to Mr Vancouver, Mrs Vancouver to Mr Engelbrecht, and Mrs Engelbrecht to Mr Moyo. Such states are called **derangements**.

So, more precisely, we are interested in the number of derangements, $D_n$, of those $n$ pairs $(A_1, a_1), (A_2, a_2), ..., (A_n, a_n)$, we talked about earlier. For starters, we see that $D_2 = 1$, since there is only one way in which two things can be muddled up: $A_1$ is attached to $a_2$ and $A_2$ to $a_1$. It is also easy to see that $D_1 = 0$.

The question now is, can we express $D_n$ as a simple function of $n$? This would provide a simple card-in-the-slot solution far superior to any attempt to produce a list of all 'the chaotic assignments'.

## Derivation of a formula for $D_n$

Assume that we have the items $A_1, A_2, ..., A_n$ which in the normal state are attached to the items $a_1, a_2, ..., a_n$ respectively.

We observe that there are $n-1$ possible ways in which $A_1$ is attached to the wrong item.

Consider the situation in which we have assigned item $a_2$ to item $A_1$ – this means we have two different ways of assigning $a_1$; we could either:

(i) attach $A_2$ to $a_1$ – leaving us with $n-2$ items $A_3, A_4, ..., A_n$ and their counterparts $a_3, a_4, ..., a_n$, which can be deranged in $D_{n-2}$ ways;
(ii) NOT attach $A_2$ to $a_1$ – leaving us with $n-1$ items $A_2, A_3, ..., A_n$ and their counterparts $a_1, a_3, ..., a_n$. These can be deranged in $D_{n-1}$ ways.

The two cases listed above are mutually exclusive and so the total number of ways of dealing with the remaining items is:

$$D_{n-1} + D_{n-2}.$$

Observe that, *for each* of the $n-1$ ways of attaching $A_1$ to the wrong counterpart, there are $D_{n-1} + D_{n-2}$ of attaching the remaining items $A_2, A_3, \ldots, A_n$ to the remaining counterparts $a_i$. Hence, using the fundamental principle of enumeration,

$$D_n = (n-1)(D_{n-1} + D_{n-2}),$$

therefore   $D_n - nD_{n-1} = (n-1)D_{n-2} - D_{n-1}$
$$= (-1)(D_{n-1} - (n-1)D_{n-2}).$$

Now we have a pattern:
if $s_n = D_n - nD_{n-1}$, then $s_n = -s_{n-1} = (-1)^2 s_{n-2} = \cdots = (-1)^{n-2} s_2$.
That is

$$D_n - nD_{n-1} = (-1)^{n-2}(D_2 - 2D_1)$$
$$= (-1)^{n-2} = (-1)^n,$$

because, as we saw earlier on, $D_2 = 1$ and $D_1 = 0$. Dividing throughout by $n!$, we have:

$$\frac{D_n}{n!} - \frac{D_{n-1}}{(n-1)!} = \frac{(-1)^n}{n!},$$

therefore   $$\frac{D_{n-1}}{(n-1)!} - \frac{D_{n-2}}{(n-2)!} = \frac{(-1)^{n-1}}{(n-1)!},$$

hence   $$\frac{D_{n-2}}{(n-2)!} - \frac{D_{n-3}}{(n-3)!} = \frac{(-1)^{n-2}}{(n-2)!}$$

$$\vdots \qquad \vdots \qquad = \qquad \vdots$$

$$\frac{D_2}{2!} - \frac{D_1}{1!} = \frac{(-1)^2}{2!}.$$

Adding these equations together gives

$$\frac{D_n}{n!} - \frac{D_1}{1!} = \frac{(-1)^n}{n!} + \frac{(-1)^{n-1}}{(n-1)!} + \cdots + \frac{(-1)^2}{2!},$$

so that, finally, the number of derangements of $n$ pairs is given by:

$$\boxed{D_n = n!\left(1 - \frac{1}{1!} + \frac{1}{2!} - \frac{1}{3!} + \cdots + \frac{(-1)^n}{n!}\right)}$$

**Example:**   Letters from the Masvingo branch of Skybank are put in a tray M, those from the Gweru branch in tray G, from Kwekwe branch in tray K and those from Bulawayo in tray B.

On a certain day, the secretary receives one letter from each of the Masvingo, Gweru, Kwekwe and Bulawayo branches. In how many ways can she put the letters in the trays M, G, K and B in such a way that no letter is in the correct tray?

(A) 24   (B) 9   (C) 6   (D) 4   (E) 3

**Solution:**   There are four pairs (letter, tray) to derange.

$$D_n = n! \left( 1 - \frac{1}{1!} + \frac{1}{2!} - \frac{1}{3!} + \cdots + \frac{(-1)^n}{n!} \right),$$

$$\text{thus} \quad D_4 = 24 \left( 1 - \frac{1}{1} + \frac{1}{2!} - \frac{1}{3!} + \frac{1}{4!} \right)$$

$$= \frac{9}{24} \times 24 = 9.$$

Hence B. You should convince yourself that this is correct, and gain some inside knowledge of how derangements work, by taking the letters B, G, K, M (in alphabetical order for convenience) and writing out (in dictionary or alphabetical / lexicographical order) the $4! = 24$ different permutations of these four, or four-letter 'words' you can make with them. Then cross off those (6) which have B in first place, and then those remaining (4) which have G in second place, and then those (3) remaining which have K in the third place, and finally those (2) remaining which have M in the fourth place. There are $24 - 15 = 9$ remaining – all derangements!

**Example:**   In the example above, in how many ways can the letters be placed in the trays if exactly one letter is placed in the correct tray?

(A) 6   (B) 3   (C) 5   (D) 9   (E) 8

**Solution:**   The letter that is placed in the correct tray can be chosen in $^4C_1 = 4$ ways. This leaves us with three letters that are to be assigned to the wrong trays and this can be done in $D_3$ ways. Using the FPE, the complete job can be done in

$$^4C_1 \times D_3 \text{ ways} = 4D_3 \text{ ways}$$

$$= 4 \times 3! \left( 1 - \frac{1}{1!} + \frac{1}{2!} - \frac{1}{3!} \right)$$

$$= 4 \times 6 \times \frac{2}{6}$$

$$= 8 \text{ ways.}  \quad \text{Hence (E).}$$

## 8.6   The exclusion–inclusion principle

The idea of derangements we have just looked at is a particular example of a broader concept called the *exclusion–inclusion principle*. It should not be surprising therefore that the discussion that follows resembles the style of the previous discussion. (In fact we could use this principle to derive the derangements formula more quickly!)

In concrete terms, take a class of $N$ students, and three properties, say, (1) female, (2) over 16 years old, and (3) loves mathematics. The class of students who are male and under 16 can be represented by the points **not** in either of the two circles in the Venn diagram on the left below, and their number is given by

$$N - (\text{number in } C_1) - (\text{number in } C_2) + (\text{number in both } C_1 \text{ and } C_2)$$

because initially we subtracted those in the intersection *twice*. We can also express this in terms of the size of the *union* of the two sets:

$$|C_1 \cup C_2| = |C_1| + |C_2| - |C_1 \cap C_2|.$$

This famous set relation, an example of the exclusion–inclusion principle, is sometimes written like this:

$$n(A \cup B) = n(A) + n(B) - n(A \cap B).$$

Now, consider three properties, and three sets of students with those properties, represented by circles $C_1, C_2, C_3$ in the Venn diagram on the right. Let $N_i$ denote the number of those having $i$th property, i.e. in circle $C_i$, and $N_{ij}$ denote the number of those having both $i$th and $j$th properties, and $N_{ijk}$ denote the number having all three properties. Then the class of students who have *none* of the three properties (younger males hating mathematics!) can be represented by the points *not* in any of the three circles in the second Venn diagram, and their number is

$$N - N_1 - N_2 - N_3 + N_{12} + N_{23} + N_{13} - N_{123}.$$

We could write this as

$$N - n(C_1 \cup C_2 \cup C_3) = N - (N_1 + N_2 + N_3) + (N_{12} + N_{23} + N_{13}) - N_{123}.$$

Generalizing further now, consider a collection of $N$ items, and $n$ properties $\alpha_1, \alpha_2, \alpha_3, \ldots, \alpha_n$, any number of which can be attached to any one of the items. Let us agree that by writing $N(\alpha_1)$ we will denote the number of all those items (in the collection of items under consideration of course) that have property $\alpha_1$. They may or may not also have any other of the remaining properties. Likewise, by $N(\alpha_1, \alpha_2)$, we denote all those items that have both properties $\alpha_1$ and $\alpha_2$, and so on. We may also think of $\alpha_i$ as the property of belonging to a set $A_i$, so that $N(\alpha_i) = |A_i|$, the size of the set, and $N(\alpha_1, \alpha_2) = |A_1 \cap A_2|$, the size of the intersection.

Also, by $N(\alpha_1')$ we denote the total number of items that do not have property $\alpha_1$, while $N(\alpha_1, \alpha_2, \alpha_3')$ denotes the total number of items that have both properties $\alpha_1$ and $\alpha_2$ but do not have property $\alpha_3$, and so on.

The exclusion–inclusion principle states that:

$$N(\alpha_1', \alpha_2', \ldots, \alpha_n') = N - N(\alpha_1) - N(\alpha_2) - \cdots - N(\alpha_n)$$
$$+ N(\alpha_1, \alpha_2) + N(\alpha_1, \alpha_3) + \cdots$$
$$- N(\alpha_1, \alpha_2, \alpha_3) - N(\alpha_1, \alpha_2, \alpha_4) - \cdots$$
$$+ (-1)^n N(\alpha_1, \alpha_2, \ldots, \alpha_n).$$

For example, if we have four properties the rule becomes:

$$N(\alpha_1', \alpha_2', \alpha_3', \alpha_4') = N - N(\alpha_1) - N(\alpha_2) - N(\alpha_3) - N(\alpha_4)$$
$$+ N(\alpha_1, \alpha_2) + N(\alpha_1, \alpha_3) + N(\alpha_1, \alpha_4)$$
$$+ N(\alpha_2, \alpha_3) + N(\alpha_2, \alpha_4)$$
$$+ N(\alpha_3, \alpha_4)$$
$$- N(\alpha_1, \alpha_2, \alpha_3) - N(\alpha_1, \alpha_2, \alpha_4)$$
$$- N(\alpha_1, \alpha_3, \alpha_4) - N(\alpha_2, \alpha_3, \alpha_4)$$
$$+ N(\alpha_1, \alpha_2, \alpha_3, \alpha_4).$$

**Proof:**  The left-hand side counts the number of objects in the collection having none of the properties $\alpha_i$. We now consider the number of times each object $x$ is counted in the right-hand side.

If $x$ has none of the properties $\alpha_i$, then it belongs to none of the sets $A_i$. So in this case $x$ is counted once in the first term and not in any other term.

Now suppose $x$ has exactly $n$ of the properties $\alpha_i$, where $n > 0$ and $n \in \mathbb{Z}$. Then:

(i) $x$ is counted once in the first term $N$;
(ii) it is counted $-n$ times in the second term;
(iii) $+^nC_2$ times in the third; and so on until we reach the $(n+1)$st term.

Thereafter, since $x$ has only $n$ of the properties, and we are intersecting more than $n$ of the sets $A_i$, $x$ is not counted any more. So, summing up, the number of times $x$ is counted is:

$$1 - {}^nC_1 + {}^nC_2 - {}^nC_3 + \cdots + (-1)^n \, {}^nC_n = (-1 + 1)^n$$
$$= 0 \text{ times}$$

by the Binomial Theorem (see Toolchest 7). It therefore contributes nothing to the total count so that only those with none of the properties $\alpha_i$ are counted!                                                                                          ◇

**Example:**   How many numbers between 1 and 1000 (inclusive) are not divisible by 2, 3 and 5?

(A) 30   (B) 300   (C) 266   (D) 356   (E) 235

**Solution:**   Let $\alpha_1$ denote divisibility by 2, $\alpha_2$ denote divisibility by 3, and $\alpha_3$ denote divisibility by 5. Further, let $(\alpha_1, \alpha_2)$ denotes divisibility by 6, $(\alpha_2, \alpha_3)$ divisibility by 15, $(\alpha_1, \alpha_3)$ divisibility by 10, and $(\alpha_1, \alpha_2, \alpha_3)$ divisibility by 30. We want to find $N(\alpha_1', \alpha_2', \alpha_3')$, which is given by the exclusion–inclusion principle as:

$$N - N(\alpha_1) - N(\alpha_2) - N(\alpha_3) + N(\alpha_1, \alpha_2) + N(\alpha_1, \alpha_3)$$
$$+ N(\alpha_2, \alpha_3) - N(\alpha_1, \alpha_2, \alpha_3),$$

where

$$N(\alpha_1) = \left[\frac{1000}{2}\right] = 500, \quad N(\alpha_2) = \left[\frac{1000}{3}\right] = 333,$$

$$N(\alpha_3) = \left[\frac{1000}{5}\right] = 200 \quad N(\alpha_1, \alpha_2) = \left[\frac{1000}{6}\right] = 166,$$

$$N(\alpha_1, \alpha_3) = \left[\frac{1000}{10}\right] = 100, \quad N(\alpha_2, \alpha_3) = \left[\frac{1000}{15}\right] = 66,$$

$$N(\alpha_1, \alpha_2, \alpha_3) = \left[\frac{1000}{30}\right] = 33.$$

Substituting gives the required number as

$$1000 - 500 - 333 - 200 + 166 + 100 + 66 - 33 = 266. \quad \text{Hence (C).}$$

**Example:**   Sweet-toothed Timothy visited his favourite sweet shop one Thursday afternoon. Timothy's Thursday afternoon policy is to buy not more than five sweets of any type and to buy at least one of each of the types available. That day, there were four varieties in the shop and Timothy wanted to buy exactly one dozen (= 12) sweets. How many distinct dozens could Timothy buy? And how many if he drops the policy of sampling all available types?

(A) 455   (B) 144   (C) 599   (D) 419   (E) 125

**Solution:**    The first problem is equivalent to finding the number of solutions in positive integers $a, b, c$ and $d$ (all less than or equal to 5) of the equation:

$$a + b + c + d = 12. \tag{8.2}$$

The second problem extends the possibilities to non-negative integers (that is, any of $a, b, c$ and $d$ may be zero). One bright idea is to use partitions: imagine 12 sticks with 11 spaces between them of which we select 3: this will create a solution $a, b, c, d$ to our equation (ignoring for the moment the upper bound of 5), and *vice versa* – each solution will select three spaces. Thus we only need to ask how many ways there are of choosing three things out of 11. And we know the answer: $^{11}C_3$. This answers the first problem, when none of $a, b, c, d$ is zero. What shall we do to include the cases where there is a zero? We could start dealing with these separately (not too hard: it involves solving equations $a + b + c = 12$, $a + b = 12$). But another bright idea is to define new variables $a + 1$, $b + 1$, $c + 1$, $d + 1$, and a new equation

$$(a + 1) + (b + 1) + (c + 1) + (d + 1) = 12 + 4 = 16,$$

which is clearly equivalent to our first equation. What we want is the number of ways of choosing three spaces out of 15, and it is $^{15}C_3 = 455$. Now we need to think of how many of these satisfy the condition $a, b, c \leq 5$. Let us defer this until we have fully appreciated what's been achieved so far.

To practise the application of our two bright ideas to more general contexts (this is training us to apply them to other problems), let us consider the more general question of finding the number of solutions in non-negative integers to the equation:

$$x_1 + x_2 + \cdots + x_n = \sum_{i=1}^{n} x_i = r, \quad r \geq n. \tag{8.3}$$

To tackle this question, we start by changing the condition on the $x_i$: we look for the number of solutions to the equation (8.3), where each $x_i$ is a *positive integer*. Consider $r$ numbered sticks lined up as below, and observe that there are $(r - 1)$ spaces between the sticks:

and that if we have a set of $n-1$ arrows, each time we choose $n-1$ positions out of the $r - 1$ in such a way that no arrows occupy the same space, we generate a solution to equation (8.3). This can be done in $^{r-1}C_{n-1}$ ways which equals the required number of solutions.

Now define $y_i = x_i - 1 \geq 0$, since $x_i \geq 1$. Consider the equations:

$$\sum_{i=1}^{n} y_i = r, \quad y_i \geq 0 \tag{8.4}$$

$$\sum_{i=1}^{n} x_i = r + n, \quad x_i > 0. \tag{8.5}$$

The number of solutions to (8.5) equals the number of solutions to (8.4), and is given by the formula above as:

$$^{r+n-1}C_{n-1} \quad (\text{write } r + n \text{ in place of } r).$$

Recall that $^nC_r = {}^nC_{n-r}$, so that $^{r+n-1}C_{n-1} = {}^{r+n-1}C_r$. Hence, the number of solutions in non-negative integers to the equation (8.3) is:

$$\boxed{^{r+n-1}C_{n-1} = {}^{r+n-1}C_r}$$

Going back to the original problem, we found that the number of integral solutions to equation (8.2), without the upper limit on the variables, is given by this formula as:

$$^{12+4-1}C_{12} = {}^{15}C_{12} = {}^{15}C_3 = 455.$$

We need the number of solutions which have none of the four properties $a > 5$, $b > 5$, $c > 5$ and $d > 5$. Let the set $A_d$ consist of solutions to (8.2) such that the variable $d$ (which we can choose in $4 = {}^4C_1$ different ways), is greater than 5. The number $r = |A_d|$ of elements of this set is the same as the number of solutions in nonnegative integers of the equation

$$a + b + c + d' = a + b + c + (d - 6) \leq 12 - 6 = 6,$$

which is given by the formula as:

$$^{6+4-1}C_6 = {}^9C_6 = {}^9C_3 = 84.$$

In a similar manner, let $A_{c,d}$ be the set of solutions to (8.2) for which precisely two of the variables, $c, d$ (which two we can choose in $6 = {}^4C_2$ different ways), are greater than 5. This is the same as the number of solutions in non-negative integers of the equation

$$a + b + c' + d' = a + b + (c - 6) + (d - 6) = 12 - 12 = 0,$$

which is 1. Also, observe that it is impossible for three or more variables to exceed 5 in a solution to equation (8.2). Hence the sets $A_a, A_b, A_c, A_d$ meet in pairs only, with each of their six overlaps containing just one element. For example, $A_c \cap A_d = A_{c,d}$. Hence, the number of distinct dozens Timothy can buy is:

$$455 - {}^4C_1 \cdot 84 + {}^4C_2 \cdot 1 = 455 - 336 + 6 = 125. \quad \text{Hence (E).}$$

It is instructive to work this problem about Timothy's sweets the routine way, observing that there are 18 solutions (like 5502) which have two fives, 48 solutions (like 5412) which have a five and a four, 24 solutions (like 5223) which have one five and no four, 18 solutions (like 4413) which have exactly two fours, 12 solutions (like 4323) which have no fives and one four, four solutions (like 4044) which have three fours, and one solution 3333 which has no fives or fours.

Here is the solution to the Appetizer Problem posed at the beginning of Section 5.

**Solution to Appetizer Problem:** Number the chairs 1–6 around the circular table as shown below. We can enumerate all the possible arrangements by considering, in turn, all the visually distinct arrangements in which the males sit in odd-numbered chairs and then consider the arrangements in which the males sit in even-numbered chairs. Let M1 denoted the male from married pair 1 and F1 his spouse. Assume the same nomenclature for pairs 2 and 3. A picture of the six solutions for males in odd numbered chairs is shown below.

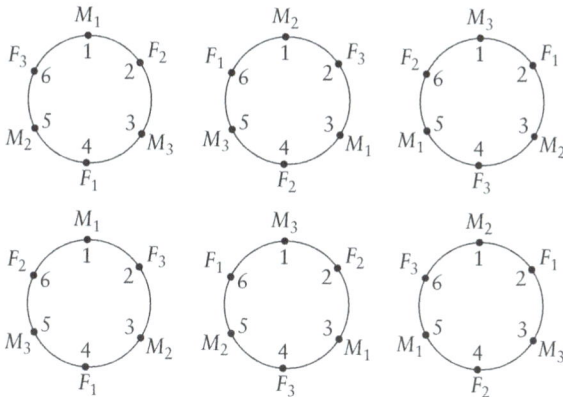

We could now picture six more, with the females in the odd-numbered chairs and the males in the even-numbered chairs, but by symmetry, we know there are six of these (just interchange the letters M and F in the six arrangements above). The answer to the problem is therefore 12; hence (A).

The generalized form of the problem above (finding the number of arrangements for $n$ couples sitting in $2n$ chairs) is an old and famous problem that was first posed by Lucas in 1891. It is known as the problème des ménages. Touchard gave a solution for the problem in 1934 but did not have a proof for his solution. The first published proof of the problem was by Kaplansky (1943) and the proof we give here is by Bogart and Doyle (described in their elegant paper, 'A non-sexist solution of the ménage problem', *American Mathematical Monthly* **43** (1986), 514–518). The problème des ménages is an example of a derangements (perfect permutations) problem.

The proof by Bogart and Doyle requires the establishment of two preliminary theorems about the placement of non-overlapping dominos (yes, those pre-school/primary school play objects!). The first concerns the placement of dominos over a linear sequence of numbers so that each domino covers two numbers and the second extends to the placement of dominos over a circular sequence of numbers.

**Theorem 1**  Let 'Line $m$, $k$' denote a configuration/placement (any) in which $k$ indistinguishable non-overlapping dominos are placed on a linear sequence of numbers $1, 2, \ldots, m$, where each domino covers two numbers. Let Dom(Line $m$, $k$) denote the number of all such configurations. Then

$$\text{Dom(Line } m, \; k) = \binom{m-k}{k}.$$

**Proof:**  Each such placement of the $k$ dominos corresponds in one-to-one fashion with a sequential arrangement of $k$ twos and $m - 2k$ ones. The example below will help you picture why the above statement is true.

| 1 2 | 3 | 4 | 5 | 6 | 7 | 8 | 9 10 | 11 12 |

Here $m = 12$ and $k = 3$ and the particular placement shown above corresponds to the sequence 211111122. Dom(Line $m$, $k$) is therefore equal to the number of partitions (see Section 8.3.2 above) of a total of $(m - 2k) + k = m - k$ ones and twos into two groups populated by $m - 2k$ ones and $k$ twos respectively. From Section 8.3.2 this number is

$$\frac{(m-k)!}{(m-2k)!k!} = \binom{m-k}{k}. \qquad \diamond$$

**Theorem 2**  Let Dom(Circle $m$, $k$) denote the number of ways to place $k$ indistinguishable and non-overlapping dominos on a circular sequence of numbers $1, 2, \ldots, m$, where each domino covers two numbers. Then,

$$\text{Dom(Circle } m, \; k) = \frac{m}{m-k} \binom{m-k}{k}.$$

**Proof:**  The strategy adopted here in enumerating the number of placements works by counting the number of placements in classes of placements which are both disjoint and exhaustive. The total number of placements is therefore be the sum of the number of placements calculated over the entire set of these disjoint (and exhaustive) classes.

Consider the following two classes of placements: (i) the class of placements where the number 1 is covered by a domino and (ii) the class of placements where the number 1 is not covered by a domino. These two classes are disjoint and exhaustive and we use them here to evaluate Dom (Circle $m, k$).

The number of placements in which $m$ and 1 are covered by a single domino (as shown above) is clearly given by Dom(Line $m - 2$, $k - 1$). By the same token, the number of placements in which 1 and 2 are covered by a single domino is Dom(Line $m - 2$, $k - 1$). The number of placements in which 1 is not covered by any domino is clearly given by Dom(Line $m - 1$, $k$). Summing the number of placements in the two classes gives:

$$\text{Dom(Circle } m, \ k) = 2\text{Dom(Line } m - 2, k - 1) + \text{Dom(Line } m - 1, k)$$

$$= 2\binom{m - k - 1}{k - 1} + \binom{m - k - 1}{k}$$

$$= \frac{2(m - k - 1)!}{(m - k - 1 - (k - 1))!(k - 1)!} + \frac{(m - k - 1)!}{(m - 2k - 1)!k!}$$

$$= \frac{(m - k)!}{(m - 2k)!k!}\left(\frac{2k}{m - k} + \frac{m - 2k}{m - k}\right)$$

$$= \frac{(m - k)!}{(m - 2k)!k!} \cdot \frac{m}{m - k} = \frac{m}{m - k}\binom{m - k}{k}.$$

◇

We now have enough tools to prove the problème des ménages. Let $M_n$ be the number of visually distinct ways of seating $n$ married couples in $2n$ chairs placed around a circular table with sexes alternating and no spouses sitting next to each other ($M_n$ is called the $n$th ménage number).

**Theorem 3** $M_1 = 0$, and, for all $n \geq 2$,

$$M_n = 2(n!) \sum_{k=0}^{n} (-1)^k \frac{2n}{2n - k}\binom{2n - k}{k}(n - k)!$$

**Proof:** First let us label the couples $1, 2, 3, \ldots, n$. Let $A$ be the set of all sex-alternating seatings of the $n$ couples. For $i = 1, 2, 3, \ldots n$, let $A_i$ be the set of seatings in $A$ where the spouses of couple $i$ are adjacent. Using the exclusion–inclusion principle, we have

$$M_n = |A - (A_1 \cup \cdots A_n)|$$

$$= |A| - \sum_{1 \leq i_1 \leq n} |A_{i_1}| + \sum_{1 \leq i_1 < i_2 \leq n} |A_{i_1} \cap A_{i_2}| - \cdots + (-1)^n |A_1 \cap \cdots \cap A_n|.$$

Now, if we let $W_k$ be the number of seatings in $A$ in which spouses of a fixed set of $k$ couples are adjacent, it follows that:

$$M_n = |A| - \binom{n}{1}W_1 + \binom{n}{2}W_2 - \cdots + (-1)^k \binom{n}{k}W_k + \cdots + (-1)^n \binom{n}{n}W_n.$$

We work out the value of $|A|$ by following through the steps one might take in building a seating in $A$. One can proceed as follows: (a) decide whether to seat the women in odd or even numbered seats (this yields two possibilities in our count); (b) place the women in those seats (this yields $n!$ possibilities); (c) place the men in the remaining seats (this yields $n!$ possibilities). Combining the counts from the independent steps (a), (b) and (c) gives $|A| = 2(n!)^2$.

We follow a similar kind of logic in calculating the value of $W_k$ as follows. In building a seating in which spouses of a fixed set of $k$ couples are adjacent one may proceed as follows: (i) decide whether to seat women in odd or even numbered seats (two ways); (ii) choose $k$ pairs of adjacent seats in which to seat the $k$ couples (Dom (Circle $2n$, $k$) $= \frac{2n}{2n-k}\binom{2n-k}{k}$ ways); (iii) assign the couples to those pairs ($k!$ ways); (iv) seat the remaining $(n-k)$ women $((n-k)!$ ways); and (v) seat the remaining $(n-k)$ men $((n-k)!$ ways). Combining the counts from these steps gives:

$$W_k = \frac{4n}{2n-k}\binom{2n-k}{k}k!\left((n-k)!\right)^2, \quad k \geq 1, \text{ and } W_0 = 2(n!)^2 = |A|.$$

It follows that:

$$M_n = \sum_{k=0}^{n}(-1)^k\binom{n}{k}W_k$$

$$= \sum_{k=0}^{n}(-1)^k\binom{n}{k}\frac{4n}{2n-k}\binom{2n-k}{k}k!\left((n-k)!\right)^2$$

$$= \sum_{k=0}^{n}(-1)^k\frac{n!}{(n-k)!k!}\cdot\frac{4n}{2n-k}\binom{2n-k}{k}k!\left((n-k)!\right)^2$$

$$= 2(n!)\sum_{k=0}^{n}(-1)^k\frac{2n}{2n-k}\binom{2n-k}{k}(n-k)!.$$

◇

## 8.7  The pigeon-hole principle

As an appetizer, here is a typical problem of the kind you should be able to solve when you have worked through this section. You are invited to try it as soon as you wish. You may find it hard for now, but by the end of the section, where its solution is given, it should not look difficult to you.

**Appetizer Problem:**  *A subset S of the set $A = \{3, 11, 19, 27, \ldots, 147, 155\}$ is said to be '158-free' if there are no two elements which add up to 158. What is the maximum size of a subset that is '158-free'?*

(A) 10   (B) 11   (C) 12   (D) 15   (E) 20

Consider the statements below:

(a) In a group of 13 people, one can find at least two born in the same month.
(b) In a group of 367 people, at least two people have birthdays on the same day of the year.
(c) There are at least two people in Harare with the same number of hairs on their heads.

The above statements are all true; (a) and (b) are quite amusingly obvious; (c) is just as amusing but lacks that flavour of 'obviousness'. In other words, (c) is a surprising result!

These three statements are simple manifestations of a more general principle known to mathematicians as the **pigeon-hole principle** (PHP). This principle was probably first formally enunciated by the German mathematician, Peter Gustav Lejeune Dirichlet (1805–1859) and it states that:

**P1:** If $n$ objects are placed in fewer than $n$ pigeon-holes (say $n - 1$), then at least two are together in the same pigeon-hole.

A more general form of this principle states that:

**P2:** If $mk + 1$ objects are distributed at random into $m$ pigeon-holes then some pigeon-hole has at least $k + 1$ objects.

P1 and P2 are informal versions of principles which, in formal mathematics, are stated more abstractly:

**P3:** If a set of $n$ elements is a union of $m < n$ of its subsets, then at least one of those subsets has more than one element.

**P4:** If a set of $mk + 1$ elements is a union of $m$ of its subsets, then at least one of those subsets has at least $k + 1$ elements.

In using this principle as a problem solving tool, the challenge is to choose the 'objects' and the 'pigeon-holes' correctly. For example, in (a) above, take the people as 'objects' and the 12 months as 'pigeon-holes'. Since there are more 'objects' than 'pigeon-holes', the PHP. tells us that 'there are at least two in the same pigeon-hole'– i.e. at least two people share the same month. Statement (c) is a bit more subtle but follows the same pattern. The result is easily established with the help of three extra facts:

(i) the population of Harare is $N$.
(ii) the set of all possible numbers of hairs $\{0, 1, 2, 3, \ldots, \text{etc.}\}$ covering a person's head is bounded with $n$ as the lowest upper bound, i.e. a person can have at most $n$ hairs. (It is a biological fact that there is an upper bound – considerably less than the population of Harare – to the number of hairs that can be growing on a human head.)
(iii) $N > n$.

Now we number pigeon-holes from 0 to $n$ and take each person in Harare to be an 'object' to be placed into a particular pigeon-hole if its number is the number of hairs on their head. The result follows easily from the PHP

**Example:**  Let $S = \{x_1, x_2, \ldots, x_{10}\}$ with $x_i \in \{1, 2, \ldots, 99, 100\}$ for $i = 1, 2, \ldots, 10$, with $x_i \neq x_j$ for $i \neq j$. Show that there are at least two subsets of $S$ for which the sum of the elements is the same.

**Solution:**  The maximum sum of elements of the subsets of $S$ is:
$$100 + 99 + 98 + 97 + 96 + 95 + 94 + 93 + 92 + 91 = 90 \times 10 + (1 + 2 + \cdots + 9) + 10 = 955.$$

Hence the set of possible values of the sum of elements of subsets of $S$ has at most 955 distinct elements. Now, $S$ has $2^{10} = 1024$ subsets. It then follows from the PHP that at least two subsets have elements adding up to the same value. Which assignment of 'objects' and 'pigeon-holes' is appropriate here?

**Example:**  25 points are marked inside the area bounded by a rectangle 6 cm long and 4 cm wide. Show that there are at least two points that are at most $\sqrt{2}$ cm apart.

**Solution:**  Mark out $1 \times 1$ (cm) squares in the area of the rectangle as shown below:

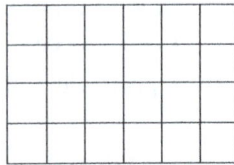

There will be 24 of these 'pigeon-holes', and we take the points to be the objects. The PHP says that there will be at least two points lying within or along the boundaries of the same $1 \times 1$ square. These can be separated by at most the length of a diagonal: $\sqrt{1^2 + 1^2} = \sqrt{2}$ cm.

**Example:**  Let $a_1, a_2, a_3, \ldots, a_{1997}$ represent an arbitrary arrangement of $1, 2, 3, \ldots, 1997$. Is the product $(a_1 - 1)(a_2 - 2)(a_3 - 3) \cdots (a_{1997} - 1997)$ an odd or even number? Justify your answer.

**Solution:**  There are 999 odd numbers and 998 even numbers between 1 and 1997. Hence there are 998 of those pairs with even number on RHS. But there are 999 pairs with odd number on the LHS, so at least one of these pairs must have an odd number on the right too, so that their difference is even. Hence the product $(a_1 - 1)(a_2 - 2)(a_3 - 3) \cdots (a_{1997} - 1997)$ has an even factor and so is even.

Finally, here is the solution to the Appetizer Problem posed at the beginning of this section.

**Solution of Appetizer Problem:** Consider pairing off the elements as follows:

$$(3, 155), (11, 147), \ldots, (75, 83).$$

Observe that we have paired them strategically so that the sum of each pair is 158, each element of A appears just once, and we have 10 such pairs. It is clear that if we take all the left-hand members, or all the right-hand members, our set will be 158-free. But, by the PHP, if we extract $10 + 1 = 11$ elements of A, we are bound to have at least two belonging to one such pair. Hence, the largest possible membership is 10, giving (A) as the correct choice.

*The pigeon principle that I know is ... you eat them!*

## 8.8 Problems

**Problem 1:** *While cleaning out the garage, John found four old single-digit house numbers: one 3, one 4 and two 5s. The number of different two-digit house numbers he can create is:*

(A) 4  (B) 5  (C) 6  (D) 7  (E) 8

**Problem 2:** *There are 10 people at a party. Each person shakes hands with the other 9 exactly once. How many handshakes will there be?*

(A) 90  (B) 19  (C) 45  (D) 38  (E) 20

**Problem 3:**   *A diagonal of a polygon is a line joining two non-consecutive vertices. Which of the following can be the number of diagonals of a polygon?*

(A) 4   (B) 164   (C) 55   (D) 35   (E) 33

**Problem 4:**   *Approximately how many digits are there in the expansion of $2^{1999}$?*

(A) 400   (B) 4000   (C) 6000   (D) 600   (E) 1999

**Problem 5:**   *A cube is cut into two pieces with a single straight cut. What is the maximum possible number of sides of a resulting cross-section?*

(A) 4    (B) 5    (C) 6    (D) 7    (E) more than 7.

**Problem 6:**   *In the first round of the Zimbabwe Maths Olympiad, there are 30 multiple choice questions with options (A)–(E). How many different answering patterns are possible?*

(A) $5^{30}$    (B) $30^5$    (C) $130^6$    (D) $6^{30}$    (E) $5 \cdot 6!$

**Problem 7:**   *The map shown is to be coloured with four colours (one of which is the colour in the X region) such that no adjacent areas have the same colour. What is the maximum number of regions, including X, that can be coloured with X's colour?*

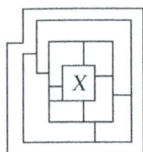

(A) 1   (B) 2   (C) 3   (D) 5   (E) Not enough information

**Problem 8:**   *How many triangles are in the figure below?*

(A) 4    (B) 8    (C) 12    (D) 16    (E) 20

**Problem 9:**   *There are 120 permutations of the letters MATHS. Suppose they are arranged in alphabetical order, from AHMST to TSMHA. What will be the 60th permutation?*

(A) *MSTHA*   (B) *MTAHS*   (C) *MHTSA*   (D) *MATHS*   (E) *MHAST*

**Problem 10:**  *I have two red, two blue and two green socks mixed up in my drawer. What is the smallest number of socks I should pick up, in the dark, so that I can be sure I have two of the same colour?*

(A) 2   (B) 3   (C) 4   (D) 5   (E) 6

**Problem 11:**  *At a party, each boy was supposed to shake hands with each of the other boys present exactly once. Each girl present was supposed to shake hands once with each of the other girls present. No members of the opposite sex were supposed to shake hands. We had more boys than girls and altogether we had seven handshakes. How many boys were at the party?*

(A) 2   (B) 3   (C) 4   (D) 5   (E) 6

**Problem 12:**  *A rectangular $4 \times 3 \times 2$ block has its surface painted red and then is cut into cubes with each edge 1 unit. The number of cubes having exactly one face painted red is:*

(A) 0   (B) 4   (C) 8   (D) 12   (E) 24

**Problem 13:**  *How many odd numbers between 0 and 1000 have distinct digits?*

(A) 375   (B) 345   (C) 350   (D) 365   (E) 375

## 8.9   Solutions

**Solution 1:**   All the two-digit numbers will have distinct digits, selected from $3, 4, 5$, except the extra possibility, 55. Considering those with distinct digits, there are three ways of choosing the first letter, then (having chosen it) there are two ways of choosing the second letter. Hence there are $3! + 1$ numbers. They are  34, 35, 43, 45, 53, 54 and 55. Hence (D).

**Solution 2:**   First try a direct count of the number of handshakes for smaller groups of people, like two, three, etc. Experimentation is a very important aspect of problem solving because, not only does it help us to become more accustomed to the problem and to discover various pitfalls lying in our path, but it reveals hidden patterns. (Mathematics is sometimes said to be a deductive science, but that is a half-truth! Many great mathematical results were first guessed at and conjectured through experimentation.) If we have two people there will be just one handshake as illustrated diagramatically below, with a dot representing a person, and a line joining two dots representing a handshake.

As seen in the second diagram, there can only be three distinct lines joining the three points. This means that when we have a group of three people, there will be three distinct handshakes.

The next two diagrams show us that for a group of four people there will be six distinct handshakes, and for group of five people there will be 10 handshakes.

This method of using diagrams to keep track of the number of handshakes works well for small groups of people, but gets quite intricate as the groups get larger. When we have enough diagrams to help us see the developing pattern, we look for a formula! We have collected enough information to be able to guess it.

| Number of people in group | Number of handshakes |
|:---:|:---:|
| 2 | $2 = 1 + 1 = \frac{(2) \times (2-1)}{2}$ |
| 3 | $3 = 1 + 2 = \frac{(3) \times (3-1)}{2}$ |
| 4 | $6 = 1 + 2 + 3 = \frac{(4) \times (4-1)}{2}$ |
| 5 | $10 = 1 + 2 + 3 + 4 = \frac{(4) \times (4-1)}{2} =$ |
| $\cdots$ | $\cdots$ |
| $\cdots$ | $\cdots$ |
| $n$ | $\frac{(n) \times (n-1)}{2}$ |

The table suggests that the number of handshakes for a group of $n$ people is $\frac{n(n-1)}{2}$.

Now let us use combinatorics, as in the example on page 266, to arrive at the same conclusion, so verifying our guess. Observe that whenever we have two people we can have a handshake so that the number of distinct handshakes equals the number of distinct pairs of people that can be chosen from the group of $n$ people. Hence the number $H_n$ of handshakes is given by

$$H_n = \binom{n}{2} = \frac{n!}{(n-2)!2!} = \frac{n(n-1)(n-2)!}{(n-2)!2!} = \frac{n(n-1)}{2}.$$

There is yet another route we could have used to get to the same formula. This results from using a different technique in keeping track of the number of handshakes.

Let the $n$ members of the group stand in a line. Let the $n$th person shake hands with the $(n-1)$th, $(n-2)$th, ..., 2nd, and lastly the 1st person, then take a seat. After this let the $(n-1)$th person repeat the procedure with each of the remaining members of the file, and then take a seat when he is through. After which the $(n-2)$th person repeats the process, then the others in turn, until the 2nd person shakes hands with the first and sits down. We make the following observations:

(a) Each person will shake the other $(n-1)$ people exactly once! – Do you believe that? Try convincing yourself by experimenting with a small group to get the physical picture of what's going on.
(b) The $n$th person is going to do a total of $(n-1)$ handshakes, then the $(n-1)$th person does $(n-2)$, the $(n-2)$th person does $(n-3)$, and so on up to just one handshake for the 2nd person.

We will then have a total of $1 + 2 + \cdots + (n-3) + (n-2) + (n-1) = H_n$ handshakes for the whole process. This is the sum of the first $(n-1)$ natural numbers. You know the formula:

$$S_n = 1 + 2 + 3 + \cdots + n = \frac{n(n+1)}{2}.$$

Writing $(n-1)$ in place of $n$ gives

$$H_n = S_{n-1} = \frac{(n-1)(n-1+1)}{2} = \frac{n(n-1)}{2} \quad \text{as before.}$$

We have still another way of reaching this conclusion. We will show that there is an *isomorphism* between this problem and the problem of the number of diagonals $D_n$ of an $n$-gon, i.e. the two problems *have the same form*. Look at the diagrams we started off with. We used these to keep track of the number of handshakes, and help us guess a formula. We now use them to derive the formula rigorously.

We will see in the next problem that $D_n = \frac{n(n-3)}{2}$. Now, the total number of lines joining $n$ points (of an $n$-gon) is the number of diagonals plus the number of sides. Hence

$$
\begin{aligned}
H_n &= D_n + n \\
&= \frac{n(n-3)}{2} + n \\
&= \frac{n^2 - 3n}{2} + \frac{2n}{2} \\
&= \frac{n(n-1)}{2}, \quad \text{as before.}
\end{aligned}
$$

The ideas you have been introduced to here are of course not restricted to handshakes. They have wide application. For example, if in a tournament each of five teams is supposed to play the other four exactly once, then we are going to have a total of $\frac{5 \times 4}{2} = 10$ matches in the tournament. Returning to the problem we started with: the number of handshakes at the 10-people party is given by

$$
H_{10} = \frac{10 \times (10-1)}{2} = 45, \quad \text{hence (C).}
$$

**Solution 3:**   Let $D_n$ be the number of diagonals of an $n$-sided polygon and let $X$, $Y$, $Z$ be three consecutive vertices of the polygon.

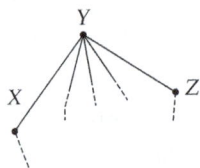

By definition, $XY$ and $YZ$ are not diagonals (in fact they are sides). Hence by joining $Y$ to the remaining $n - 3$ vertices we can draw all the diagonals emanating from $Y$. Doing this for all the $n$ vertices gives us a number $n(n - 3)$. However $n(n-3)$ is twice as large as $D_n$ because a line joining two vertices can be thought of as emanating from any one of the two vertices so that each diagonal $Y$ has in fact been counted *twice*. So

$$
2D_n = n(n-3), \quad \text{so that } D_n = \frac{n(n-3)}{2} \quad \text{thus}
$$

$$
D_{10} = \frac{10 \times 7}{2} = 35. \text{ Hence (D).}
$$

We can easily verify that no positive integral values of $n$ greater than 3 exist for which $D_n = 4$, 164, 55 or 33.

**Solution 4:**   Observe that $10^3 = 1000$ and $2^{10} = 1024$, so that $10^3 \approx 2^{10}$. Hence

$$2^{1999} = (2^{10})^{199.9}$$
$$\approx (10^3)^{199.9}$$
$$= 10^{599.7}.$$

Hence there are approximately 600 digits, which is (D).

Notice that by using $10^3 \approx 2^{10}$ we can calculate an approximate value for $\log_{10}(2)$: it is about $3/10$, since $10^{\frac{3}{10}} \approx 2$. Thus $\log 2^{1999} = 1999 \log 2 \approx 3 \times 199.9 = 599.7$, so that $2^{1999} \approx 10^{599.7}$.

**Solution 5:**   Consider a cube as shown with vertices labelled $A, B, C, D, E, F, G, H$. The plane cut passing through the midpoints of each of $BC, CD, DH, HE, EF, FB$ leaves a hexagonal shaped area. To show that this is the best possible solution, note that each edge of the cross-section corresponds to a face of the cube.

Since there are six faces, there can be at most six sides in any cross-section. Hence (C).

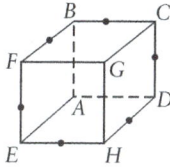

**Solution 6:**   There are six ways of responding to each of the 30 questions: either choose $A, B, C, D, E$ or leave the question unanswered. Hence we have

$$6 \times 6 \times 6 \times 6 \times \cdots \times 6 \ \ \text{(thirty factors)} = 6^{30}$$

ways in which one can answer the whole paper. Hence (D).

**Solution 7:**   No region adjacent to $X$ can be coloured with the colour $X$. If we try to colour the five regions bordering $X$ with just two colours $C_1, C_2$, then we have a clockwise sequence $C_1, C_2, C_1, C_2, C_1$, and the fifth region borders the first, both having the same colour! Therefore either $A$ or $B$ is adjacent to regions bounded with three colours. So either $A$ or $B$ is coloured with the colour of $X$. No other region (in particular $I$) can be coloured with the colour of $X$. Hence (B).

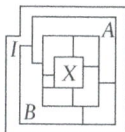

**Solution 8:**   Using the four-fold symmetry of the figure, there are 16 triangles, in four groups of four. A representative of each group is shown below.

Group I    Group II    Group III    Group IV

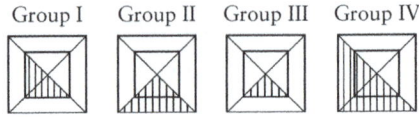

Hence (D).

**Solution 9:**   There are 4! = 24 permutations beginning with $A$, and 24 beginning with $H$. Thus arrangements 49 – 72 must begin with $M$. There are 3! = 6 permutations beginning with $MA$ and thus arrangements 55 – 60 must begin with $MH$. The 60th arrangement will be the last in the sequence, i.e $MHTSA$. Hence (C).

**Solution 10:**   In any selection of three socks, we might choose a red (R), a blue (B) and a green (G). Now the fourth has to be one of R, B and G, and so it's inevitable that we get a matching pair once we choose the fourth one. Hence the smallest number of socks we should choose so that we are sure of getting at least two of the same colour is 4. Hence (C).

**Solution 11:**   (See Problem 2 above.) Let there be $b$ boys and $g$ girls at the party with $b > g$. So $\frac{b(b-1)}{2} + \frac{g(g-1)}{2} = 7$, or $b(b-1) + a(a-1) = 14$. Now $b(b-1)$ and $a(a-1)$ are always even (products of consecutive integers) and they add up to 14. Hence we consider the sums

$$12 + 2 = 14, \quad \text{or } (4 \times 3) + (2 \times 1) = 14$$
$$8 + 6 = 14$$
$$10 + 4 = 14.$$

Check that only the first can be expressed as $b(b-1) + g(g-1)$. Since $b > g$, we have $b = 4$ and $g = 2$. There were four boys at the party, so that (C) is the correct choice.

**Solution 12:**   The dimensions of the sides of the large cube are $4 \times 3$, $4 \times 2$, and $3 \times 2$. The only blocks which will have exactly one face painted red will be the blocks which did not touch an edge of the original cube, but were on its surface. Clearly, any block on a face which had a length of 2 will have to touch one of the edges, so we can thus eliminate all the faces that had 2 as one of their dimensions. This only leaves the two faces that measured $4 \times 3$, and they each have two centre blocks. There are thus four new cubes which have exactly one face painted red. Hence (B).

  A visual approach to the solution is as follows, using the sketch of the $4 \times 3 \times 2$ block below.

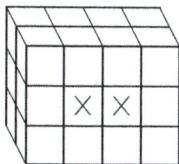

We can see that there are four cubes with exactly one face painted red (two marked with X's, and the other two on the opposite side).

**Solution 13:**   A number between 0 and 1000 has either 1, 2 or 3 digits. Consider these cases separately:

*1-digit numbers:*  We have five of these, namely 1, 3, 5, 7 and 9.

*2-digit numbers:*   Since the number is odd, the last digit is either 1, 3, 5, 7 or 9. So there are five different ways of choosing the last digit. Having chosen the last digit, there are then eight different digits that can be used in writing the first digit of the 2-digit number (zero cannot be used, and the digit must be distinct from the digit already used). Hence we have a total of $8 \times 5 = 40$  2-digit odd numbers with distinct digits.

*3-digit numbers:*  Here we have two types that have to be considered separately, namely those that contain zero as the middle number and those that do not.

*Those with zero in middle:*  As before, for such a number, there are five different ways of choosing the last digit, and having done that (and observed that the middle digit is fixed already), there are eight different ways of writing the first digit. Hence we have a total of $5 \times 1 \times 8 = 40$ such numbers.

*Those without zero in middle:* We have five different ways of writing the last digit, and for each of these we have eight choices for the middle digit, and then seven ways for the first (do you see why?). Hence we have total of $5 \times 8 \times 7 = 280$  such numbers.

   Taking stock of our findings, we have a total of $280 + 40 + 40 + 5 = 365$ numbers between 0 and 1000 for which the digits are different. Hence (D).

# 9 Miscellaneous problems and solutions

"Hmm, it looks like your vehicle is fraught with miscellaneous problems Mr Bond. The number plate is incorrect as well"

## 9.1 Problems

The problems in this last Toolchest can fairly be called a 'miscellany', in that they vary greatly, both in the tools needed for solution, and in difficulty. They are only roughly ordered, in that you will generally find the easier problems before the more difficult ones, but there is no guarantee of that: in any case, different people will find different problems easier. We also make no claim to have supplied all the necessary tools, or similar examples, in the foregoing Toolchests. Eight important areas have been discussed in detail. But in this ninth Toolchest you will be challenged to discover and apply a wider variety of methods – some you meet in standard curricula, like algebra and coordinate geometry, as well as some you don't, like functional equations.

**Problem 1:** *A circular sheet of paper of radius 6 cm is cut into six equal sectors. Each sector is formed into the curved surface of a cone, with no overlap. What is the height (in cm) of each cone?*

(A) $\sqrt{27}$   (B) $\sqrt{32}$   (C) $\sqrt{35}$   (D) 6   (E) 7

**Problem 2:**   *The minimum value of the function $f(x) = 2^{x^2-2x}$ is*

(A) $\frac{1}{2}$   (B) $-2$   (C) 2   (D) 0   (E) 1

**Problem 3:**   *The number of points common to the graphs of $y = |x|$ and $y = |x^2 - 4|$ is*

(A) 0   (B) 1   (C)2   (D) 3   (E) 4

**Problem 4:**   *When doing a series of additions on a calculator a student noted that she added 35 095 instead of 35.95. In order to correct this error in a single step, she should now*

(A) add 35.95   (B) subtract 35 059.05   (C) subtract 35 130.95
(D) add 35 130.95   (E) subtract 35 095

**Problem 5:**   *If $R = 3^x + 3^{-x}$ and $S = 3^x - 3^{-x}$, then $R^2 - S^2$ is*

(A) $2(3^{2x})$   (B) $2(3^{-2x})$   (C) 0   (D) 4   (E) 12

**Problem 6:**   *If $f(x) = x^2 - 7x + k$ and $f(k) = -9$, find $f(-1)$.*

(A) $-9$   (B) $-3$   (C) 3   (D) 5   (E) 11

**Problem 7:**   *The volume of a cylinder with radius $r$ and height $h$ is given by the expression $\pi r^2 h$. What is the volume of sand in the cylindrical hole with radius 2 m and depth 5 m?*

(A) $20\pi$   (B) $40\pi$   (C) $10\pi$   (D) None of these   (E) $5\pi$

**Problem 8:**   *If $f(2x - 1) = 4x^2 - 10x + 16$, find $f(x)$.*

(A) $x^2 + 3x + 18$   (B) $x^2 - 3x + 18$   (C) $x^2 - 6x + 12$   (D) $x^2 - 7x + 18$
(E) $x^2 - 3x + 12$

**Problem 9:**   *When $x^3 - kx^2 - 10kx + 25$ is divided by $x - 2$, the remainder is 9. The value of $k$ is*

(A) 25   (B) $-\frac{1}{2}$   (C) $\frac{7}{4}$   (D) 1   (E) $-\frac{13}{8}$

**Problem 10:**   *What is the value of $2^{\log_a(a^5)}$?*

(A) $2a^5$   (B) $(2^a)^5$   (C) $a^5$   (D) $5^2$   (E) $2^5$

**Problem 11:**   *If $a^x = c^q = b$ and $c^y = a^z = d$, then:*

(A) $xy = qz$   (B) $\frac{x}{y} = \frac{q}{z}$   (C) $x^y = q^z$   (D) $x-y = q-z$   (E) $x+y = q+z$

**Problem 12:**   *If $c^d = 3$, then $c^{4d} - 5$ equals*

(A) 76   (B) 7   (C) 22   (D) 86   (E) None of these

**Problem 13:**   *If $xy = 6$, $yz = 9$ and $zx = 24$, then a value of $xyz$ is*

(A) 648   (B) 1296   (C) 48   (D) 1.5   (E) 36

**Problem 14:**   *The points $A(3,4)$, $B(4,3)$, $C(-3,-4)$, $D(3,-4)$, $E(4,-3)$ are marked on a coordinate grid. The line segment that is horizontal is:*

(A) *AD*   (B) *BE*   (C) *BC*   (D) *CD*   (E) *AB*

**Problem 15:**  *If* $a + b + c = 10$ *and* $ab + bc + ac = 20$, *what is the value of* $a^2 + b^2 + c^2$?

(A) 60   (B) 70   (C) 40   (D) 20   (E) 10

**Problem 16:**  *A bus passes through stations A, B and C in that order. Before passing through the station A, the only person in the bus is the bus driver. At B and C, it picks up twice as many people as it picked up from the previous stop. Which of the following can be the number of people in the bus after it has passed station C if no-one alighted from the bus at each of the three stations?*

(A) 24   (B) 8   (C) 20   (D) 18   (E) 14

**Problem 17:**   *I define a function* $\Gamma(r)$ *such that* $\Gamma(r+1) = r\Gamma(r)$ *and* $\Gamma(1) = 1$ *whenever r is a positive whole number. Find* $\Gamma(7)$.

(A) Cannot be determined   (B) 120   (C) 720   (D) 5040   (E) 24

**Problem 18:**  *All pupils in two schools take an exam. The table below shows the average scores for boys and for girls at each school. Also shown is the overall average for each school and the overall average for all boys at the two schools. What is the overall average for all the girls at the two schools?*

(A) 81   (B) 82   (C) 83   (D) 84   (E) 85

|  | Zamunya School | Wakiti School | Both Schools |
|---|---|---|---|
| Boys | 71 | 81 | 79 |
| Girls | 76 | 90 | ? |
| Boys and Girls | 74 | 84 | |

**Problem 19:**   *'As I was going to St Ives I met a man with seven wives; each wife had seven sacks, each sack had seven hens, each hen had seven chicks.' How many were going to St Ives?*

(A) 2 401   (B) 2 801   (C) 1   (D) 2 802   (E) 16 807

**Problem 20:**   *'As I was going to St Ives I met a man with seven wives; each wife had seven sacks, each sack had seven hens, each hen had seven chicks.' Suppose it costs the man 10c per day to feed a chick and 20c per day to feed a hen. What is the daily cost of hen and chick feed?*

(A) $2.10   (B) $6.30   (C) $44.10   (D) $88.20   (E) $308.70

**Problem 21:**   *Consider distinct points A and B, with x-coordinates* $x = a$ *and* $x = b$, *respectively, on the curve* $y = x^2 + x + 1$. *At what point will the line AB (produced) cut the y-axis?*

(A) $y = \frac{a+b+1}{ab-1}$    (B) $y = a^2+a+1$    (C) $y = b^2+b+1$    (D) $y = a^2-b^2+1$
(E) $y = 1 - ab$

**Problem 22:**    *The school choir has 100 female and 80 male members. The church choir has 80 female and 100 male members. There are 60 females who are members of both choirs and altogether there are 230 people who are in at least one of the choirs. How many males are in the school choir but not in the church choir?*

(A) 10    (B) 20    (C) 30    (D) 50    (E) 70

**Problem 23:**    *A box is filled with coins and beads, all of which are coloured either silver or gold. Twenty percent of the objects in the box are beads, and 40 percent of the coins are silver. One item is randomly chosen from the box. What is the probability that it is a gold coin?*

(A) 0.8    (B) 0.48    (C) 0.6    (D) 0.4    (E) 0.52

**Problem 24:**    *The diagram shows a maze used for testing and training mice. A mouse begins at point A and at each junction as represented by a dot has equal chances of choosing any of the new paths available. The mouse can move in any direction but cannot travel along any path more than once. What is the probability of the mouse reaching point C after passing through more than three junctions?*

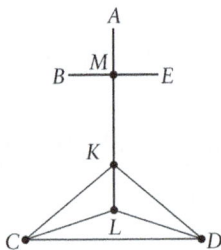

(A) 1    (B) $\frac{1}{6}$    (C) $\frac{2}{9}$    (D) $\frac{1}{9}$    (E) 0

**Problem 25:**    *What is the value of*

$$\sqrt{1 + 2\sqrt{1 + 2\sqrt{1 + 2\sqrt{1 + 2\sqrt{1 + \cdots}}}}}?$$

(A) $1 - \sqrt{2}$    (B) $1 + \sqrt{3}$    (C) $1 + \sqrt{2}$    (D) $\sqrt{2} - 1$    (E) $\sqrt{3}$

**Problem 26:**    *Suppose we have the system of equations*

$$a + b + c = 10$$
$$a^2 + b^2 = 65$$
$$a + c = 3.$$

*Find all possible values of c.*

(A) −1 and 7   (B) 1 and −7   (C) −1 and −7   (D) 4 and −4   (E) 7 and −4

**Problem 27:**  There are 14 people in a room. What is the probability that at least two of them were born in the same calendar month?

(A) 0   (B) $\frac{1}{12}$   (C) $\frac{14}{12}$   (D) 1   (E) $\frac{1}{14}$

**Problem 28:**  There are 14 people in a room. What is the probability that at least two of them share the same birthday?

(A) 0   (B) $\frac{2}{365}$   (C) 0.14   (D) 0.22   (E) 0.02

**Problem 29:**  How many people should be gathered in a room before the probability that at least two of them share the same birthday is greater than even (i.e. probability 0.5)?

(A) 48   (B) 100   (C) 23   (D) 183   (E) 36

**Problem 30:**  Mercy goes shopping with a $50 note. She buys meat for $23.99, onions for $15.45, and one tomato. The till operator gives her $9.27 change. When she gets home she looks at her till slip and notices that she received 45c change too little. How much did the tomato cost?

(A) $1.84   (B) $0.84   (C) $1.29   (D) $0.39   (E) $1.74

**Problem 31:**  If $\log_2(\log_3 a) = 2$ what is the value of a?

(A) 3   (B) 9   (C) 27   (D) 81   (E) 243

**Problem 32:**  If $a = \log_{10} 12$ and $b = \log_2 12$, express $\log_6 10$ in terms of a and b.

(A) $\frac{b}{ab-a}$   (B) $ab - a$   (C) $\frac{a}{b}$   (D) $\frac{ab-a}{b}$   (E) $\frac{b}{a}$

**Problem 33:**  There are 300 boys who represent their school in sports in summer and in winter. In summer 60 play cricket and the remainder play tennis. In winter they must play soccer or hockey but not both. 56% of the hockey players play cricket in summer and 30% of the cricket players play soccer in winter. How many boys play tennis and soccer?

(A) 21   (B) 30   (C) 54   (D) 99   (E) 120

**Problem 34:**  An electric device contains two components W and Z. If the device fails, the probability that W will need replacement is 0.5. If W has to be replaced, the probability that Z will also have to be replaced is 0.7. If it is not necessary to replace W, the probability that Z will have to be replaced is 0.1. What proportion of all the failures will require the replacement of W and Z?

(A) 0.035   (B) 0.05   (C) 0.15   (D) 0.35   (E) 0.7

**Problem 35:**   *If we define $i = \sqrt{-1}$ what is the value of $i^{62}$?*

(A) $i$   (B) 1   (C) $-1$   (D) $-i$   (E) $i^{-1}$

**Problem 36:**   *If we define $(n) = n^n$, what is the value of $((2))$?*

(A) 64   (B) 256   (C) 8   (D) 2   (E) 16

**Problem 37:**   *Mr Gore drives at 60 km/hr and leaves Masvingo at 0800 hours. Mr Muzvanya drives at 100 km/hr and leaves Masvingo at 0900 hrs, driving on the the same road as Mr Gore. At what time will Muzvanya catch up with Gore?*

(A) 1 030 hrs   (B) 1 200 hrs   (C) 1 100 hrs   (D) 1 145 hrs   (E) 1 015 hrs

**Problem 38:**   *Ackerman's function is defined for non-negative integers n and k by the following three equations. Evaluate $f(2,2)$.*

(i) $f(0,n) = n + 1$;
(ii) $f(k,0) = f(k-1,1)$;
(iii) $f(k+1,n+1) = f(k, f(k+1,n))$.

(A) 6   (B) 7   (C) 5   (D) 9   (E) 8

**Problem 39:**   *Two forgetful friends agree to meet one afternoon. Each one remembers that the meeting time was between 2 pm and 5 pm but each has forgotten the exact time agreed upon. Each one decided to go to the meeting place at a random time between 2 pm and 5 pm, wait for half an hour, and then leave if the other one does not arrive. What is the probability that the two friends will meet?*

(A) $\frac{11}{36}$   (B) $\frac{1}{3}$   (C) $\frac{1}{6}$   (D) $\frac{25}{36}$   (E) $\frac{1}{36}$

**Problem 40:**   *In the sum below, each letter represents a different nonzero digit. If $R = 8$ and $F = 1$ what is $T$?*

$$
\begin{array}{r}
TWO \\
+TWO \\
\hline
FOUR
\end{array}
$$

(A) 5   (B) 6   (C) 7   (D) 8   (E) 9

**Problem 41:**   *Four of the vertices of a cube form a regular tetrahedron, as shown in the figure. What fraction of the volume of the cube is occupied by the tetrahedron?*

(A) $\frac{2}{3}$   (B) $\frac{1}{6}$   (C) $\frac{1}{3}$   (D) $\frac{1}{2}$   (E) $\frac{5}{6}$

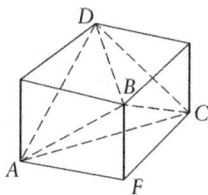

**Problem 42:**   *Consider the four statements below:*

- *All idiots are fools.*
- *No twits are idiots.*
- *Everyone is a twit or a nut.*
- *Fools are not nuts.*

*Which of the following additional statements must be true?*

1. *No nuts are fools.*
2. *All twits are fools.*
3. *There are no idiots.*

(A) 1 and 3   (B) 1 and 2   (C) 2 and 3   (D) 1 only   (E) 3 only

**Problem 43:**   *The 'stem and leaf' table below gives the number of goals scored by each of 22 football clubs in one season. Find the median of the distribution.*

| Stems | Leaves | Cumulative count |
|-------|--------|------------------|
| 2 | 9 | 1 |
| 3 | | 1 |
| 4 | 18 | 3 |
| 5 | 0001112379 | 13 |
| 6 | 0013 | 17 |
| 7 | 11349 | 22 |

(A) 50   (B) 53   (C) 55   (D) 57   (E) 59

**Problem 44:**   *Consider the equation $ax^2 + bx + 1 = 0$. If a and b are randomly chosen real numbers, each lying between 0 and 4, what is the probability that the equation has no real roots?*

(A) $\frac{2}{3}$   (B) $\frac{1}{3}$   (C) $\frac{3}{16}$   (D) $\frac{13}{16}$   (E) $\frac{1}{16}$

**Problem 45:**   *A shop has 600 shirts for sale. Of these 100 are red, 200 are blue and the remainder are yellow. Not all of the shirts have collars and/or pockets; 400 have collars and 150 have pockets. What is the maximum possible number of blue shirts with pockets and without collars?*

(A) 50   (B) 100   (C) 150   (D) 200   (E) None of these

**Problem 46:**  *Last Sunday's newspaper had its pages numbered 1 to 16. There was a fierce wind blowing and along the pavement I found a double page with the number 5 on one of the pages. What were the numbers on the other three pages? Can you find an algorithm to determine the three numbers sharing a sheet with the page-number k, if there are n sheets (4n pages)?*

(A) 4, 9, 10   (B) 4, 11, 12   (C) 6, 11, 12   (D) 6, 9, 10   (E) 6, 13, 14

**Problem 47:**  *If $a + b = 10$ and $b + c = 20$ and $a + c = 30$, find the value of $a^2 + b^2 + c^2$.*

(A) 500   (B) 350   (C) 400   (D) 3600   (E) 1100

**Problem 48:**  *A 7 by 7 'magic square' is constructed, in which each of the 49 squares contains one of the integers in the set $S = \{1, 2, 3, \ldots, 49\}$ and all squares have different numbers, so that the 49 integers are distributed through the small squares. The numbers in each column and those in each row have the same sum. What is this sum?*

(A) 125   (B) 175   (C) 49   (D) 25   (E) 50

**Problem 49:**  *Find the value of $\sqrt{5 + \sqrt{5}} - \sqrt{5 - \sqrt{5}}$.*

(A) $\sqrt{10 - 4\sqrt{5}}$   (B) $-10$   (C) $\sqrt{30}$   (D) $\sqrt{10 - 14\sqrt{5}}$   (E) $\sqrt{5 - 6\sqrt{5}}$

**Problem 50:**  *On a die the numbers on opposite faces add up to 7. The die shown is rolled edge over edge along the path until it rests on the square labeled X. What will be the number on the uppermost face when the die is in square X?*

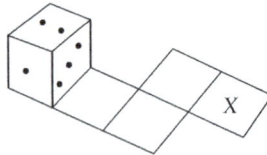

(A) 1   (B) 2   (C) 3   (D) 4   (E) 5

**Problem 51:**  *Jonah is in a race. He has to start at point X, run to the fence AB and touch it and then run to finish at point Y. He has calculated correctly that to run the shortest distance he should touch the fence at point R where $A\hat{R}X = 30°$. The distance from X to the fence is 400 m whilst that from Y to the fence is 500 m Find the total distance run by Jonah.*

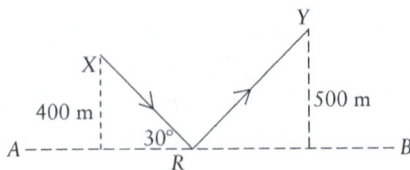

(A) 1800 m   (B) 1600 m   (C) 2000 m   (D) 900 m   (E) 1000 m

**Problem 52:**   *The figure below is called a network. It represents connections between three airports $B_1, B_2, B_3$ in one country and three airports $C_1, C_2, C_3$ in another country. The number on each linking line gives the number of different airlines flying on the route. Which of the matrices below does not give correct information on flights between the two countries?*

(A) $\begin{bmatrix} 0 & 1 & 2 \\ 2 & 3 & 0 \\ 0 & 0 & 1 \end{bmatrix}$  (B) $\begin{bmatrix} 0 & 1 & 2 \\ 1 & 0 & 1 \\ 0 & 3 & 1 \end{bmatrix}$  (C) $\begin{bmatrix} 3 & 0 & 2 \\ 0 & 1 & 0 \\ 1 & 2 & 0 \end{bmatrix}$

(D) $\begin{bmatrix} 2 & 3 & 0 \\ 0 & 0 & 1 \\ 0 & 1 & 2 \end{bmatrix}$  (E) $\begin{bmatrix} 2 & 3 & 0 \\ 0 & 1 & 2 \\ 0 & 0 & 1 \end{bmatrix}$

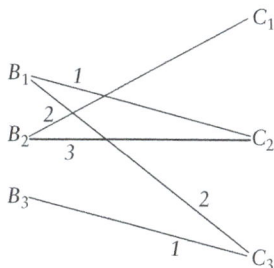

**Problem 53:**   *At a practice session, each of three players was required to hit a ball which was thrown to him. The balls were thrown once to each player, independently. The probabilities of each player hitting the ball were $\frac{1}{6}, \frac{1}{4}, \frac{1}{3}$ respectively. What is the probability that exactly one of the players hits his ball?*

(A) $\frac{31}{72}$  (B) $\frac{1}{72}$  (C) $\frac{30}{72}$  (D) $\frac{9}{72}$  (E) $\frac{51}{72}$

**Problem 54:**   *Two fair dice are thrown. Let the numbers showing be $a, b$ respectively, and let $s = a + b$. Let $X = P(s \leq 4)$, $Y = P(s = 7)$, and $Z = P(s > 9)$. (This means $X$ is the probability that $s \leq 4$, etc.) Then the relationship between $X, Y$ and $Z$ is*

(A) $X < Y = Z$  (B) $X = Y < Z$  (C) $X > Y > Z$  (D) $X = Y = Z$  (E) $X < Y < Z$

**Problem 55:**   *Evaluate*

$$\lim_{x \to \infty} \frac{8 + 2^x}{8 - 2^x}$$

(A) 0  (B) 1  (C) $-1$  (D) 2  (E) $-2$

**Problem 56:**   *Each of Tapina, Suwisa and Gift buy pens and pencils, each one buying 10 items. Gift buys twice as many pens as Suwisa buys pencils,* *and Suwisa buys twice as many pens as Tapina buys pencils. Each bought an even number of pencils. How many pencils were bought altogether?*

(A) 16   (B) 12   (C) 18   (D) 6   (E) 10

**Problem 57:**   *Patricia walks to visit her friend Charles, and then returns home by exactly the same route; all in all, she takes a total of 6 hours walking time. Patricia walks at a speed of 2 km/h when going uphill, at 6 km/h when going downhill, and at 3 km/h when walking on a flat ground. What is the total distance she walked?*

(A) 9 km   (B) 12 km   (C) 18 km   (D) 22 km   (E) 36 km

**Problem 58:**   *If $f(x - y) = f(x)f(y)$ for all x and y, and $f(x)$ never equals zero, then $f(3)$ equals:*

(A) −3   (B) 3   (C) 4   (D) $\pm\sqrt{7}$   (E) None of these

**Problem 59:**   *If $2^{2\log_2 x} = \left(2^{(\log_2 x)+1}\right) - 1$, find x.*

(A) 1   (B) 2   (C) 3   (D) 4   (E) −1

**Problem 60:**   *If $\log_{(a^2)} b + \log_{(b^2)} a = \frac{5}{2}$, $a, b > 0$, then the number of values of b, for a given value of a is*

(A) 1   (B) 2   (C) 3   (D) 4   (E) 5

**Problem 61:**   *If $x^2 + \dfrac{1}{x^2} = 14$ and $x > 0$ what is the value of $x + \dfrac{1}{x}$?*

(A) 2   (B) 16   (C) 3   (D) 8   (E) 4

**Problem 62:**   *How many values of $x \in \mathbb{R}$ are there satisfying the equation*

$$x^2 - \sqrt{(x - 1)^3} - \sqrt{(x - 1)} + 2(1 - x) = 0?$$

(A) 1   (B) 2   (C) 3   (D) 4   (E) 0

**Problem 63:**   *A right-angled triangle ABC has perimeter 624 cm and area 6864 cm².   Find the length of its hypotenuse.*

(A) 265   (B) 290   (C) 269   (D) 303
(E) Cannot be determined using given information

**Problem 64:**   *If I live in Utopia and have n coins, and can buy anything that costs between 1 cent and $1, inclusive, without needing any change, what is the smallest possible value of n?*

(A) 8   (B) 9   (C) 10   (D) 11   (E) 7

(The Utopian coinage system includes a 1c piece, 5c piece, 10c piece, 20c piece, 50c piece and a $1 piece.)

**Problem 65:**   *If x, y and z are positive, and $\frac{xy}{x+y} = a$, $\frac{xz}{x+z} = b$, and $\frac{yz}{y+z} = c$, then x is:*

(A) $\frac{abc}{ab+ac+bc}$  (B) $\frac{2abc}{ab+ac+bc}$  (C) $\frac{abc}{ac+bc-ab}$  (D) $\frac{2abc}{bc+ac-ab}$  (E) None of these

**Problem 66:**  The letters $A, B, D$ and $M$ represent the numbers $1, 8, 9$ and $9$, not necessarily in that order. The largest possible sum of the three digit numbers $BAD, DAM$ and $MAD$ is:

(A) 2159  (B) 2655  (C) 2656  (D) 2657  (E) 2958

**Problem 67:**  A man travels $100$ km at $x$ km/h, $400$ km at $2x$ km/h, and $600$ km at $3x$ km/h. For the entire trip, his average speed in km/h is:

(A) 1100  (B) $11x$  (C) $27x$  (D) $2x$  (E) $\frac{11x}{5}$

**Problem 68:**  Consider the equation

$$2x^4 - 11x^3 + 16x^2 - 11x + 2 = 0.$$

Which of the following pairs shows the correct values of $x + \frac{1}{x}$?

(A) $(4,2)$  (B) $(\frac{1}{2},2)$  (C) $(4,\frac{3}{2})$  (D) $(11,16)$  (E) $(-11,2)$

**Problem 69:**  In 'the traveller's game' a person is supposed to move from top left corner $A$ to opposite corner $B$ of a seven-by-seven 'gridded' square shown below, along one of the 'shortest routes' – that is, observing the following rule:

The traveller moves along the edges of the unit squares of the grid, in two directions only: in a right-hand direction or downwards (or we could put as: moving eastwards or southwards).

How many such routes there are from the top left-hand corner to the bottom right-hand corner?

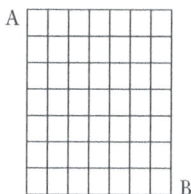

(A) 3432  (B) 1626  (C) 924  (D) 64  (E) 16

**Problem 70:**  The diameter $CD$ of a circle with centre $G$ is produced to a point $K$ so that $DK$ equals the radius of the circle $CD$. A circle with diameter $GK$ is drawn. Find the area common to the two circles if $GK = 8$ units.

(A) 16  (B) $(\frac{16\pi}{3} - 4\sqrt{3})$  (C) $16\pi$  (D) $(\frac{8\pi}{3} - \sqrt{3})$  (E) $(\frac{32\pi}{3} - 8\sqrt{3})$

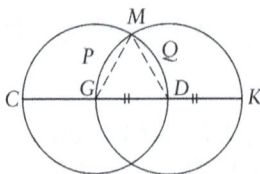

**Problem 71:**   *A mathematics student was required to solve the problem below in an examination:*
*How many triplets $(a, b, c)$ between 0 and 100 satisfy the system of equations*
*(i) $7^a = b$   (ii) $7^b = c$   (iii) $14^c = 7^{ab}$?*
*He solved the problem in four stages:*
*Stage I (i) $\times$ (ii) $\implies 7^{a+b} = bc$.*

*Stage II $\left(7^{(a+b)}\right)^2 = (bc)^2$.*

*Stage III*

$$\text{By (II)} \quad 7^{a^2 + b^2 + 2ab} = b^2 c^2$$
$$\implies \quad (7^a)^2 \cdot \left(7^b\right)^2 \cdot \left(7^{ab}\right)^2 = b^2 c^2$$
$$\implies \quad b^2 \cdot c^2 \cdot \left(7^{ab}\right)^2 = b^2 c^2$$
$$\implies \quad ab = 0.$$
$$\text{So} \quad 14^c = 7^0 = 1 \implies c = 0.$$

*Stage IV This is a contradiction, because (ii) now reads $7^b = 0$ and we know that $a^x > 0$, for real numbers $a, x$ when $a > 0$. Hence, the system has no solutions. There is an error in the student's reasoning. At what stage did the student make the error?*

*(A) I   (B) II   (C) III   (D) IV*

**Problem 72:**   *The equation*

$$\lim_{x \to \infty} \left(\frac{1}{x}\right) = 0,$$

*which is to be read as 'the limit of $\frac{1}{x}$ as x approaches infinity equals zero', is a convenient symbolic way of representing the statement that, as x takes on progressively large values, the value of $\frac{1}{x}$ gets closer and closer to zero. Given that*

$$\lim_{x \to \infty} \left(1 + \frac{a}{x}\right)^x = e^a$$

*(where e is a rather special number ≈ 2.718 . . .), find in terms of e the value of*

$$\lim_{n \to \infty} \left(\frac{n+1}{n+2}\right)^{3n}.$$

(A) $e^{-3}$  (B) $e^{-2}$  (C) $e^{-1}$  (D) cannot be determined  (E) $e$

**Problem 73:**  *Two lovers agree to meet at the time between 1 o'clock and 2 o'clock when the minute hand is exactly above the hour hand. At what time will they meet?*

(A) 5 minutes after 1  (B) $5\frac{3}{11}$ minutes after 1  (C) $5\frac{5}{11}$ minutes after 1
(D) $5\frac{4}{11}$ minutes after 1  (E) $5\frac{6}{11}$ minutes after 1

**Problem 74:**  *Using the Principle of Mathematical Induction to prove that if n is a positive integer then $2^{3n} - 1$ is divisible by 7.*

**Problem 75:**  *Show that $10^{2n-1}$ is divisible by 11 for all positive integers n.*

**Problem 76:**  *Find a number less than 3000 which, when divided by 1 leaves remainder 0, when divided by 2 leaves remainder 1, when divided by 3 leaves remainder 2, . . ., when divided by 9 leaves remainder 8, and when divided by 10 leaves remainder 9.*

**Problem 77:**  *The process of 'geometrical squaring' starts with a square of a given size within which a circle of maximum area is drawn. A second square of maximum area is drawn within the circle and the process repeated an arbitrary but finite number of times. Let the area of the square we start with be $A_1$, that of the second $A_2$, the third $A_3$ and so on.*

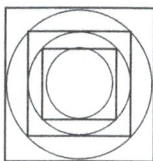

*Let $S_n = A_1 + A_2 + A_3 + \ldots + A_n$ and let $M_n = kA_1 - S_n$ where k is a positive whole number. What is the smallest value of k for which $M_n$ is always strictly positive?*

(A) 2  (B) 3  (C) 1  (D) 4  (E) 5

**Problem 78:**  *If $a, b, c$ are positive integers such that $a^2 + b^2 = c^2$, show that there exist three positive integers $p, q, r$, where $p, q$ have opposite parity, such that*

$$(a, b, c) = (r(p^2 - q^2),\ 2pqr,\ r(p^2 + q^2)).$$

**To start you off:** Such triples $(a, b, c)$ are called Pythagorean triples (see Toolchest 1, page 52). There are a number of ways of proving the required result. One is by number theory, assuming first that $a, b, c$ are relatively prime, and using congruence arithmetic (Toolchest 4) to show that $c$ must be odd and $a$ and $b$ must have opposite parity; then use the factorization of a difference of squares to prove that $a^2$ is a product of two relatively prime numbers which must therefore be squares of odd numbers, etc. Other methods use coordinate geometry, observing that Pythagorean triples correspond to points $P(\frac{a}{c}, \frac{b}{c})$ on the unit circle centred at the origin. You can parametrize such points with the gradient $t$ of the line joining $P$ to the point $(-1, 0)$.

## 9.2   Solutions

**Solution 1:**   The diagram shows one sector of radius 6.

The length of curved arc of this sector is one-sixth of the circular circumference, that is $\frac{2\pi \cdot 6}{6} = \frac{12\pi}{6} = 2\pi$. This must be equal to the circumference of the base of the cone, so we have: $2\pi r = 2\pi$, hence $r = 1$. By Pythagoras, then, the height $h$ of each cone is $h = \sqrt{6^2 - 1^2} = \sqrt{35}$. Hence (C).

**Solution 2:**   $f(x)$ will be a minimum for $x^2 - 2x$ a minimum, but $x^2 - 2x = (x - 1)^2 - 1$ is a minimum for $x = 1$, with $f(1) = 2^{1-2} = \frac{1}{2}$. Hence (A).

**Solution 3:**   The sketch illustrates the answer. Hence (E).

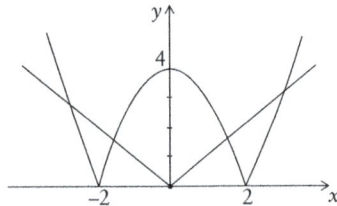

**Solution 4:**   If she subtracts 35 095, 'things will go back to where they were' just before she made the mistake (i.e. she 'isolates' the mistake) so that she can add the number she meant to add correctly, i.e. 35.95. However all this can be done in a single step by adding the number given by

$$-35\,095 + 35.95 = -35059.05$$

i.e. subtracting 35059.05. Hence (B).

**Solution 5:**  $R + S = 2(3^x)$ and $R - S = 2(3^{-x})$ giving $R^2 - S^2 = (R + S)(R - S) = 2(3^x)2(3^{-x}) = 4$. Hence (D).

**Solution 6:**  $f(k) = -9 = k^2 - 7k + k$, giving $k^2 - 6k + 9 = 0 = (k - 3)^3$, so that we have $k = 3$ and hence $f(x) = x^2 - 7x + 3$. Hence $f(-1) = 1 + 7 + 3 = 11$, so that (E) is the correct choice.

**Solution 7:**  Sorry – this is a deliberately deceptive question, to train you to scrutinize the data you are given carefully, and take nothing for granted! Observe that there is no sand in the hole, so that, although the volume of the cylindrical hole is $\pi r^2 h = 20\pi$, the volume of the sand is 0. Hence (D).

**Solution 8:**  If we let $g(x) = f(2x - 1) = 4x^2 - 10x + 16$, then

$$g\left(\frac{x + 1}{2}\right) = f\left(\frac{2(x + 1)}{2} - 1\right)$$

$$= f(x)$$

$$\text{hence } f(x) = 4\left(\frac{x + 1}{2}\right)^2 - \frac{10(x + 1)}{2} + 16$$

$$= (x + 1)^2 - 5(x + 1) + 16$$

$$= x^2 - 3x + 12. \text{ Hence (E).}$$

The equation $f(2x - 1) = 4x^2 - 10x + 16$ is an example of an important class of equations called functional equations, and what we have just done is to solve a functional equation. The question you may well be asking is, how did we find the function $m(x) = \frac{x+1}{2}$ such that $g(m(x)) = f(x)$? The answer is in the fact that, if $m^{-1}(x)$ is the inverse function of $m$, then:

$$g(x) = g(m(m^{-1}(x))) = f(m^{-1}(x)).$$

Therefore the function we want has inverse $m^{-1}(x) = 2x - 1$, from which we deduce $m(x) = \frac{x+1}{2}$.

There is an alternative method, expressing the function in terms of $2x - 1$ instead of $x$:

$f(2x - 1) = (2x - 1)^2 - 3(2x - 1) + 12$, so that $f(x) = x^2 - 3x + 12$.

**Solution 9:**  Using the remainder theorem:

$$f(2) = 9 = 8 - 4k - 20k = 25, \text{ so that } k = 1. \text{ Hence (D).}$$

**Solution 10:**  By the basic property of logarithms that 'log of product is sum of logs' (which follows from the index laws and the definition of the logarithm), we have

$$\log_a(a^5) = 5\log_a a = 5 \times 1 = 5, \text{ and so } 2^{\log_a(a^5)} = 2^5. \text{ Hence (E).}$$

**Solution 11:**

$$a^x = c^q \quad \text{yields} \quad a = c^{\frac{q}{x}}, \tag{9.1}$$

$$\text{and} \quad a^z = c^y \quad \text{yields} \quad a = c^{\frac{y}{z}}. \tag{9.2}$$

Therefore, equating the indices in (9.1) and (9.2),

$$\frac{q}{x} = \frac{y}{z},$$

so that $qz = xy$. Hence (A).

**Solution 12:**

$$c^{4d} - 5 = (c^d)^4 - 5 = 3^4 - 5 = 81 - 5 = 76. \quad \text{Hence (A).}$$

**Solution 13:**   Multiplying the equations together gives

$$(xy)(yz)(zx) = 6 \times 9 \times 6 \times 4,$$
$$\text{therefore} \quad x^2 y^2 z^2 = 6^2 \cdot 3^2 \cdot 2^2,$$
$$\text{so that} \quad xyz = 6 \cdot 3 \cdot 2 \quad (\text{or} \; -6 \cdot 3 \cdot 2)$$
$$= 36. \quad \text{Hence (E).}$$

**Solution 14:**   Two points on the same horizontal level will have the same $y$-coordinate. The only two such coordinates here are $C$ and $D$. Hence (D).

**Solution 15:**

$$100 = (a + b + c)^2$$
$$= a^2 + b^2 + c^2 + 2ab + 2bc + 2ac$$
$$= a^2 + b^2 + c^2 + 2(ab + bc + ac)$$
$$= a^2 + b^2 + c^2 + 2(20),$$
$$\text{therefore} \quad (a + b + c)^2 = 100 - 40 = 60. \quad \text{Hence (A).}$$

**Solution 16:**   Suppose it picked up $x$ people at station A. This implies that $2x$ were picked at B and $4x$ at C. Hence, after passing C there was a total of $x + 2x + 4x + 1 = 7x + 1$ people. We need to find a number which is of the form $7x + 1$; but there is only such number among the given choices and it is 8, given by $x = 1$. Hence (B).

**Solution 17:**   From the first defining equation

$$\Gamma(r + 1) = r\Gamma(r) = r\Gamma\left((r - 1) + 1\right) = r(r - 1)\Gamma(r - 1)$$
$$= r(r - 1)(r - 2)\Gamma(r - 2) \quad \text{(proceeding similarly)}$$
$$= r(r - 1)(r - 2)(r - 3)\Gamma(r - 3)$$
$$= r(r - 1)(r - 2) \cdots 4 \cdot 3 \cdot 2 \cdot \Gamma(1)$$

$$= r(r-1)(r-2)\cdots 4\cdot 3\cdot 2\cdot 1 \quad \text{(since } \Gamma(1)=1\text{)}$$
$$= r!.$$

Hence $\Gamma(7) = (7-1)! = 6! = 1\times 2\times 3\times 4\times 5\times 6 = 720$. Hence (C).

**Solution 18:**   Let $k$ be the unknown average, and denote the total numbers of pupils by $a, b, c, d$ as in the table below:

|  | Zamunya | Wakiti | Both Schools |
|---|---|---|---|
| Boys | $a$ | $b$ | $a+b$ |
| Girls | $c$ | $d$ | $c+d$ |
| Boys and Girls | $a+c$ | $b+d$ | |

Looking at the total marks scored, by all boys, by all girls, by all Zamunya pupils and by all Wakiti pupils, respectively, we have four equations:

$$71a + 81b = 79(a+b) \tag{9.3}$$
$$76c + 90d = k(c+d) \tag{9.4}$$
$$71a + 76c = 74(a+c) \tag{9.5}$$
$$81b + 90d = 84(b+d). \tag{9.6}$$

From equation (9.3), we have $\frac{b}{a} = 4$;
from equation (9.5), we have $\frac{a}{c} = \frac{2}{3}$;
from equation (9.6), we have $\frac{b}{d} = 2$.
From equation (9.4), we have:

$$k = \frac{76c + 90d}{c+d} = \frac{76(c+d) + 14d}{c+d} = 76 + \frac{d}{c+d}.$$

But

$$\frac{c+d}{d} = \frac{c}{d} + 1 = \frac{c}{a}\cdot\frac{a}{b}\cdot\frac{b}{d} + 1 = \frac{3}{2}\cdot\frac{1}{4}\cdot\frac{2}{1} + 1 = \frac{7}{4}.$$

Hence $k = 76 + 14\frac{4}{7} = 76 + 8 = 84$, which is (D).

**Solution 19:**   This is an age-old catch-question: only one is going to St Ives, for all the others are coming *from* St Ives! But if you want to know how many objects (man, wives, sacks, cats, kittens) were on the road, in addition to 'me', the answer is given as the sum of a geometric series:

$$1 + 7 + 7^2 + 7^3 + 7^4 = \frac{7^5 - 1}{7 - 1} = \frac{16\,806}{6} = 2\,801.$$

Of course, if you include 'me' the answer is 2 802 objects. Take your pick!

**Solution 20:**  No catch here! The cost of feeding each sack-full will be:

$$7 \times 20c + 49 \times 10c = \$6.30.$$

Thus total costs will be $7 \times 7 \times \$6.30 = \$308.70$. Hence (E).

**Solution 21:**  $A$ and $B$ are the points $(a, f(a))$, $(b, f(b))$, i.e. $(a, a^2 + a + 1)$ and $(b, b^2 + b + 1)$

Now, assuming $x_1 \neq x_2$, the equation of a straight line through the points $(x_1, y_1)$ and $(x_2, y_2)$ is given by

$$\frac{y - y_1}{x - x_1} = \frac{y_2 - y_1}{x_2 - x_1}.$$

So the formula gives the equation of the line as:

$$\frac{y - a^2 - a - 1}{x - a} = \frac{b^2 + b + 1 - a^2 - a - 1}{b - a}$$
$$= \frac{b^2 - a^2 + b - a}{b - a}$$
$$= \frac{(b - a)(b + a) + (b - a)}{b - a}$$
$$= \frac{(b - a)(b + a + 1)}{b - a}$$
$$= b + a + 1,$$

since distinct points ensure $a \neq b$. Now this line will cut the y-axis where $x = 0$, that is:

$$\frac{y - a^2 - a - 1}{-a} = a + b + 1,$$

therefore  $y = -(a^2 + ab + a) + a^2 + a + 1$
$$= 1 - ab. \quad \text{Hence (E).}$$

**Solution 22:**  Total number in both choirs is $180 + 180 - 230 = 130$. Total number of males in both choirs is $130 - 60 = 70$. Thus, number of males in school choir only is $80 - 70 = 10$. Hence (A).

**Solution 23:**  80% of objects are coins while 60% of the coins are gold. Hence

$$P(\text{gold coin selected}) = 0 \cdot 6 \times 0 \cdot 8 = 0 \cdot 48. \quad \text{Hence (B).}$$

**Solution 24:**   The various ways in which the mouse can reach $C$ after passing through more than three junctions are shown below together with the probabilities that the mouse uses that path.

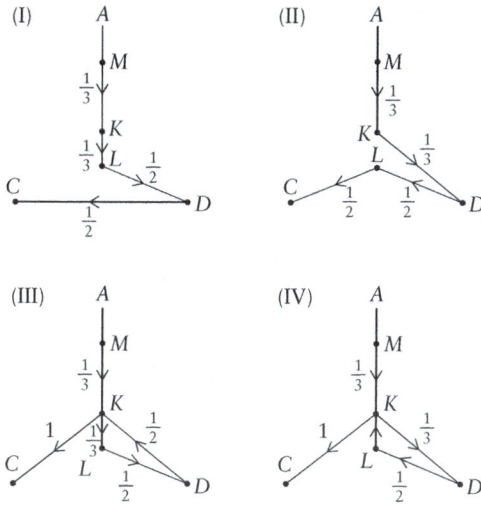

As an illustration of how the probabilities were obtained consider diagram (I). The probability that the mouse chooses $MK$ when it's at junction $M$ is given by $\frac{1}{3}$, since there is a total of three ways of proceeding from $M$ all of which are equally likely to get chosen. The same sort of reasoning is used to obtain the remaining probabilities here and in all figures.

Hence the required probability

$$= \left(\tfrac{1}{3} \times \tfrac{1}{3} \times \tfrac{1}{2} \times \tfrac{1}{2}\right) + \left(\tfrac{1}{3} \times \tfrac{1}{3} \times \tfrac{1}{2} \times \tfrac{1}{2}\right) + \left(\tfrac{1}{3} \times \tfrac{1}{3} \times \tfrac{1}{2} \times \tfrac{1}{2} \times 1\right)$$
$$+ \left(\tfrac{1}{3} \times \tfrac{1}{3} \times \tfrac{1}{2} \times \tfrac{1}{2} \times 1\right) = \tfrac{1}{9}. \quad \text{Hence (D).}$$

**Solution 25:**   Let $x$ denote the expression. Then $x = \sqrt{(1 + 2x)}$, so that $x^2 - 2x - 1 = 0$. Solving gives $x = \frac{2 \pm \sqrt{4+4}}{2}$, and since $x \geq 0$ the result follows immediately. Hence (C).

**Solution 26:**   We have $a + b + c = b + (a + c) = b + 3 = 10$, so that $b = 7$ and hence $a^2 = 65 - 49 = 16$, giving $a = \pm 4$.
Finally, then

$$c = 10 - (a + b)$$
$$= 10 - 7 - a$$
$$= 3 - a$$
$$= 3 \pm 4.$$

Thus $c = -1$ or $7$. Hence (A).

**Solution 27:** Since there are only 12 months in the year, at least two people must have a birthday in the same month, by the pigeon-hole principle (PHP in Toolchest 8). Hence (D).

**Solution 28:** Take the people in order, selecting a first person. Whatever day is his birthday, since there are 365 days in the year (we suppose for simplicity that there are no leap year birthdays on the 29th February), the probability that the second person has a different birthday is $\frac{364}{365}$. Then the probability that the third person does not share birthdays with either of the first two is $\frac{363}{365}$. Proceeding similarly, the probability that the fourteenth person does not share any of the previous 13 birthdays is $\frac{365-13}{365} = \frac{352}{365}$. The total probability that each of these thirteen events takes place is the product of thirteen terms: $P = \frac{364}{365} \cdot \frac{363}{365} \cdots \frac{352}{365} \approx 0.7769$. Therefore the probability that two of the group do share a birthday is $1 - 0.7769 \approx 0.2231$. Hence (D).

**Solution 29:** Continuing the process of the previous solution beyond 13, you will find that the product of 23 of these diminishing fractions (all less than 1) will fall to less than 0.5, hence any group of at least 23 people will have more than even chance of two individuals sharing a birthday. Hence (C). This is a well-known *counter-intuitive* result: most of us would want the group to be much larger before we would bet on it!

**Solution 30:** Costs are $\$50 - \$9.27 - \$0.45 = \$40.28$. Thus the tomato costs $\$40.28 - \$15.45 - \$23.99 = \$0.84$. Hence (B).

**Solution 31:** Let $\log_3 a = x$ so that

$$\log_2(\log_3 a) = \log_2 x = 2, \text{ so that } x = 2^2 = 4;$$
$$\log_3 a = 4 \text{ so that } a = 3^4 = 81. \text{ Hence (D).}$$

**Solution 32:** We need to use the 'base conversion' result that if $a, b, c > 0$ then

$$\log_a b = \frac{\log_c b}{\log_c a}.$$

Let $\log_a b = x$, so that $b = a^x$. Taking logs to base $c$ gives

$$\log_c b = \log_c (a^x) = x \log_c a,$$
$$\text{therefore} \quad x = \log_a b = \frac{\log_c b}{\log_c a}. \tag{9.7}$$

Setting $c = b$, we get another important result:

$$\log_a b = \frac{\log_b b}{\log_b a} = \frac{1}{\log_b a}. \tag{9.8}$$

Now, to solve our problem, we try using equation (9.7):

$$\log_6 10 = \frac{\log_{12} 10}{\log_{12} 6} = \frac{\log_{12} 10}{\log_{12}\left(\frac{12}{2}\right)}$$

$$= \frac{\log_{12} 10}{\log_{12} 12 - \log_{12} 2} = \frac{\log_{12} 10}{1 - \log_{12} 2}.$$

It is clear that our second equation (9.8) will finish the job:

$$\log_6 10 = \frac{\frac{1}{a}}{1 - \frac{1}{b}} = \frac{b}{ab - a}. \quad \text{Hence (A).}$$

**Solution 33:**   We start by listing the two categories of sport:

*Summer sports*: cricket, tennis        *Winter sports*: soccer, hockey.

It has been clearly indicated that no individual plays both cricket and tennis, and the same applies for soccer and hockey. Also we are told that every boy plays a summer sport and a winter sport. This leads us to the Venn diagram below, where the empty spaces are indeed empty, and all 300 boys are distributed among the four intersections (whose sizes still have to be found, although the fruits of the investigation below are already recorded in the diagram). You may prefer to use the Venn diagram on the right, which makes clearer the distributions:

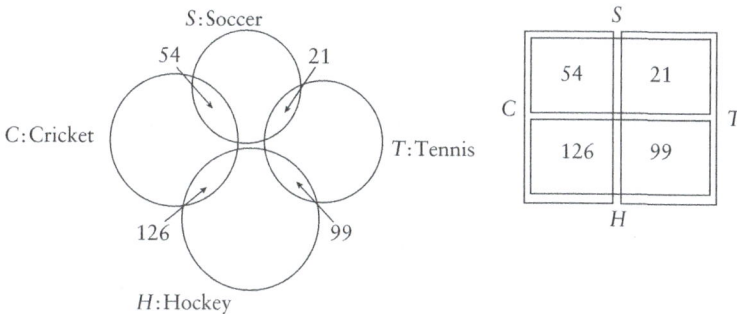

Let's establish our goal: we aim to find $n(T \cap S)$. The given information: 'In summer 60% play cricket and the remainder play tennis' tells us:

$$n(C) = \frac{60}{100} \times 300 = 180, \quad n(T) = 300 - 180 = 120,$$

that is, the total number of cricket players is 180 and the total number of tennis players is 120.

The diagram indicates that the statement '30% of the cricket players play soccer in winter' delivers enough information to allow us to find very easily $n(S \cap C)$ and $n(S \cap H)$ since $n(C)$ is already known:

$$n(S \cap C) = \frac{30}{100} \times 180 = 54, \quad \text{and} \quad n(S \cap H) = 180 - 54 = 126.$$

Further, the statement that '56% of the hockey players play cricket in summer' shows us that

$$n(C \cap H) = (0.56)n(H), \text{ so that } n(H) = \frac{n(C \cap H)}{0.56} = \frac{100 \times 126}{56}.$$

$$n(H \cap T) = n(H) - n(H \cap C) = \frac{100 \times 126}{56} - 126$$

$$= 126\left(\frac{100}{56} - 1\right) = \frac{44}{56} \times 126 = 99.$$

so   $n(S \cap T) = n(T) - n(H \cap T) = 120 - 99 = 21.$ Hence (D).

**Solution 34:**  First let's write down, using convenient notation, the given information:

$P(\text{replace } W/\text{failure}) = 0.5;$
$P(\text{replace } Z/\text{replace } W) = 0.7;$
$P(\text{replace } Z/\text{not replace } W) = 0.1.$

Now let $A$ be the event 'replace $Z$' and $B$ be the event 'replace $W$'. Then

$$P(\text{replace } W \text{ and replace } Z) = P(A \cap B)$$

$$= P(A/B)P(B) = 0.7 \times 0.5 = 0.35. \text{ Hence (D).}$$

**Solution 35:**   If $i = \sqrt{-1}$ then $i^2 = -1$, and $i^4 = i^2 \cdot i^2 = -1 \times -1 = 1$. Now, $i^{62} = (i^4)^{15} \cdot i^2 = (1)^{15} \cdot (-1) = -1$. Hence (C).

**Solution 36:**

$$(2) = 2^2 = 4, \text{ so that } ((2)) = (4) = 4^4 = 256. \text{ Hence (B).}$$

**Solution 37:**   Draw a picture to aid imagination.

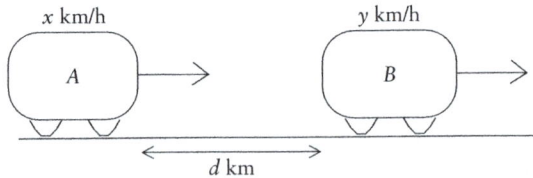

If we have an object $B$ moving at a speed of $y$ km/h and behind it another object $A$ going at a speed of $x$ km/h, then, depending on the speeds of $A$ and $B$, one of three things may be happening to the distance $d$ between the two moving objects.

*Case (I)*, $x < y$ the distance $d$ is increasing and $A$ will never overtake $B$.

*Case (II)*, $x = y$ the distance $d$ is constant and $A$ will never overtake $B$.

*Case (III)*, $x > y$ the distance $d$ between $A$ and $B$ is decreasing and there shall come a time when $A$ overtakes $B$. The question now is, at what rate is the distance between the two decreasing?

Simple reasoning reveals that the distance decreases at a rate of $(x - y)$ km/hr: For in one hour $A$ moves a distance $x$ km, and at the same time $B$ has moved $y$ km, hence in an hour's time, the distance between them has decreased by $(x - y)$ km.

This means that, if at a certain moment the distance between $A$ and $B$ is $d_1$, then the time it will take for $B$ to overtake $A$ is

$$\frac{d_1}{x - y}.$$

We have another interesting case to consider: what happens when $A$ and $B$ are approaching each other?

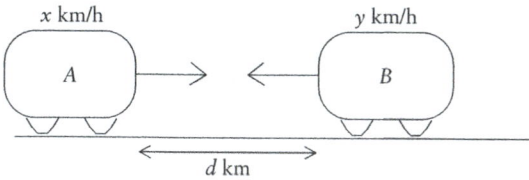

At what rate is $d$, the distance between the two, decreasing? A method of reasoning akin to the one illustrated above shows that the distance between the two is decreasing at a rate of $(x + y)$ km/h.

Now, coming to the given problem, we note that at the time Muzvanya takes off, Gore is at a distance of $60 \times 1 = 60$ km. And the distance between them diminishes at $(x - y) = (100 - 60) = 40$ km/h. Hence the time it will take for Muzvanya to overtake Gore is

$$\frac{60}{40} = \frac{3}{2} = 1\frac{1}{2} \text{ hrs.}$$

Hence Muzvanya will catch up with Gore at 1030 hrs, which is answer (A).

**Solution 38:**   Using (iii) twice:

$$f(2,2) = f(1+1, 1+1) = f(1, f(1+1, 1)) = f(1, f(1+1, 0+1))$$
$$= f(1, f(1, f(2,0))). \tag{9.9}$$

Now using (ii), and (iii):

$$f(2,0) = f(1,1) = f(0, f(1,0)). \tag{9.10}$$

Applying (ii) and (i) then gives:

$$f(1,0) = f(0,1) = 1 + 1 = 2. \tag{9.11}$$

By (9.10) and (9.11), $f(2,0) = f(0,2)$, and by (i), $f(0,2) = 2 + 1 = 3$, hence:

$$f(2,0) = f(1,1) = 3. \tag{9.12}$$

Now (9.9) and (9.12) give:

$$f(2,2) = f(1, f(1,3)).  \tag{9.13}$$

A little experiment will show you that to get $f(1,3)$ (using (iii) and (i)) you first need $f(1,2)$, for which you first need $f(1,1)$, and we can summarize the steps in one general equation:

$$f(1,m) = f(0+1, (m-1)+1) = f(0, f(1, m-1)) = f(1, m-1) + 1.$$

We already have $f(1,1) = 3$, so $f(1,2) = 4$, $f(1,3) = 5$, $f(1,4) = 6$, so returning to equation (9.13), and using (iii) and (i) once more, gives us

$$f(2,2) = f(1, f(1,3)) = f(1,5) = f(0+1, 4+1) = f(0, f(1,4))$$
$$= f(0,6) = 7. \text{ Hence (B)}.$$

**Solution 39:**  If we measure time in hours with 2 pm as our reference point, and we let the times they go to the meeting place be $x$ and $y$ respectively, taking values between 0 and 3, then

$$0 \le y \le 3$$

$$\text{and}\quad 0 \le x \le 3.$$

Therefore they meet if $|x - y| \le \frac{1}{2}$. This inequality can be represented as the shaded region in the diagram:

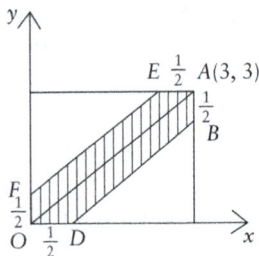

because

$$|x - y| \le \frac{1}{2} \equiv \left( x \ge y \text{ and } x - y \le \frac{1}{2} \right) \text{ or } \left( x < y \text{ and } y - x \le \frac{1}{2} \right)$$

$$\equiv \left( x \ge y \text{ and } y \ge x - \frac{1}{2} \right) \text{ or } \left( x < y \text{ and } y \le x + \frac{1}{2} \right).$$

If $B, D, E$ and $F$ are points as shown in the diagram, such that $OD = AB = EA = OF = \frac{1}{2}$, then the shaded area $ODBAEF$ is the set of pairs $(x,y)$ for which the two friends meet, while the unshaded area of the whole $4 \times 4$ square is the set of pairs for which they don't meet. Therefore

$$P(\text{meeting}) = \frac{\text{shaded area}}{\text{area of square}} = \frac{3^2 - (\frac{5}{2})^2}{3^2} = \frac{11}{36}. \text{ Hence (A)}.$$

**Solution 40:**   Sums such as the one given in the question in which each letter represents a different digit (sometimes zeros included!) are called *alphametrics*. The normal base 10 addition rules of carrying and all the arithmetic rules (commutativety, associativity, etc.) are retained.

Since $R = 8$ and $4 + 4 = 8$ and $9 + 9 = 18$ (so that 8 units remain and 1 ten is carried over) $O$ is four or 9.

However, in the latter case $T$ will also have to be 9 (since $F = 1$ and in $9 + 9 = 18$, 8 remains and 1 is carried over, yet in $4 + 4 = 8$ nothing is carried over) which cannot be. So, $O = 4$ and $T = 7$. (That the entire sum works can be confirmed by further substitutions $U = 6$, $W = 3$.) Hence (C).

**Solution 41:**   Let the side of the cube be $l$ and the vertices of the tetrahedron be $A, B, C, D$. As a general rule, the volume of any solid which rises to a point, whatever the shape of the base, is one-third the product of the base area and the perpendicular height. If we have two cones as in the diagram with congruent circular bases and equal vertical heights (the left hand cone is drawn right circular, i.e. upright) then their volumes are both $\frac{1}{3}\pi r^2 h$.

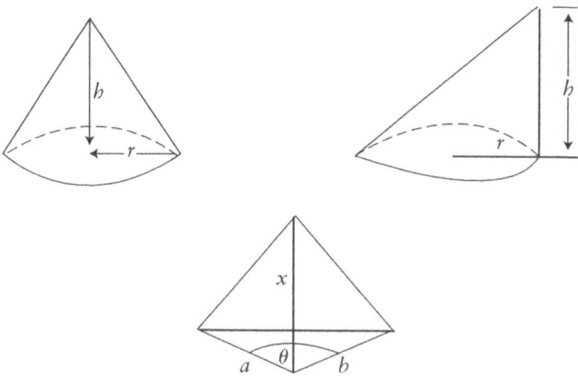

Similarly the volume of the tetrahedron in the diagram is $\frac{1}{3}\left(\frac{1}{2}ab\sin\theta\right)x = \frac{1}{6}abx\sin\theta$.

In the given problem, the volume not occupied by the tetrahedron is easily seen to be made up of four tetrahedrons at the four unlabelled corners. hence the volume is four times the volume of the tetrahedron $ABCF$. To find this volume, consider the triangle $AFC$ as the base. Its area is $\frac{1}{2}AF \cdot FC \sin A\hat{F}C = \frac{1}{2}l^2 \sin 90° = \frac{l^2}{2}$ (or just half the square).

Now the volume of the tetrahedron $ABCF$ is $\frac{1}{3}(\frac{l^2}{2})l = \frac{l^3}{6}$. Thus

$$\text{volume of tetrahedon } ABCD = l^3 - 4\left(\frac{l^3}{6}\right) = \frac{V}{3},$$

where $V = l^3$ is the volume of the cube. Hence (C).

**Solution 42:**   We give labels to our statements:

(a) All idiots are fools.
(b) No twits are idiots.
(c) Everyone is a twit or a nut.
(d) Fools are not nuts.

That no nuts are fools follows from (d). And while all fools are twits, there is nothing to imply the reverse. From (c), (d), all fools are twits. If there are any idiots, then they would be fools by (a) and thus twits, contradicting (b). Thus there are no idiots so that only 1 and 3 are true. Hence (A).

**Solution 43:**   Interpreting the table, we have 22 numbers:

$$29, 41, 48, 50, 50, 50, 51, 51, 51, 52, 53, 57, \ldots 79,$$

for which we want the median. The 11th and the 12th numbers are 53 and 57, and their average $\frac{53+57}{2} = 55$ is the median. Hence (C).

**Solution 44:**   The best way to approach this is via what's called a 'geometric probability' argument, using the diagram below:

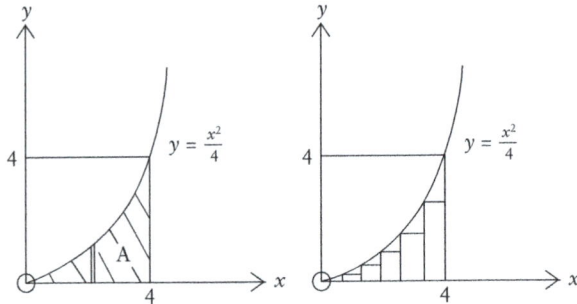

The probability that the equation has no real roots is the probability that $b^2 < 4a$ when $a$ and $b$ are random numbers on the real line between 0 and 4. This situation corresponds to choosing randomly a pair $(a, b)$ in the $4 \times 4$ square, and those pairs that are above the curve $4y = x^2$ correspond to no real roots. The shaded area $A$ *under* the graph is found by integrating:

$$\int_0^4 \frac{x^2}{4} dx = \frac{x^3}{12} \Big|_0^4 = \frac{16}{3}.$$

The required probability is therefore given by

$$\frac{\text{unshaded area}}{\text{area of square}} = \frac{16 - \frac{16}{3}}{4^2} = \frac{2}{3}. \quad \text{Hence (A)}.$$

If you haven't learnt any calculus you can still find that area in the manner it was first done historically! Divide the interval from 0 to 4 into $n$ equal subintervals of length $4/n$, and draw the stepped rectangles under the curve as in the diagram on the right, which is done for $n = 6$. Work out the heights of the $n$ rectangles (each of width $4/n$), and sum their areas to get the total area $A_n$ under the $n$ steps, using the formula for the sum of squares given in Toolchest 6:

$$A_n = \left(\frac{4}{n}\right)\left(0 + \frac{1}{4}\left(\frac{4 \cdot 1}{n}\right)^2 + \frac{1}{4}\left(\frac{4 \cdot 2}{n}\right)^2 + \cdots + \frac{1}{4}\left(\frac{4 \cdot (n-1)}{n}\right)^2\right)$$

$$= \frac{16}{n^3}\left(1^2 + 2^2 + \cdots + (n-1)^2\right)$$

$$= \frac{16}{n^3} \cdot \left(\frac{n-1}{6}\right) \cdot n \cdot (2n-1)$$

$$= \frac{16}{6}\left(1 - \frac{1}{n}\right)\left(2 - \frac{1}{n}\right).$$

Now we find the limiting value of this area as $n$ grows larger and larger. Since the product of the two brackets approaches $1 \cdot 2 = 2$, the limiting area, which will be the area under the parabola, is $\frac{16}{3}$ as we found before by integrating.

**Solution 45:**   Common sense is all you need to apply here! There are 200 shirts without collars. Since we have a total of 200 blue shirts all these could be blue, and 150 of them could have pockets. Hence (C).

**Solution 46:**   Shown in the diagram is a cross-section (magnified) of a newspaper and the page numbers as indicated on each side of each sheet of paper. Using the sheet labelled X, we see that the other numbers were $6, 11, 12$, and hence (C). By a similar process of thinking about the more general $n$-sheet newspaper, observing that numbers of adjacent pages sum to $4n + 1$ and numbers of backed pages differ by 1, the algorithm is:

if $k$ is odd, $k \leq 2n$, the other three page-numbers will be: $k+1$, $4n-k$, $4n-k+1$;

if $k$ is even, $k \geq 2n + 1$, the same as previous;

if $k$ is odd, $k \geq 2n + 1$, the other three page-numbers will be: $k - 1$, $4n - k + 2$, $4n - k + 1$;

if $k$ is even, $k \leq 2n$, the same as previous.

For $k = 5$, $n = 4$, we have the first category, and the algorithm can be checked against (C).

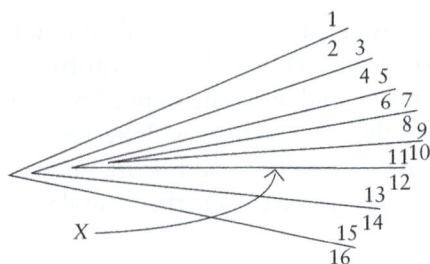

## Solution 47:

$$\text{Let (i) } a + b = 10, \quad \text{(ii) } b + c = 20, \quad \text{(iii) } a + c = 30$$

$$\text{Adding: } 2(a + b + c) = 60, \text{ so that } a + b + c = 30 \qquad (9.14)$$

$$\text{and so } (9.14) - \text{(i) gives } c = 20; \quad (9.14) - \text{(ii) gives } a = 10;$$

$$(9.14) - \text{(iii) gives } b = 0$$

$$\text{so that } a^2 + b^2 + c^2 = 20^2 + 10^2 + 0^2 = 500. \quad \text{Hence (A)}.$$

Alternatively we could observe that:

$$\text{squaring both sides of (i): } a^2 + b^2 + 2ab = 100$$

$$\text{squaring both sides of (ii): } b^2 + c^2 + 2bc = 400$$

$$\text{squaring both sides of (iii): } a^2 + c^2 + 2ac = 900$$

$$\text{and adding } 2a^2 + 2b^2 + 2c^2 + 2ab + 2bc + 2ac = 1400.$$
$$(9.15)$$

$$\text{Squaring both sides of (9.14): } a^2 + b^2 + c^2 + 2ab + 2bc + 2ac = 900.$$
$$(9.16)$$

$$\text{Now, (9.15)-(9.16) gives } a^2 + b^2 + c^2 = 1400 - 900$$
$$= 500 \text{ as before.}$$

**Solution 48:** Adding the seven columns gives the sum of all the numbers from 1 up to 49, which is $\sum_{i=1}^{49} i = (49)\left(\dfrac{50}{2}\right) = 1225$, so that any one column must add up to $\dfrac{1225}{7} = 175$. Hence (B).

**Solution 49:** Let $X = \sqrt{5+\sqrt{5}} - \sqrt{5-\sqrt{5}}$, $a = \sqrt{5+\sqrt{5}}$, $b = \sqrt{5-\sqrt{5}}$, so that $X = a - b$, and hence

$$X^2 = a^2 + b^2 - 2ab$$

$$= 5 + \sqrt{5} + 5 - \sqrt{5} - 2\sqrt{\{(5+\sqrt{5})(5-\sqrt{5})\}}$$

$$= 10 - 2\sqrt{20}$$

$$= 10 - 4\sqrt{5}$$

hence $X = \sqrt{10 - 4\sqrt{5}}$.  Hence (A).

**Solution 50:** The figures show the successive moves of the die.

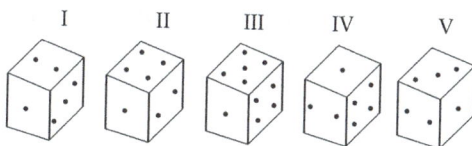

After the first move the bottom face is 3, then 2, then 6 (opposite 1), then 4, which is opposite 3, so the top face at the end will be 3. Hence (C).

Alternatively, it is perhaps easier to follow the mystery face, working backwards, using the six directions *top, bottom, left, right, front, back*: whatever the top face is at the end will (on reversing the moves) be left, then left still, then bottom, then right – but this is 3!

**Solution 51:** Using the property (think of a light ray) that the shortest distance is obtained when angles $XRA$ and $YRB$ are equal, we have $XR = \frac{400}{\sin 30^\circ} = 800$ and $YR = \frac{500}{\sin 30^\circ} = 1000$ to give a total distance of 1800 m. Hence (A).

You can verify the equal angles property by drawing the line $RY'$ to the 'mirror image' $Y'$ of the point $Y$ in the line $AB$. Clearly the route $XRY'$ will be shortest when it is a straight line! It follows from similar triangles that the two angles $XRA$ and $YRB$ are equal.

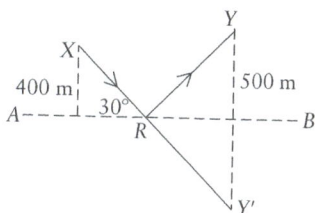

**Solution 52:** Let 0(zero)-connection denote the collection of airports (paired) in different countries with no airlines flying on the route that joins

them, one-connection denote the collection of those with one airline and so on. We will then have

| 0-connection | 1-connection | 2-connections | 3-connections |
|---|---|---|---|
| $B_1 - C_1$ | $B_1 - C_2$ | $B_2 - C_1$ | $B_2 - C_2$ |
| $B_2 - C_3$ | $B_3 - C_3$ | $B_1 - C_3$ | |
| $B_3 - C_2$ | | | |
| $B_3 - C_1$ | | | |

So that whenever we see a 3, we know that the airports involved are $B_2$ and $C_2$. Using this and the other results above, we can easily find out the order of airports each matrix represents. Take matrix (A) for example. You should be able to use the above findings to discover that it represents the following order

|       | $C_1$ | $C_2$ | $C_3$ |
|-------|-------|-------|-------|
| $B_1$ | 0 | 1 | 2 |
| $B_2$ | 2 | 3 | 0 |
| $B_3$ | 0 | 0 | 1 |

where the 0 at the top-left corner, say, represents the number of lines on the route between airport $B_1$ and $C_1$, and the 1 at the bottom-right corner represents the number of airlines flying on the route between airport $C_3$ and $B_3$. Continuing this way for the matrix, we see that it does correctly represent the information on flights between the two countries. If the same treatment is applied to the remaining four matrices, we that $C, D, E$ make good sense for appropriate orderings of the airports, but rows 2 and 3 in matrix $B$ do not correspond to any network connection. Hence (B).

**Solution 53:** Let the players be $A, B$, and $C$ and let $P(A) = \frac{1}{6}$ be the probability that player $A$ hits the ball, while $P(\bar{A}) = 1 - \frac{1}{6} = \frac{5}{6}$ is the probability that player $A$ does not hit the ball. Similarly,

$P(B) = \frac{1}{4}, P(\bar{B}) = 1 - \frac{1}{4} = \frac{3}{4}, P(C) = \frac{1}{3}, P(\bar{C}) = 1 - \frac{1}{3} = \frac{2}{3}.$

Now, $P(\text{one}) = P(\text{exactly one of the players hits his ball})$ is given by:

$$P(\text{one}) = P(A_H)P(\bar{B}_H)P(\bar{C}_H) + P(B_H)P(\bar{A}_H)P(\bar{C}_H) + P(C_H)P(\bar{A}_H)P(\bar{B}_H)$$

$$= \frac{1}{6} \cdot \frac{3}{4} \cdot \frac{2}{3} + \frac{1}{4} \cdot \frac{5}{6} \cdot \frac{2}{3} + \frac{1}{3} \cdot \frac{5}{6} \cdot \frac{3}{4}$$

$$= \frac{6 + 10 + 15}{72} = \frac{31}{72}. \text{ Hence (A).}$$

**Solution 54:**   Let the possible values of $a + b$ be shown by the table below:

|  |  | Die 1 |  |  |  |  |  |
|---|---|---|---|---|---|---|---|
|  |  | 1 | 2 | 3 | 4 | 5 | 6 |
|  | 1 | 2 | 3 | 4 | 5 | 6 | 7 |
|  | 2 | 3 | 4 | 5 | 6 | 7 | 8 |
| Die 2 | 3 | 4 | 5 | 6 | 7 | 8 | 9 |
|  | 4 | 5 | 6 | 7 | 8 | 9 | 10 |
|  | 5 | 6 | 7 | 8 | 9 | 10 | 11 |
|  | 6 | 7 | 8 | 9 | 10 | 11 | 12 |

We can see from the table that:

$$P(s \leq 4) = \frac{6}{36} \quad \text{i.e.} \quad X = \frac{6}{36}.$$

$$P(s = 7) = \frac{6}{36} \quad \text{i.e.} \quad Y = \frac{6}{36}.$$

$$P(s > 9) = \frac{6}{36} \quad \text{i.e.} \quad Z = \frac{6}{36}.$$

Therefore $X = Y = Z = \frac{6}{36}$. Hence (D).

**Solution 55:**   The trick is to divide through by $2^x$, and use the fact that, as $x$ gets larger and larger, $\frac{8}{2^x}$ gets closer and closer to zero.

$$\frac{8 + 2^x}{8 - 2^x} = \frac{\frac{1}{2^x}(8 + 2^x)}{\frac{1}{2^x}(8 - 2^x)}$$

$$= \frac{\frac{8}{2^x} + 1}{\frac{8}{2^x} - 1}.$$

Hence

$$\lim_{x \to \infty} \frac{8 + 2^x}{8 - 2^x} = \lim_{x \to \infty} \frac{8 \cdot 2^{-x} + 1}{8 \cdot 2^{-x} - 1} = \frac{1}{-1} = -1. \text{ Hence (C)}.$$

**Solution 56:**   Let subscript 1 denote pens and subscript 2 denote pencils for each of Gift, $G$, Suwisa, $S$ and Tapina, $T$. Then $G_1 = 2S_2$ and $S_1 = 2T_2$ so that $S_2$, $T_2$ have possible values 2 and 4 only. If $S_2 = 2$, then $G_1 = 4$, $S_1 = 10 - S_2 = 8$ and $T_2 = 4$. If $S_2 = 4$ then $G_1 = 8$, $S_1 = 10 - S_1 = 6$ and $T_2 = 3$ which is impossible. Thus $S_2 = 2$, $T_2 = 4$ and $G_2 = 6$ making (B) the correct answer.

**Solution 57:**   Observe that the distances that are uphill on the journey to Charles's home become downhill on the journey back home and *vice versa*. The flat ground remains flat irrespective of whether she is going home or going to her friend's house. So if we suppose that Patricia walks $x$ km uphill,

$y$ km downhill and $z$ km on the flat going to her friend's; then the time taken
$6 = \frac{x}{2}+\frac{y}{6}+\frac{z}{3}+\frac{y}{2}+\frac{x}{6}+\frac{z}{3}$ giving $36 = 3x+y+2z+3y+x+2z = 4(x+y+z)$.
Hence total distance (to and fro) is $2(x+y+z) = 18$ km. Hence (C).

**Solution 58:**   We need to think of a relationship between $x$ and $y$ that
allows us to solve this functional equation, i.e. to find an expression for
$f(x)$ in terms of $x$. If we let $y = \frac{1}{2}x$, then $f(x-y) = f(\frac{x}{2}) = f(x)f(\frac{x}{2})$ giving
$f(x) = 1$ since $f(\frac{x}{2}) \neq 0$. This is for all $x$, so that $f(3) = 1$. Hence (E).

**Solution 59:**   Convince yourself that $p = a^{\log_a p}$, because $\log_a x$ and $a^x$ are
inverse functions. Therefore $2^{2\log_2 x} = \left(2^{\log_2 x}\right)^2 = (x)^2$ and $2^{(\log_2 x)+1} =$
$2 \cdot 2^{(\log_2 x)} = 2x$. Therefore $x^2 = 2x - 1$, so that $x^2 - 2x + 1 = 0$ and hence
$(x-1)^2 = 0$, giving $x = 1$. Hence (A).

**Solution 60:**   We need to express the LHS as a sum/difference of logarith-
mic terms taken to a common base:

$$\log_{a^2} b = \frac{\log_a b}{\log_a a^2} = \frac{\log_a b}{2\log_a a} = \frac{\log_a b}{2}.$$

Similarly, $\log_{b^2} a = \frac{\log_b a}{2}$. Therefore

$$\log_{(a^2)} b + \log_{(b^2)} a = \frac{\log_a b}{2} + \frac{\log_b a}{2}$$
$$= \frac{\log_a b}{2} + \frac{1}{2\log_a b} = \frac{5}{2},$$

therefore   $(\log_a b)^2 - (5\log_a b) + 1 = 0$,

which is quadratic. Put $x = \log_a b$, giving $x^2 - 5x + 1 = 0$ and $x = \frac{5\pm\sqrt{25-4}}{2} = \frac{5\pm\sqrt{21}}{2}$. Hence $\log_a b = \frac{5+\sqrt{21}}{2}$ or $\log_a b = \frac{5-\sqrt{21}}{2}$, so that $b = a^{\frac{5+\sqrt{21}}{2}}$ or $b = a^{\frac{5-\sqrt{21}}{2}}$. For a given value of $a$ we therefore have two values
of $b$ for which the equation is satisfied. Hence (B).

**Solution 61:**

$$\left(x+\frac{1}{x}\right)^2 = x^2 + \frac{1}{x^2} + 2$$
$$= 14 + 2 = 16$$
$$= 4^2,$$

therefore   $x + \frac{1}{x} = \pm 4.$

Since it is given that $x > 0$, we have (E).

**Solution 62:**   The insight you need is: try expressing all in terms of $x - 1$, which suggests 'completing the square':

$$x^2 - \sqrt{(x-1)^3} - \sqrt{(x-1)} + 2(1-x) = 0,$$

giving   $x^2 - 2x + 1 - \sqrt{(x-1)^3} - \sqrt{(x-1)} + 1 = 0,$

hence   $(x-1)^2 - \sqrt{(x-1)^3} + \sqrt{(x-1)} + 1 = 0.$

Now let $\sqrt{(x-1)} = y$, which gives the biquadratic (or quartic) equation

$$y^4 - y^3 - y + 1 = 0.$$

Ignoring the signs and writing down the four coefficients (including zero for $x^2$) gives

$$\overset{\longleftarrow}{\underset{\longrightarrow}{1 \quad 1 \quad 0 \quad 1 \quad 1}}$$

The number above is *palindromic*, i.e. the digits read the same in both directions shown by the arrows. Biquadratic equations with palindromic coefficients are solved easily by suitable substitutions as follows (see also Problem 68): in $y^4 - y^3 - y + 1 = 0$, divide throughout by $y^2$ to give

$$y^2 - y - \frac{1}{y} + \frac{1}{y^2} = 0, \quad \text{hence} \quad y^2 + \frac{1}{y^2} - 1\left(y + \frac{1}{y}\right) = 0.$$

Now put $k = y + \frac{1}{y}$, so that $y^2 + \frac{1}{y^2} = k^2 - 2$ (check). Therefore

$$0 = (k^2 - 2) - k = k^2 - k - 2 = (k-2)(k+1),$$

$$\text{giving} \quad k = y + \frac{1}{y} = 2 \quad \text{or} \quad -1,$$

$$\text{hence} \quad y^2 - 2y + 1 = 0 \quad \text{or} \quad y^2 + y + 1 = 0,$$

$$\text{and so} \quad (y-1)^2 = 0 \quad \text{or} \quad y = \frac{-1 \pm \sqrt{1-4}}{2}.$$

Therefore   $y = 1$   or   $y$ is not real since $b^2 - 4ac < 0$,

giving   $\sqrt{x-1} = 1$, so   $x - 1 = 1$, hence   $x = 2$.

Clearly, the only value of $x$ is 2. Hence (A).
    To be sure you understand why this works you should try the same procedure on the general palindromic quartic equation: $ax^4 + bx^3 + cx^2 + bx + a = 0$.

*Wash your sins, not only your face!*

**Solution 63:**   Label the right-angled triangle as in the diagram.

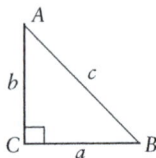

Then $p = a + b + c$ is the perimeter of the triangle, and $A = \frac{ab}{2}$ is the area of the triangle. Thus we have $2ab = 4A$. By Pythagoras we also have $a^2 + b^2 = c^2$. Hence

$$p^2 = (a+b+c)^2 = a^2 + b^2 + c^2 + 2ac + 2ab + 2bc$$

$$= c^2 + c^2 + 2ac + 2bc + 4A$$

$$= 2c(a+b+c) + 4A$$

$$= 2cp + 4A,$$

therefore   $c = \dfrac{p^2 - 4A}{2p}.$

Hence, putting $p = 624$ and $A = 6864$:

$$c = \frac{624^2 - 4 \times 6\,864}{2 \times 624}$$

$$= \frac{389\,376 - 27\,456}{1\,248}$$

$$= \frac{361\,920}{1\,248} = 290. \text{ Hence (B).}$$

**Solution 64:**   The solution procedure is almost entirely by common sense: first you'll need four 1c coins to get any number of cents up to 4, then a 5c coin gives you every number up to 9, a 10c coin up to 19, a 20c coin up to 39, and you'll need another 10c or 20c coin to get up to 49; then a 50c coin will take you up to 99. If you chose (at the previous step) to have another 10c coin, you'll need either a dollar coin or one more 1c coin to pay one dollar. But if you chose to have another 20c coin, you don't need any more coins. Hence the minimum number of coins is $n = 9$, or solution (B). The following coins do the job:

$$1c,\ 1c,\ 1c,\ 1c,\ 5c,\ 10c,\ 20c,\ 20c,\ 50c$$

**Solution 65:**   A reasonable strategy is to use the given equations to get each of $x$, $y$, $z$ as subjects, and then solve the three equations for $x$ in terms of $a$, $b$, $c$ – that is, eliminate $y$ and $z$.

$$\frac{xy}{x+y} = a,$$

$$\text{giving}\quad xy = a(x+y) = ax + ay,$$

$$\text{so that}\quad xy - ax = ay,$$

$$\text{hence}\quad x = \frac{ay}{y-a}. \tag{9.17}$$

Similar treatment of the other two expressions yields:

$$x = \frac{zb}{z-b} \tag{9.18}$$

$$\text{and}\quad y = \frac{zc}{z-c}. \tag{9.19}$$

Setting out to eliminate $y$ from (9.19) and (9.17) gives

$$x = \frac{a\left(\frac{zc}{z-c}\right)}{\frac{zc}{z-c} - a}$$

$$= \frac{azc}{z-c} \div \frac{zc - az + ac}{z-c}$$

$$= \frac{azc}{zc - az + ac}. \tag{9.20}$$

Finally, (9.18) and (9.20) allow us to eliminate $z$:

$$x = \frac{zb}{z-b},$$

$$\text{hence}\quad zx - bx = zb,$$

$$\text{therefore}\quad z = \frac{bx}{x-b}.$$

$$\text{Now} \quad x = \frac{azc}{zc - az + ac} = \frac{\frac{abxc}{x-b}}{\frac{bcx}{x-b} - \frac{abx}{x-b} + \frac{ac(x-b)}{x-b}},$$

$$= \frac{abcx}{x-b} \times \frac{x-b}{bcx + abx + ac(x-b)},$$

hence $\quad bcx + abx + ac(x-b) = abc,$

so that $\quad x(bc + ab + ac) = 2abc,$

and finally $\quad x = \dfrac{2abc}{ab + bc + ac}.$ Hence (B).

**Solution 66:**

$$BAD + DAM + MAD = (100B + 10A + D) + (100D + 10A + M)$$
$$+ (100M + 10A + D)$$
$$= 102D + 101M + 100B + 30A.$$

Setting $D = 9$, $M = 9$, $B = 8$, $A = 1$, the largest sum is $918 + 909 + 800 + 30 = 2657$. Hence (D).

**Solution 67:**

$$\text{Total distance travelled} = 100 + 400 + 600 = 1100 \text{ km.}$$
$$\text{Total time taken} = \frac{100}{x} + \frac{400}{2x} + \frac{600}{3x}$$
$$= \frac{500}{x} \text{ hr.}$$
$$\text{So, average speed} = \frac{\text{total distance}}{\text{total time}}$$
$$= 1100 \div \frac{500}{x}$$
$$= \frac{11x}{5} \text{ km/h.} \quad \text{Hence (E).}$$

**Solution 68:** Write out the coefficients of the equation as below:

2    11    16    16    11    2

The coefficients are said to be *palindromic* if the numbers are the same whichever direction you read the list in (see also Problem 62). Such equations are analysed using the following device. First, divide the equation throughout by $x^2$:

$$0 = 2x^2 - 11x + 16 - \frac{11}{x} + \frac{2}{x^2}, \quad (x^2 \neq 0)$$

$$= 2x^2 + \frac{2}{x^2} - 11x - \frac{11}{x} + 16$$

$$= 2\left(x^2 + \frac{1}{x^2}\right) - 11\left(x + \frac{1}{x}\right) + 16 = 0$$

$$= 2\left(\left(x + \frac{1}{x}\right)^2 - 2\right) - 11\left(x + \frac{1}{x}\right) + 16 = 0$$

$$= 2\left(x + \frac{1}{x}\right)^2 - 11\left(x + \frac{1}{x}\right) + 12 = 0.$$

Next, let $x + \frac{1}{x} = y$, to obtain a quadratic equation in $y$:

$$0 = 2y^2 - 11y + 12$$

$$= (2y - 3)(y - 4),$$

therefore     $y = 4$ or $\frac{3}{2}$. Hence (C).

**Solution 69:**

| 1 | 1 | 1 | 1 | 1 | 1 | 1 | 1 |
|---|---|---|---|---|---|---|---|
| 1 | 2 | 3 | 4 | 5 | 6 | 7 | 8 |
| 1 | 3 | 6 | 10 | 15 | 21 | 28 | 36 |
| 1 | 4 | 10 | 20 | $^F$35 | 56 | 84 | 120 |
| 1 | 5 | 15 | $^G$35 | $^P$70 | 126 | 210 | 330 |
| 1 | 6 | 21 | 56 | 126 | 252 | 462 | 792 |
| 1 | 7 | 28 | 84 | 210 | 462 | 924 | 1716 |
| 1 | 8 | 36 | 120 | 330 | 792 | 1716 | 3432 |

Observe that any node (intersection of edges) on the top edge of the grid can be reached from the top-left corner in exactly one way. The same holds for any node along the left-hand edge of the grid. We place the number 1 at each of these nodes to record this fact, or (as in the diagram) place the number inside the square to the bottom-right of the corresponding node. Consider the square $P$, for example (at $(5, 5)$ in the grid, with '70' in it). Its top-left node can be reached either from the left ($G$), or from above ($F$). Hence, if we know the number of pathways from the top-left corner to $F$, and the number to $G$, we can calculate the number of pathways from the top-left corner to $P$, as $p = g + f$. Using this rule and starting from top-left corner, we fill in the numbers in the square. The number attached to the bottom-right node is 3432. Hence (A).

Notice that these numbers are also entries in Pascal's triangle!

It is interesting to consider the number of paths for the more general $n \times n$ grid. Place it on a coordinate system as shown below:

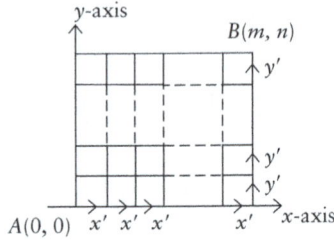

Let $x'$ denote a unit step along or parallel to the $x$-axis in the positive direction, and let $y'$ denote a unit step along or parallel to the $y$-axis in the positive direction.

Observe that each way of moving from $A$ to $B$ consists of $m$ steps $x'$ and $n$ steps $y'$, and that each unique permutation of these items defines a unique route from $A$ to $B$. The number of routes from $A$ to $B$ therefore equals the number of permutations of the $m$ steps $x'$ and $n$ steps $y'$. This is:

$$\binom{m+n}{n} = \binom{m+n}{m} = \frac{(m+n)!}{m!n!}.$$

For example, in the problem above, $m = n = 7$, giving the number of paths as:

$$\frac{(7+7)!}{7!7!} = \frac{14!}{7!7!} = 3432.$$

**Solution 70:**   The diagram illustrates the given information. Let the second circle cut the first at $M$ as shown and let $GD = DK = r$. Hence $GM = GD = r$ (radii). Therefore triangle $GMD$ is equilateral. The required area is therefore $2A$ where $A$ is area of sector $GMQD$ of the left-hand circle added to area of segment $GMP$ of the right-hand circle. But the latter area (of the segment $GMP$) is equal to the area of the triangle subtracted from the area of sector $GDMP$. Thus

$$A = 2(\text{area of sector}) - \text{area of triangle}$$
$$= 2\frac{60}{360} \times \pi r^2 - \frac{1}{2}r^2 \sin 60°$$
$$= \frac{\pi r^2}{3} - \frac{r^2 \sqrt{3}}{4}.$$

But $GK = 2r = 8$, so that $r = 4\,\text{cm}$, giving the required area as $2A = 2\left(\frac{16\pi}{3} - 4\sqrt{3}\right)\,\text{cm}^2$. Hence (E).

**Solution 71:** The stage is (III), hence (C) is correct, because:

$$\left(7^{a+b}\right)^2 = 7^{2a+2b} \neq 7^{a^2+b^2+2ab}.$$

**Solution 72:** This question is really a question about the laws of indices.

$$\left(\frac{n+1}{n+2}\right)^{3n} = \left(\frac{n+1}{n+2}\right)^{n \times 3} = \left(\left(\frac{n+1}{n+2}\right)^{n}\right)^{3}.$$

Now, $n + 1 = n\left(1 + \frac{1}{n}\right)$ and $n + 2 = n\left(1 + \frac{2}{n}\right)$ so that

$$\frac{n+1}{n+2} = \frac{n\left(1+\frac{1}{n}\right)}{n\left(1+\frac{2}{n}\right)} = \frac{1+\frac{1}{n}}{1+\frac{2}{n}}$$

Hence

$$\left(\frac{n+1}{n+2}\right)^{3n} = \left(\left(\frac{1+\frac{1}{n}}{1+\frac{2}{n}}\right)^{n}\right)^{3}$$

$$= \left(\frac{\left(1+\frac{1}{n}\right)^{n}}{\left(1+\frac{2}{n}\right)^{n}}\right)^{3}$$

$$\to \left(\frac{e}{e^2}\right)^{3} = e^{-3}. \text{ Hence (A).}$$

**Solution 73:** To answer this question, we need to know what happens to both hands in an hour.

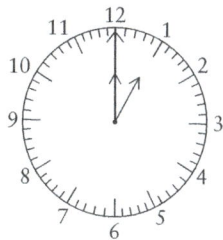

In 1 hour, the minute hand moves through all 60 of the small divisions shown above while the hour hand moves through 5 out of the 60. Hence, in an hour's time, the minute hand gains 55 divisions on the hour hand. The statement '55 divisions gained in 1 hour' allows us to solve our problem by employing simple proportion. At 1 o'clock, there are five divisions that keep the hands apart. The question is, how long will it take for the minute hand to gain exactly five divisions on the hour hand (for when this happens the two hands will coincide) if we know that the minute hand gains on the hour hand at a rate of 55 divisions per hour? This is a simple proportion problem and we see that it will take the minute hand $\frac{5}{55} \times 1$ hours to gain five divisions on the hour hand. This time is $\frac{1}{11}$ hours, or $\frac{60}{11}$ minutes. That is, the required time is $5\frac{5}{11}$ minutes after 1, making (C) the correct choice.

## Solution 74

**Method 1:** Let $P(n)$ be the statement '$2^{3n} - 1$ is divisible by 7'. For $n = 1$, $2^{3(1)} - 1 = 8 - 1 = 7$, so $P(1)$ is true. Assume inductively that $P(k)$ is true for some $k \geq 1$. That is, we assume $2^{3k} - 1$ is divisible by 7, so that there exists an integer $A$ such that $2^{3k} - 1 = 7A$.

Now consider the situation when $n = k + 1$. We have

$$
\begin{aligned}
2^{3(k+1)} - 1 &= 2^{3k+3} - 1 \\
&= 8 \cdot 2^{3k} - 1 \\
&= 8(2^{3k} - 1) + 7 \\
&= 8 \cdot 7A + 7 \quad \text{by the inductive hypothesis} \\
&= 7(8A + 1).
\end{aligned}
$$

Since $8A + 1$ is an integer, $2^{3(k+1)} - 1$ is divisible by 7. We have deduced $P(k+1)$ from $P(k)$. But we already showed that $P(1)$ holds. Therefore, by the PMI, the statement $P(n)$ is true for all $n$, that is, $2^{3n} - 1$ is divisible by 7 for all positive integers $n$. The proof is complete.

**Method 2:** For any real number $a$ and any positive integer $n$, we have the identity

$$
a^n - 1 = (a - 1)(a^{n-1} + \cdots + a^2 + a + 1),
$$

which follows by cancellation of the terms in pairs when multiplying out the RHS, or else using the formula for the sum of the GP in the second bracket. Put $a = 2^3 = 8$ and we have the required result:

$$
2^{3n} - 1 = \left(2^3\right)^n - 1 = (8 - 1)(\cdots) = 7(\cdots).
$$

**Method 3:** Using 'arithmetic modulo 7' as in Toolchest 4, which means dealing with the seven classes of integers determined by the seven different remainders on division by 7. Sums, products and powers of integers in the same class will also be in the same class. Now $8 \equiv 1 \pmod{7}$, which means that 8 and 1 are in the same class. Therefore

$$
2^{3n} = \left(2^3\right)^n = 8^n \equiv 1^n \pmod{7} = 1 \pmod{7}
$$

so that $8^n - 1 \equiv (1 - 1) \pmod{7} \equiv 0 \pmod{7}$.

See how powerful this method is by showing that the remainder when $2^{100}$ is divided by 7 is 2: use the fact that $100 = 3 \times 33 + 1$ and $2^3 \equiv 1 \pmod{7}$.

**Solution 75:** Let $P(n)$ be the statement '$10^{2n} - 1$ is divisible by 11'. For $n = 1$, $10^{2(1)} - 1 = 100 - 1 = 99$, so $P(1)$ is true.

Assume inductively that $\mathcal{P}(k)$ is true for some $k \geq 1$. That is, we assume $10^{2k} - 1$ is divisible by 11, so that there exists an integer $A$ such that $10^{2k} - 1 = 11A$.

Now consider the situation when $n = k + 1$. We have

$$10^{2(k+1)} - 1 = 10^{2k+2} - 1$$
$$= 10^2 \cdot 2^{3k} - 1$$
$$= 10^2(10^{2k} - 1) + 99$$
$$= 10^2 \cdot 11A + 99 \quad \text{by the inductive hypothesis}$$
$$= 11(100A + 9).$$

Since $100A + 9$ is an integer, we have deduced the truth of $\mathcal{P}(k+1)$ from the truth of $\mathcal{P}(k)$. Therefore, by the PMI, the statement $\mathcal{P}(n)$ is true for all $n$. The result is proved.

Now see if you can apply **Methods 2** and **3** as in the previous problem to get this result!

**Solution 76:** Observe that if our unknown number $x$ leaves remainder $k - 1$ on division by $k$ then $x + 1$ will be divisible by $k$. This can be neatly expressed in modulo arithmetic:

$$x \equiv (k - 1) \pmod{k} \Rightarrow x + 1 \equiv k \pmod{k} \Rightarrow x + 1 \equiv 0 \pmod{k}.$$

Thus the number $x + 1$ is divisible by each of the numbers from 2 to 9. The least common multiple of these is $2^3 \cdot 3^2 \cdot 5 \cdot 7 = 2\,520$, and no further multiple will be in the given range (less than 3 000). Therefore $x = 2\,520 - 1 = 2\,519$.

**Solution 77:** To get a feel for the problem, try $k = 1, 2, \cdots$:

For $k = 1$ : $M_n = A_1 - S_n = -A_2 - A_3 - \ldots - A_n < 0.$

For $k = 2$ : $M_n = A_1 - (A_2 + A_3 + \ldots + A_n).$ Will this be positive? For all $n$?

For $k = 3$ : $M_n = 2A_1 - (A_2 + A_3 + \ldots + A_n).$ This will have more chance of being positive for all $n$.

It becomes clear that we want to compare the size of $A_1$ with the sum of the other areas, taken for larger and larger $n$. Consider the first two of the squares obtained by 'geometrical squaring'. Let the side of the first (larger) square be $a_1$ and that of the second square be $a_2$.

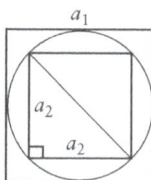

Referring to the right-angled triangle in the figure,

$$a_2^2 + a_2^2 = (\text{diameter of circle})^2 \quad (\text{by Pythagoras' theorem})$$

$$= a_1^2,$$

therefore $\quad a_2^2 = \frac{1}{2}a_1^2,$

so that $\quad A_2 = \frac{1}{2}A_1.$

Applying the same sort of reasoning to any pair of consecutive squares, an inductive argument (see Toolchest 2) leads us to the general result

$$A_{n+1} = \frac{1}{2}A_n, \quad n \geq 1.$$

In words, the equation tells us that the area of each square is half the area of its larger neighbour square. This allows us to express the area of each square generated by this process in terms of the area of the first square. We have:

$$A_2 = \frac{1}{2}A_1, \quad A_3 = \frac{1}{2}A_2 = \frac{1}{2} \cdot \frac{1}{2}A_1 = \left(\frac{1}{2}\right)^2 A_1$$

$$A_4 = \frac{1}{2}A_3 = \frac{1}{2}\left(\frac{1}{2}\right)^2 A_1 = \left(\frac{1}{2}\right)^3 A_1$$

$$\vdots \qquad \vdots$$

$$A_n = \left(\frac{1}{2}\right)^{n-1} A_1.$$

Hence

$$S_n = A_1 + A_2 + A_3 + \cdots + A_n$$

$$= A_1 + \frac{1}{2}A_1 + \left(\frac{1}{2}\right)^2 A_1 + \cdots + \left(\frac{1}{2}\right)^{n-1} A_1$$

$$= A_1 \left(1 + \frac{1}{2} + \left(\frac{1}{2}\right)^2 + \left(\frac{1}{2}\right)^3 + \cdots \left(\frac{1}{2}\right)^{n-1}\right).$$

Now, as more and more squares are drawn, the sum $1 + \frac{1}{2} + \left(\frac{1}{2}\right)^2 + \left(\frac{1}{2}\right)^3 + \cdots$ gets closer and closer to 1. Or, more precisely, the 'sum to infinity' (see Toolchest 6) of the above geometric series is

$$\frac{a}{1-r} = \frac{1}{1-(\frac{1}{2})} = 2.$$

Hence, $S_n$ approaches $2A_1$ as more and more squares are drawn. We see from the definition of $M_n$ that $k = 2$ ensures that $M_n > 0$ for any number

$n$ of squares drawn and that this is the smallest whole number value of $k$ to do so. Hence the answer is (A).

**Remarks:** Observe that, if the problem had required the process to be repeated an infinite number of times, that is, $M = kA_1 - S$, where $S$ is the 'sum to infinity of the infinite series', then

$$S = \sum_{n=1}^{\infty} A_n = 2A_1, \text{ so that } M = kA_1 - S = (k - 2)A_1,$$

so that $k = 2$ would now give $M = 0$, which is not positive. The answer in this case would be (B).

**Solution 78:**    First, we give a number-theoretic solution. Suppose we are given any three positive integers satisfying $a^2 + b^2 = c^2$. We may assume that the greatest common factor $r$ has been cancelled, so that $(a, b, c)$ is a *reduced* Pythagorian triple, while $(ra, rb, rc)$ was the original.

**Step 1.**    We deduce that $c$ is odd, and one of $a$, $b$ is odd (say $a$) and the other ($b$) is even. If $a$, $b$ were both even, then their squares would be divisible by 4, so their sum, $c^2$, would be divisible by 4, hence $c$ would be even too – and so $(a, b, c)$ would not be *reduced*, as we insisted at the start. And if $a$, $b$ were both odd, then their squares (of form $(2s + 1)^2 = 4s^2 + 4s + 1$) would give remainder 1 on division by 4 – we write $a^2 \equiv 1 \pmod 4$, $b^2 \equiv 1 \pmod 4$. Hence $c^2 = a^2 + b^2$ would give remainder 2 on division by 4, or: $c^2 \equiv (1 + 1) \pmod 4 \equiv 2 \pmod 4$. But, whether $c$ is odd (so $c^2 \equiv 1 \pmod 4$)) or $c$ is even (so $c^2 \equiv 0 \pmod 4$)), this is impossible.

**Step 2.**    Now $c - b$ and $c + b$ are both odd (difference and sum of odd and even). And they cannot have a common factor, because anything that divides both is not 2, and would also divide their sum and difference, which are $2c$ and $2b$, so would divide both $c$ and $b$, hence both of their squares too, hence their difference $a^2$ too, again contradicting the assumption about the triple being *reduced*.

**Step 3.**    The product $a^2 = c^2 - b^2 = (c - b)(c + b)$ is a square, that is, its proper factors occur in pairs. But these pairs cannot be split up between the two factors $(c - b)$, $(c + b)$, because these were shown in step 2 to have no common factor. Therefore each of these has factors occurring in pairs, so must be a square number too. Put $c - b = n^2$, $c + b = m^2$.

**Step 4.**    Both $n$ and $m$ must be odd, because their squares are odd (see step 2), and the square of an even number is even. Therefore their sum and difference are even numbers, and so we can define two new integers: $p = \frac{m+n}{2}$, $q = \frac{m-n}{2}$.

**Step 5.**    Clearly, $m = p + q$ and $n = p - q$, so $c + b = m^2 = p^2 + 2pq + q^2$ and $c - b = n^2 = p^2 - 2pq + q^2$. Adding and subtracting these two equations gives us: $c = p^2 + q^2$, $b = 2pq$ and hence $a = c^2 - b^2 = p^2 - q^2$. Now, since $c = p^2 + q^2$ is odd, one of $p, q$ must be odd and the other even. Also, they are relatively prime, for if they had a common factor $s$, it would divide $a$, $b$ and $c$.

**Conclusion:** Every reduced Pythagorian triple $(a, b, c)$ is given by $(p^2 - q^2, 2pq, p^2 + q^2)$, where $p, q$ have opposite triple parity and are relatively prime. Hence every Pythagorian triple has the required form.

The second approach uses analytic geometry. If $a, b, c$ are positive integers such that $a^2 + b^2 = c^2$, then $\left(\frac{a}{c}\right)^2 + \left(\frac{b}{c}\right)^2 = 1$, so the point $P : \left(\frac{a}{c}, \frac{b}{c}\right)$ lies on the unit circle in the first quadrant. In fact, every Pythagorean triple corresponds to such a point with rational coordinates, and conversely. There are two methods for proceeding:

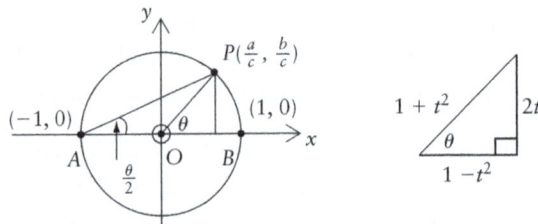

**Method 1:** Let $\theta$ be the angle $P\hat{O}B$, so that angle $P\hat{A}O = \frac{\theta}{2}$, and we choose parameter $t = \tan\frac{\theta}{2} = \frac{b/c}{1 + a/c} = \frac{b}{c+a}$, a rational number. Express this rational number in its lowest terms, $\frac{q}{p}$. Now we can use the standard expressions for $\cos\theta$ and $\sin\theta$ in terms of $t = \tan\frac{\theta}{2}$. (These can be derived from the half angle formula, $\tan\theta = \frac{2\tan\frac{\theta}{2}}{1+\tan^2\frac{\theta}{2}}$, and Pythagoras' Theorem applied to the right triangle in diagram.)

$$\frac{a}{c} = \frac{OB}{OP} = \cos\theta = \frac{1 - t^2}{1 + t^2} = \frac{p^2 - q^2}{p^2 + q^2}.$$
$$\frac{b}{c} = \frac{PB}{OP} = \sin\theta = \frac{2t}{1 + t^2} = \frac{2pq}{p^2 + q^2}.$$

Hence, $a : b : c = \frac{a}{c} : \frac{b}{c} : 1 = p^2 - q^2 : 2pq : p^2 + q^2$, so that $(a, b, c) = r(p^2 - q^2, 2pq, p^2 + q^2)$, where $r$ is some positive integer.

**Method 2:** Let the line $AP$, passing through $A(-1, 0)$, have gradient $t$, so that its equation is $y = t(x + 1)$. This line meets the circle $x^2 + y^2 = 1$ at points where $x^2 + (t(x + 1))^2 = 1$. This is a quadratic equation for $x$, which is easily shown to have roots $x = -1$, (as expected) and $x = \frac{1-t^2}{1+t^2}$, with corresponding values $y = 0$ and $y = \frac{2t}{1+t^2}$. Therefore the point $P :$ $\left(\frac{a}{c}, \frac{b}{c}\right) = \left(\frac{1-t^2}{1+t^2}, \frac{2t}{1+t^2}\right)$, and, letting $t = q/p$, we conclude as in method 1.

# Further Training Resources

## Training books

1. Barbeau, E. J., Klamkin, M. S., Moser, W. O. J. *Five Hundred Mathematical Challenges*. The Mathematical Association of America. 24 July 1997. (236 pages).
2. Burns, J. C. *Seeking Solutions: Discussion and Solution of the Problems from the International Mathematical Olympiads 1988–1990*. Australian Mathematics Trust. (89 pages).
   (To order go to: http://www.amt.canberra.edu.au/amtshop.html)
3. Bradley, C. J. Introduction to Number Theory and Inequalities. United Kingdom Mathematic Trust. 2006.
4. Engel, A. *Problem Solving Strategies*, Springer. 1st edn. 1998; Corr. 2nd printing edition 11 May 1999. (420 pages).
5. Gardiner, A. D. and Bradley, C. J. Plane Euclidean Geometry. UKMT. 2005.
6. Gardiner, A. D. *The Mathematical Olympiad Handbook: An introduction to Problem Solving based on the first 32 British Mathematical Olympiads 1965–1996*. Oxford University Press. October 1997. (248 pages).
7. Griffiths, M. The Backbone of Pascal's Triangle. UKMT. 2007 (First Published 2008).
8. Halmos, P. R. *Problems for Mathematicians, Young and Old*. The Mathematical Association of America. December 1991. (328 pages).
9. Jacobs, D. A. *Mathlete's Training Guide: Introductory Problem Solving Skills for Grade 9 and 10 pupils*. South African Mathematical Society.
   (To order go to: http://www.mth.uct.ac.za/imo/imopub.html) or http://www.mth.uct.ac.za/imo/mathlete.html)
10. Rusczyk, R. and Lehoczky, S. *The Art of Problem Solving: Volume 1: The BASICS Solutions*. Mu Alpha Theta, National High School Mathematics. 4th edition 30 June 2002. (174 pages).
11. Lehoczky, S. and Rusczyk, R. *The Art of Problem Solving Volume 2: And Beyond, Solutions*. Greater Testing Concepts. April 1994. (211 pages).
12. Plank, A. W. and Williams, N. H. *Mathematical Toolchest*. Australian Mathematics Trust. (120 pages).
13. Reiman, I. *International Mathematical Olympiad 1959–1999*. Anthem Press. May 2002. (400 pages).
14. Smith, G. A Mathematical Olympiad Primer. UKMT. 2007 (First Published 2008).
15. Tabov, J. B. and Taylor, P. J. *Methods of Problem Solving, Book 1*. Australian Mathematics Trust. 1996.
    (To order go to: http://www.amt.canberra.edu.au/book09.html)
16. Tabov, J. B. and Taylor, P. J. *Methods of Problem Solving, Book 2*. Australian Mathematics Trust. 1996.
    (To order go to: http://www.amt.canberra.edu.au/book19.html)

## Journals, magazines, problem collections and training booklet series

1. Derek Holton has for many years been at the forefront of mathematical olympiad training, in Australia, New Zealand (where he is professor of Pure Mathematics at the University

of Otago), and internationally. His problem solving training booklets are world famous. A good starting-point when looking for his work is his homepage hosted by the University of Otago: http://www.maths.otago.ac.nz/home/department/staff/derek_holton.html

2. The Australian Mathematics Trust's shop has much useful material: http://www.amt.canberra.edu.au/amtshop.html

3. The University of Capetown Mathematics Competition compilation is one of seven items listed in the UCT creative mathematics accessories shop: http://www.mth.uct.ac.za/digest/.

4. The United Kingdom Mathematics Trust produces good material and programmes related to olympiads and mathematics enrichment. See, in particular, their publication: Ten years of Mathematical Challenges (1997–2006): http://www.mathcomp.leeds.ac.uk/publications

5. The Canadian Mathematical Society publications and training resources for mathematics competitions: www.math.ca/competitions/

6. Zimaths Magazine online (when working): www.uz.ac.zw/science/maths/zimaths

## Books on the history of mathematics

1. Boyer, C. B. and Merzbach, U. C. *A History of Mathematics*. Wiley; 2nd edition 6 March 1991. (736 pages).
   (For details go to: http://www.amazon.com/History-Mathematics-Carl-B-Boyer/dp/0471543977)

2. Eves, H. *An Introduction to the History of Mathematics*. Brooks Cole; 6th edition 2 January 1990. (800 pages).
   (For details go to: http://www.amazon.com/Introduction-History-Mathematics-Saunders/dp/0030295580)

3. Katz, V. J. *A History of Mathematics: An Introduction*. Addison Wesley; 2nd edition 6 March 1998. (880 pages).
   (For details go to: http://www.amazon.com/History-Mathematics-Introduction-2nd-Katz/dp/0321016181)

## Some useful web-sites

www.uccs.edu/asoifer (Abraham Soifer's Centre for Excellence in Mathematics Education)

http://pass.maths.org.uk/index.html (Plus Magazine)

www.turnbull.mcs.st-and.ac.uk/~history or www-history.mcs.st-and.ac.uk/. (The MacTutor History of Mathematics Archives)

www.greylabyrinth.com (A good puzzle site)

www.artofproblemsolving.com (Another good puzzle site)

http://stimulus.ucam.org (Resources for mathematical enrichment, including NRich online mathematics club)

http://motivate.maths.org

http://thesaurus.maths.org

www.amt.canberra.edu.au/wfnmc.html (World Federation of National Mathematics Competitions)

# Index

Note: page numbers in *italics* refer to figures, whilst those in **bold** refer to examples of the application of theorems and concepts.

The manufacturer?s authorised representative in the EU for product
safety is Oxford University Press España S.A. of El Parque Empresarial
San Fernando de Henares, Avenida de Castilla, 2 ? 28830 Madrid
(www.oup.es/en or product.safety@oup.com). OUP España S.A. also acts
as importer into Spain of products made by the manufacturer.
Printed and bound by CPI Group (UK) Ltd, Croydon, CR0 4YY

08/01/2025

01817157-0008